Optimization and Dynamics with Their Applications

Akio Matsumoto
Editor

Optimization and Dynamics with Their Applications

Essays in Honor of Ferenc Szidarovszky

Springer

Editor
Akio Matsumoto
Department of Economics
Chuo University
Hachioji, Tokyo
Japan

ISBN 978-981-10-4213-3   ISBN 978-981-10-4214-0 (eBook)
DOI 10.1007/978-981-10-4214-0

Library of Congress Control Number: 2017934862

Printed on acid-free paper

This Springer imprint is published by Springer Nature
The registered company is Springer Nature Singapore Pte Ltd.
The registered company address is: 152 Beach Road, #21-01/04 Gateway East, Singapore 189721, Singapore

Preface

This book contains a collection of research papers on optimization and dynamics with their applications celebrating academic achievements of Ferenc Szidarovszky for half of a century from 1968 to Present. He made outstanding contributions to various fields including decision theory, numerical analysis, system dynamics in economics, game theory, optimal maintenance and replacement policies, to name only a few. The authors of the chapters of this book are his colleagues, ex-students and friends and have been collaborated in the past. They are, like other researchers who read his papers, inspired by his deep insight and wisdom and agree to contribute their new findings to a special book.

Ferenc Szidarovszky, currently Full Professor at Department of Applied Mathematics, University of Pécs in Hungary, has been a distinguished researcher, a best teacher and big buddy who loves jokes. Born as the only son of intellectual family where his mother as well as his three sisters were school teachers and his father was a civil engineer, he found himself to shine at mathematics during high school years when he won two second prizes in the International Student Mathematics Olympics. Without an entrance examination, he was admitted as a mathematics student at the Eötvös Loránd University of Sciences in Budapest and obtained his B.Sc. degree and Master's degree in 1966 and 1968. At the same university, he continued the graduate program in mathematics and completed his Doctorate degree in 1970 by writing a thesis on numerical methods. After graduation, he became a faculty member at the Department of Geometry and then the newly founded Department of Numerical and Computer Methods at the Eötvös Loránd University of Sciences. After spending 9 years, he moved to University of Horticulture in 1977 and spent 9 years teaching numerical analysis, operations research and computer science while he served as the acting head of Department of Computer Science. During these years, he, as a young assistant professor, attended the game theory seminar organized by Prof. Jenő Szép at the Karl Marx University of Economics and encountered with one of his life-long research subjects, game theory, in particular, oligopoly theory. He obtained another Doctorate degree in Economics from Karl Marx University of Economics in 1977 and joined its Institute of Mathematics and Computer Science in 1986. He became a member of the Department of Systems and

Industrial Engineering, University of Arizona in the US, first as a visiting professor between 1988 and 1990, then as tenured full professor in 1990 and stayed there until he retired in 2011.

Looking back his academic life after graduation, Szidar encountered five special events that were critical for his future. The first one occurred just after graduating at the Eötvös University of Sciences in June 1968 when he was offered a faculty position. During his student years he was a member of the violin session of the University Symphony Orchestra, so did not have time and willingness to be involved in any political activity. Based on this, the Communist Party leadership was against his hiring. The department head offered him to go to Soviet Union with a group of Hungarian students during the summer to an International Student Camp, what he did. So there was no more objection for hiring him, so he became an assistant professor at the Department of Geometry teaching graphical and numerical methods. In 1972, with one of his collaborators he could develop a good personal relation with a couple of professors from the University of Arizona. Based on their scientific discussions, the American partners submitted a research proposal to NSF for collaborative research with them. They got the research fund, so as a result, from 1973 until 1986 he was able to visit Arizona in almost every year and in two occasions he was also invited to be a visiting professor for one and three semesters. This connection was very helpful for him to join the faculty of the Systems and Industrial Engineering Department of the University of Arizona in 1988. Third, at the end of the 70s, the famous Japanese economist, Koji Okuguchi, was a referee of one of his papers. After submitting the report he contacted Szidar, and after exchanging several letters he visited Hungary, and this visit was then repeated in almost every year. Professor Okuguchi even learned the Hungarian language. This cooperation became very successful resulting in a great number of joint papers on different aspects of oligopoly theory and a book, *The Theory of Oligopoly with Multi-Product Firms* (Springer-Verlag, 1990), which had a second edition as well. In 1991, Prof. Okuguchi organized a special session in a conference in Dublin, where Szidar met Carl Chiarella, a professor with the University of Technology, Sydney. Their meeting was followed by a more than a decade long cooperation, dealing mainly with continuously distributed time delays in oligopolies and in other dynamic economic models. They also coauthored a book with two other scientists, *Nonlinear Oligopolies: Stability and Bifurcations* (Springer-Verlag, 2010). In the late 90s, Szidar met the editor of this volume in a conference held in Odense, Denmark. They immediately found many common research areas and a very successful cooperation started, which continues even today. Their joint work on revisiting dynamic monopolies, oligopolies and a great number of classical economic models resulted in a large number of papers, in addition to a monograph on game theory, *Game Theory and its Applications* (Springer-Verlag, 2015) as well as to editing a conference volume of the 9th International Conference on Nonlinear Economic Dynamics held in Tokyo, *Essays in Economic Dynamics* (Springer Science, 2016).

Szidar published 24 books not including second editions, more than 400 journal papers, 37 book chapters and 114 papers in conference proceedings, advised 67 MS students and supervised 12 Ph.D. students. He was involved in many scientific assignments, some of them are an Associate Editor of *Pure Mathematics and Applications*, an Editorial Board Member of *International Review of Pure and Applied Mathematics*, an Area Editor (North America) of *International Journal of Internet and Enterprise Management*, an Advisory Editorial Board Member of *Scientia Iranica.* He received various professorial awards including *National Award for Outstanding Academic Merit* from Ministry of Education of Hungary in 1969, *Candidate in Mathematical Science* from Hungarian Academy of Science in 1975, *Doctor of Engineering Science* from Hungarian Academy of Science in 1986, *Doctor Habil in Engineering* from Budapest Technical University in 1998 and *Dr. Honoris Causa* from University of Pécs in 2014.

During those days, he managed to balance his two loved ones, mathematics and classical music; in day time, he conducted mathematical research with enthusiasms, taught undergraduate and graduate students in a passionate way, while in the evening, he took a music trip, relaxed himself and sat back listening to classical music, which relieves stress and improves vigor. He also shared his love to classical music with friends and colleagues, as well as his house any time when they visited Tucson.

We appreciate financial supports: Graduate School of Economics of Chuo University with the MEXT-supported Program for the Strategic Research Foundation at Private University 2013–2017, the Japan Society for the Promotion of Science (Grant-in-Aid for Scientific Research (C) 24530202, 25380238, 26380316 and 16K03556) and Joint Research Grant of Chuo University.

We academically and socially owe much to Ferenc Szidarovszky. The present collection of papers expresses a modest sign of our gratitude.

Tokyo, Japan Akio Matsumoto

Contents

Contributors

Mohammad AmirRahmani Faculty of Civil Engineering, University of Tabriz, Tabriz, Iran

Ioannis K. Argyros Department of Mathematics Sciences, Cameron University, Lawton, USA

A. Terry Bahill Systems and Industrial Engineering, University of Arizona, Tucson, AZ, USA

Gian-Italo Bischi DESP-Department of Economics, Society, Politics, University of Urbino, Urbino, Italy

Suna Cinar Department of Industrial and Manufacturing Engineering, Wichita State University, Wichita, KS, USA

Emery A. Coppola NOAH LLC, Lawrenceville, NJ, USA

R.E. Fitz Universidad Autónoma Chapingo, México-Texcoco, Mexico

Ferenc Forgó Corvinus University of Budapest, Budapest, Hungary

Laura Gardini DESP-Department of Economics, Society, Politics, University of Urbino, Urbino, Italy

Huairui Guo ReliaSoft Corporation, Tucson, AZ, USA

Maryam Hamidi Department of Industrial Engineering, Lamar University, Beaumont, TX, USA

Zsuzsanna Hauck Faculty of Business and Economics, University of Pécs, Pécs, Hungary

Haitao Liao Department of Industrial Engineering, University of Arkansas, Fayetteville, AR, USA

Yi Luo Center for Sustainable Systems, School of Natural Resources and Environment, The University of Michigan, Ann Arbor, MI, USA

Á. Alberto Magreñán Universidad Internacional de La Rioja (UNIR), Logroño, La Rioja, Spain

Akio Matsumoto Department of Economics, Senior Researcher, International Center for further Development of Dynamic Economic Research, Chuo University, Hachioji, Tokyo, Japan

Mark Molnar Faculty of Economics and Business Management, Department of Macroeconomics, Szent Istvan University, Gödöllő, Hungary

Sandor Molnar Faculty of Mechanical Engineering, Department of Mathematics and Informatics, Szent Istvan University, Gödöllő, Hungary

S.F. Pérez Universidad Autónoma Chapingo, México-Texcoco, Mexico

Haiyan Qiao School of Computer Science and Engineering, California State University San Bernardino, San Bernardino, CA, USA

Gábor Rappai Faculty of Business and Economics, University of Pécs, Pécs, Hungary

M.R. Salazar Universidad Autónoma Chapingo, México-Texcoco, Mexico

Juan Antonio Sicilia Universidad Internacional de La Rioja (UNIR), Logroño, La Rioja, Spain

Iryna Sushko Institute of Mathematics NASU, Kiev, Ukraine

Miklos Szidarovszky ReliaSoft Corporation, Tucson, AZ, USA

József Vörös Faculty of Business and Economics, University of Pécs, Pécs, Hungary

Mehmet Bayram Yildirim Department of Industrial and Manufacturing Engineering, Wichita State University, Wichita, KS, USA

Qiuze Yu Wuhan University, Wuhan, China

Mahdi Zarghami Faculty of Civil Engineering, University of Tabriz, Tabriz, Iran

Jijun Zhao Institute of Complexity Science, Qingdao University, Qingdao, Shandong, China

Introduction

Summary of Szidar's Works

Ferenc Szidarovszky deals with a wide spectrum of scientific topics ranging from pure mathematics to applications for industrial problems, and without doubt, has made outstanding contributions in many research areas. He is a great team player having collaborators, students and coauthors from all continents. Knowing that it is almost impossible, we classified his main contributions into the six fields in which widely cited papers, highly innovative papers and influential papers are published: Numerical Analysis, Optimization, Dynamics, Game Theory and Oligopoly, Industrial Applications and Economic Dynamics and briefly outlined each of them.

Numerical Analysis

Numerical analysis is the main subject in his earlier years. Szidar developed a general scheme for matrices with nonnegative inverses including M-matrices. He presented convergence conditions for algorithms modeled by point-to-set mappings and also gave convergence conditions for several types of Newton-type methods, for nonstationary multistep iterations and for modified contractions. He expanded special methods for solving utilization equations, solved a special matrix equation in microelectronics, introduced iteration technics for nonlinear systems of circuit equations and estimated errors in solving polynomial equations. The earlier contributions are summarized in a book with S. Yakowitz, *Principles and Procedures of Numerical Analysis* (1978) and the one with I.K. Argyros, *The Theory and Applications of Iteration Methods* (1993).

Optimization

Developing a method for model choice in multiobjective optimization problems, Szidar showed the relation between ordered weighted average (OWA) operator and compromise programming and applied OWA operator for water resources problems. Further, he made several applications: he examined optimal control of invasive species in Arizona, developed a model for conflict analysis in forest management, constructed a multiobjective optimization model for wine production, for optimal product structure in food industry, for the electric power industry and in natural resources management. His book with M. Zarghami, *Multicriteria Analysis. Applications to Water and Environment Management* (2011) contains some of his contributions.

Dynamic Systems Including Simulations

Szidar developed stability conditions for adaptive control systems, presented a general technique for finite memories data manipulation and smooth prediction as well as for validating recursive filters. Delays are important in various fields of natural and social sciences. He developed a simple, elementary method for stability of dynamic systems with one or two delays. Numerical simulation is a complement of analytical considerations, especially in the case when analytical methods are limited. Szidar examined convergence properties of perturbation analysis estimate and introduced correlated sampling in integration. He performed agent-based simulations for public radio membership campaign, for social dilemma games, for finite neighborhood binary games and for battle of sexes game. Further, he worked out a method for weighted Monte Carlo integration. He gave general characterization of two-person binary games, and examined multi-agent learning models.

Game Theory and Oligopoly Theory

Attending the seminars given by Prof. Jenö Szép of Karl Marx University, Budapest, Szidar came across game theory and was under his tutelage just after he assumed his position. He has been working on game theory since then. Four books on game theory were written under joint authorship with his Hungarian colleagues and the editor of this book. In addition, many papers on game theory and oligopoly were published in academic journals ranging from mathematics to economics. Among them, concerning the bargaining game, Szidar examined it under uncertainty and under fuzzy environment, gave a new classification of the Nash bargaining solution with a modified version of alternating offer method and applied it to train scheduling. He developed a dynamic game for computer network security, made equilibrium analysis in asymmetric contests with endogenous prices. He also gave stability conditions for general quadratic games and examined rewards and costs in strategic interaction. Concerning conflict, he first yielded sufficient and necessary conditions of equilibria in general conflict models and analyzed intergroup-conflicts in utilizing public goods. Oligopoly theory that can be traced back to the middle of nineteenth century has been augmented with game theory. Szidar jointly wrote two books on oligopoly theory, the latest one is *Nonlinear Oligopolies: Stability and Bifurcations* (2010) with G-I Bischi, C. Chiarella and M. Kopel while the earlier one is given in section Economics below. Independently of Reinhard Selten who was a giant in Game Theory and won Nobel Memorial Prize in Economic Science in 1994, he introduced best responses as functions of the total industry output and gave a constructive proof of existence and uniqueness of equilibrium in concave oligopolies with a simple computer method finding it. For nondifferentiable concave oligopolies, he proved that industry output is unique and set of equilibria is either a single point or a polyhedron. The following subjects are thoroughly and rigorously examined in various oligopoly frameworks: existence and uniqueness of equilibria with stability issues for modified oligopolies with capacity constraints, market saturation, product differentiation, output adjustment costs, oligopoly-oligopsony, labor managed oligopolies, rent seeking games, considering pollution treatments, and R&D. Multi-market models and leader-follower models were constructed and examined under various environments: with uncertainty, with partially cooperative firms, with intertemporal demand interaction, with misspecified demand functions, with antitrust thresholds, with cost subsidies, with advertisements, with cost externalities, with cartelizing groups, with socially concerned firms, with contingent workforce and unemployment insurance system, with discontinuous payoffs, with production adjustment and investment costs, with incomplete information. He gave conditions for stability and controllability and also for successful learning with different learning schemes. Comparing equilibrium prices in Cournot and Bertrand oligopolies, he discovered the differences and similarities between them. Furthermore he applied oligopoly theory for international fisheries and examined some model types with one or two time delays.

Industrial Applications

Mainly working as a consultant for the Ministry of Industry, Hungary and for several industrial organizations, Szidar has a lot of industrial experience and accumulates special knowledge on water resources systems, decision making in energy policy, modeling in hydrology, modeling earth quake protection devices and using neural networks for prediction and optimization in water resources management. As a natural consequence, he has papers dealing with real problems in society. He expanded a Bayesian method for analyzing underground flooding in mines and constructed optimal observation network in the mining industry. He developed mathematical models for optimal inspection, repair and preventive replacement strategies, and for optimal utilization system of mineral resources whereas he proved the convergence of the geostatistical Kriging method, examined optimal strategies in lease contracts with non-cooperative, cooperative game theories and conflict resolution and estimated component reliability with missing failure data. In connection with water resources, Szidar developed method for optimal design of flood levees, multiobjective models for optimal water allocation in agriculture and between several users. Further he applied oligopoly theory in water management and leader-follower games in water allocation problem. He described the motion of ice sheets and glaciers by a numerical method solving the associated partial differential equations and used neural networks to predict transient underground water levels, to analyze conflict between water supply and environmental health risk, to estimate saltwater upcoming and to forecast algae counts in surface waters.

Economics

One of his mentors, Koji Okuguchi was a well-known economist and specialized in oligopoly theory. Working with him, Szidar solved many challenging problems, resulting in a large number of papers and a book, *The Theory of Oligopoly with Multi-Product Firms* (1999). He gave a condition for solvability of nonlinear input-output models and examined stability of dynamic consumer-producer markets. Moreover, cooperating with the editor of this book who is also an economist, he expanded his range of work to considering more variety of economic problems including dynamic analysis of subsidy games, duopolies with advertisements, neoclassical growth model, multiplier-accelerator model, nonlinear cobweb model, classic IS-LM model with tax collection, heterogeneous agent model of asset price, Goodwin's business cycle model, Kaldor–Kalecki model and Hicksian trade cycles.

Contributions

It may be convenient to group the contributions according to major points of emphasis. Part I is concerned with operations research, the papers of Part II are organized around dynamic analysis and Part III contains a collection of various application results.

Let us start with Part I that is opened by Ioannis Konstantinos Argyros, a professor of Cameron University and Szidar's coauthor in many papers and a book. His paper with Ángel Alberto Magreñán and Juan Antonio Sicilia is entitled, Developments on the Convergence of Some Iterative Methods. It is well known that nonlinear equations are important tools in solving optimization problems based on the first order conditions. It is also well known that there are many different types of iteration schemes for their solutions and among them the Newton-type methods got the largest attention in the literature. This paper reconsiders the classical Newton–Kantorovich method and presents several improvements to the existing literature in terms of the convergence domain and convergence ratio. Both can decrease computational effort and time significantly, so in addition to the theoretical interest the results have practical importance as well. The paper is clearly and well written, and two numerical examples show the applicability of the new results when the classical approach does not work.

The second paper, The Non-symmetric L-Nash Bargaining Solution is presented by Ferenc Forgó, an old friend and a former colleague and coauthor of Szidar. In an earlier paper, Forgó and Szidarovszky (2003, *European Journal of Operational Research*), they examined a limiting property of the two-person Nash bargaining solution if the disagreement vector converges to negative infinity in a give direction. In this paper, he presents a nice generalization of that result in the more general case of the non-symmetric Nash solution. The convergence of the non-symmetric Nash solution is proved and a simple algorithm is presented to define and compute the limiting vector. It is also proved that in the case of polyhedral feasible sets there is a finite threshold such that the limit is reached if the disagreement point is finite and below the threshold. As an example the well-known firm-union bargaining problem is selected to illustrate the theoretical results. It is also shown that in most cases the limit vector can be obtained as the solution of a multiobjective optimization problem, and in addition the relation of the result of the paper with the alternating offer method of Rubinstein (1982, *Econometrica*) is briefly outlined.

The third paper, Analyzing the Impact of Process Improvement on Lot Sizes in JIT Environment When Capacity Utilization Follows Beta Distribution, is by József Vörös, a former student and coauthor of Szidar and a professor at the University of Pécs, collaborating with Gábor Rappai and Zsuzsanna Hauck. It addresses one of the most important issues in management science. Their model and method is closely related to a former article of the first two authors, Vörös and Rappai (2016, *Applied Mathematical Modelling*). In JIT environment the employees have the obligation to report all quality problems in an assembly line resulting in frequent stoppages, and therefore the output is random, and based on the literature, Beta distribution is assumed. Its parameters are related to process quality. Better quality means that higher percent of the product can go to the market directly while the rest has to go through repairs. The optimum lot size therefore depends on these factors. This paper calculates the inventory cost, when several cases have to be considered based on the capacity of the repair shop and the lot size. Closed form expressions are derived; however, the optimum lot size depends on several random components. Therefore the authors used simulation and examined how the optimum lot size and the expected optimal annual cost depend on model parameters, especially on the parameters of the Beta distribution. The variance of the annual total cost gives a measure of economic risk to the management.

Yi Luo, a former Ph.D. student of Szidar who is currently a Research Lab Specialists with the University of Michigan, Ann Arbor studies a behavioral game theoretical model in the fourth paper Exploring Efficient Reward Strategies to

Encourage Large-Scale Cooperation Among Boundedly Rational Players with the Risk and Impact of the Public Good. It describes the players' decision making processes incorporating the risk and the impact of the public good. It is demonstrated that the conventional reward to achieve large-scale cooperation can be significantly reduced along the process and as the number of players increases, the increased interaction among them makes their decisions more rational.

The first paper of Part II is Periodicity Induced by Production Constraints in Cournot Duopoly Models with Unimodal Reaction Curves by Gian-Italo Bischi, Laura Gardini and Iryna Sushko, the first two are professors at University of Urbino (Italy) that has been a research center for nonlinear economic dynamics in Europe and the third is a mathematician from Institute of Mathematics NASU in Ukraine. They organize the Urbino group in which senior and junior researchers are involved and take the lead role for developments of nonlinear dynamics. Szidar is one of the most active members. It is well known that Rand (1978, *Journal of Mathematical Economics*) augments a classical Cournot model with tent-shaped reaction functions and shows the occurrence of robust chaos. They impose maximum production constraints yielding flat-top shaped reaction functions and demonstrate that the presence of such constraints can be a source of superstable cycles. Further they analyze the appearance of border collision bifurcations and global bifurcation that have become a focus topic in the literature on piecewise-smooth dynamic systems.

The second paper is developed by Haiyan Qiao, an ex Ph.D. student of Szidar and now an associate professor at California State University San Bernardino. In her paper, An Adaptive Learning Model for Competing Firms in an Industry, she considers an N-firm oligopoly in which the firms know the marginal cost of each firm as well as the marginal price. However, the reservation price is uncertain for them and they develop an adaptive scheme to learn about it based on repeated market observation comparing their believed prices with the actually received market price. First, Haiyan proves the asymptotic stability of the learning process. Then it is assumed that the firms get delayed price information from the market. Second, fixed delay is assumed and it is shown that system remains stable if the delay is below a certain threshold. It is also derived in the paper that beyond this threshold, stability is lost forever. Then continuously distributed delay is assumed with exponential kernel function and it is shown that the system remains stable regardless of the expected length of the delay. This is an interesting but understandable result, since with exponential kernel function, small delays have the largest weights.

The work of Jijun Zhao, an ex Ph.D. student of Szidar and now a full professor at Qingdao University in China, is the third paper, The Coordination and Dynamic Analysis of Industrial Clusters: A Multi-agent Simulation Study. She examines the dynamic evolution of industrial clusters using oligopoly theory and agent-based simulation. The literature review summarizes the main references on oligopoly theory as well as on agent-based industrial cluster models. The cluster contains supplier and producer agents. The production functions, prices of supplies and products are assumed to be linear. The model also considers the innovation development and spillover effects, the influence of the technology level on the final product prices, the needed labor level for both the suppliers and the producers, the price of labor, innovation costs and the profit functions of all firms. The dynamic system is based on discrete gradient adjustment with both linear and nonlinear adjustment schemes. This complex system is analyzed by using agent-based simulation showing the time series of the outputs of the agents, average prices of the products and labor, and total labor usage. The effect of the parameters in the adjustment schemes as well as in the selection between lincar and nonlinear gradient adjustments is examined by showing their influence on the dynamic behavior of the entire system.

The fourth paper is presented by Sándor Molnár, the director of the Institute of Mathematics and Computer Science of the Szent István University in Hungary, an old colleague, friend and coauthor of Szidar in several papers and four books, *Introduction to Matrix Theory* (2002) and the other three are in Hungarian. In a joint work with his son, Márk, Approximation of LPV-Systems with Constant-Parametric Switching Systems, they introduce a new approximation of nonlinear systems by special linear time varying systems. The approximation method provides the basis for a variety of switching-type systems. The optimization of linear control systems shows similarities with linear programming, which is well established by the main approximation theorem of this paper. The theory is applied to the Buck-Boost converter circuit.

The last paper of Part II, Love Affairs Dynamics with One Delay in Losing Memory or Gaining Affection by Akio Matsumoto who has been working with Szidar from the beginning of 2000. He constructs a Romeo–Juliette model in which the delayed time evolution of a love affair between two individuals called Romeo and Juliette are examined. There exist multiple steady states and it is shown first that the nonzero steady states are always stable and the stability of the zero steady state depends on model parameters in a no-delay case and then that introducing delay gives rise to stability switch under which the nonzero steady state can be destabilized and bifurcated to a limit cycle.

Part III is started by Terry Bahill, a former colleague of Szidar at the University of Arizona, who jointly worked with Szidar in the book, *Linear Systems Theory* (1992). He is an internationally recognized expert of introducing mathematical models in the science of baseball and his picture is in the Baseball Hall of Fames exhibition. His paper, Optimizing Baseball and Softball Bats, first introduces a mathematical model to describe the collisions between baseballs, softballs and bats focusing on the speed and spin of balls and bats by using simple Newtonian principles. The batted-ball speed is maximized first and then recommendations are given to the batter to achieve optimal bat performance. Since the optimization problem has no analytic solution, professional software is used.

The second paper of Part III, Reverse Logistic Network Design for End-of-Life Wind Turbines, is presented by Suna Cinar, a Ph.D. student in Wichita State University whom Szidar helped as an outside advisor and Mehmet Bayram Yildirim. It is an example of the recovery of valuable material that can be recycled/recovered or remanufactured at the end of wind turbines useful life by designing an effective reverse logistics network. Clean energy is one of the most important issues in energy research today. Wind turbine is an important energy source, it needs significant investment and at its end-of-life its disposal creates a huge problem for the owner, who can benefit from a well-designed reverse logistic network of best disposal alternative since it has a significant amount of reusable material. The paper introduces a mixed integer linear programming model to minimize the transportation and operating costs and also to find the best locations for recycling and remanufacturing facilities. Since most readers are not familiar with the relevant issues, the paper first gives a brief overview of the wind turbine supply chain and then a mathematical model is formulated to optimize the reverse logistic network. A case study illustrates the model and its solution is based on actual data, where several scenarios are examined and compared. This model is cost-minimizing which is then extended to a total profit maximizing model that is also illustrated based on the same data. Since during the last three years with the University of Arizona, Szidar was also involved with clean energy modeling, mainly solar energy, we are sure that he would like this paper.

Maryam Hamidi, ex Ph.D. student of Szidar and Haitao Liao, ex-colleague in the University of Arizona are coauthors in the third paper, Maintenance Outsourcing Contracts Based on Bargaining Theory. They discuss a two-person game between an owner of equipment and a servicing agent. The subject is to find a mutually agreeable contract. Two models are introduced, both are based on Nash's bargaining solution, which is a good choice since it is the limiting outcome of a real dynamic bargaining process. In the first model the strategies of the players are the scheduled preventive maintenance times and the time when the agent orders the spare parts. They affect the aging process of the equipment as well as the cost of the servicing agent in possible inventory costs and penalties for late services. The Nash bargaining solution is determined which is Pareto optimal but does not give the total maximal benefit for the players. In the second model this bargaining solution is selected as the disagreement point and the failure and preventive replacement costs are also added as decision variables and the Nash bargaining solution is obtained with the additional constraint that the total profit of the two players is on its maximal level. They also relate this solution to the Shapley values if the game is considered as a classical cooperative game. Both models are illustrated with well-selected numerical examples.

The fourth paper is given by M.R. Salazar, a professor with the Universidad Autónoma Chapingo, Mexico and a coauthor of Szidar in several papers. Her paper with R.E. Fitz and S.F. Pérez, Agricultural Production Planning in a Fuzzy Environment, deals with determining optimal cropping patterns in a region of Mexico which usually faces with water shortage. Because of uncertainty of future crop prices several price predictions are used. The objective function is the total profit by selling the products, and the constraints include area limitations for all seasons by available irrigation water amounts, annual water supply, the types of water needed for certain crops and a minimal area for each crop by the minimum possible demands. The objective functions are different for the different price predictions, so the problem is modeled as a multiobjective programming problem. The satisfaction level of each objective is modeled by transforming them into the unit interval [0, 1] which gives the satisfaction level between 0 and 100%. The lowest satisfaction level is maximized based on Itoh's fuzzy approach, which reduces the problem to an LP model. A particular case study in the Alto Lerma Irrigation District of Mexico illustrates the model and solution methodology.

The fifth paper by Qiuze Yu, Huairui Guo and Miklos Szidarovszky, Optimal Replacement Decisions with Mound-Shaped Failure Rates, addresses one of the most important problems of reliability engineering. The optimal timing of preventive replacement is critical in minimizing life-time costs. In the literature mostly normal, Weibull, Gamma or Gumble distribution of time to failure is assumed, where the failure rate function is monotonic. However, in many applications, especially if fatigue is involved lognormal distribution is assumed and used for which the failure rate is mound-shaped. The paper examines the existence of finite optimum in such cases. In addition to the lognormal distribution log-logistic, log-Gamma and log-Weibull variables are also considered. Conditions are derived for the existence of finite optimum which gives the optimal replacement schedule otherwise the item should be replaced upon the first failure. The mathematical derivations are correct and the results might be very helpful in the industry. Nice illustrative examples show how the methodology is used, and show the different possible scenarios.

The sixth paper of Mahdi Zarghami, a professor of University of Tabriz, Iran, and Mohammad AmirRahmani, A System Dynamics Approach to Simulate the Restoration Plans for Urmia Lake, Iran, deals with the restoration plan of the Urmia Lake in Iran, which supplies water for a large region of the country for agriculture, industrial and domestic users. Because of several reasons and their combinations the water level is shrinking continuously. To stop this tendency and start the recovery of the lake, several long-term strategies were developed and this paper examines the consequences of six restoration plans. The authors are not looking for optimal plan, instead using simulation of the dynamic system describing the state of health of the lake, they evaluate all of the selected six options and make recommendations for long-term water management policies. The topic of the paper is very important, shortage of water is expected to grow all over the world and might result even in serious international conflicts.

Part III is concluded by Emery Coppola, a consultant in hydrology and water resources management, a former Ph.D. student of Szidar. In his dissertation he developed a special neural network-based methodology to control the quantity and quality of groundwater resources. In his paper, A Decision Support System for Managing Water Resources in Real-Time Under Uncertainty, coauthored with Suna Cinar, they introduce a new model of watershed management including groundwater and surface water resources. Groundwater elevations and surface water flows are predicted by using artificial neural networks. The management objective is to maximize total groundwater pumping while minimizing the potential negative impacts including excessive drawdown in the unconfined and confined aquifers as well as in the riparian corridor in addition to streamflow depletion of the river.

Part I
Operations Research

Developments on the Convergence of Some Iterative Methods

Ioannis K. Argyros, Á. Alberto Magreñán and Juan Antonio Sicilia

Iterative methods, play an important role in computational sciences. In this chapter, we present new semilocal and local convergence results for the Newton-Kantorovich method. These new results extend the applicability of the Newton-Kantorovich method on approximate zeros by improving the convergence domain and ratio given in earlier studies. These advantages are also obtained under the same computational cost. Numerical examples where the old sufficient convergence criteria are not satisfied but the new convergence criteria are satisfied are also presented in this chapter.

1 Introduction

Let $\mathcal{X}$ and $\mathcal{Y}$ be Banach spaces. Let $U(x_0, R)$ stand for the open ball centered at $x_0 \in \mathcal{X}$ and of radius $R > 0$ and let $\bar{U}(x_0, R)$ stand for its closure. We shall also denote by $Ł(\mathcal{X}, \mathcal{Y})$ the space of bounded linear operators from $\mathcal{X}$ to $\mathcal{Y}$.

In this chapter we are concerned with the problem of approximating a locally unique zero $x^\star$ of F, where F is a Fréchet-differentiable operator defined on $\bar{U}(x_0, R)$ and with values in $\mathcal{Y}$. Many problems, including finding optimal solutions by solving for the first order conditions, are reduced to finding zeros of operators using Mathematical Modelling (Argyros 2007; Argyros et al. 2012; Smale 1986). The zeros

I.K. Argyros (✉)
Department of Mathematics Sciences, Cameron University,
Lawton 73505, USA
e-mail: iargyros@cameron.edu

Á.A. Magreñán · J.A. Sicilia
Universidad Internacional de La Rioja (UNIR),
Av. de La Paz, 137, 26002 Logroño, La Rioja, Spain
e-mail: alberto.magrenan@unir.net

J.A. Sicilia
e-mail: juanantonio.sicilia@unir.net

A. Matsumoto (ed.), *Optimization and Dynamics with Their Applications*,
DOI 10.1007/978-981-10-4214-0_1

of these operators can be found in closed form only in special cases. That is why most solution methods for these problems are iterative. In Computational Sciences the practice of Numerical Functional Analysis is essentially connected to variants of Newton's method (Amat et al. 2004; Argyros and Szidarovszky 1993; Argyros 2004, 2007; Argyros and Hilout 2010a, b, 2012; Argyros et al. 2012; Argyros and George 2015; Cianciaruso 2007; Ezquerro et al. 2010; Kantorovich and Akilov 1982; Magreñán and Argyros 2015; Magreñán 2014a, b; Potra, and Pták 1984; Proinov 2010; Rheinboldt 1988; Smale 1986; Wang 1999; Zabrejko and Nguen 1987).

The Newton-Kantorovich method defined by

$$x_n = x_{n-1} - F'(x_{n-1})^{-1}F(x_{n-1}) \text{ for } x_0 \in \mathcal{X} \text{ and each } n \in \mathbb{N} \tag{1.1}$$

is undoubtedly the most popular method for generating a sequence $\{x_n\}$ approximating the solution $x^\star$. The convergence analysis of iterative methods is usually divided into two categories: semilocal and local convergence analysis. In the semilocal convergence analysis one derives convergence criteria from the information around an initial point whereas in the local analysis one finds estimates of the radii of convergence balls from the information around a solution. There is a plethora of local as well as semilocal convergence results for Newton's method defined above. We refer the reader to Amat et al. (2004), Argyros and Szidarovszky (1993), Argyros (2004, 2007), Argyros and Hilout (2010a, b, 2012), Argyros et al. (2012), Cianciaruso (2007), Ezquerro et al. (2010), Kantorovich and Akilov (1982), Potra, and Pták (1984), Proinov (2010), Rheinboldt (1988), Smale (1986), Wang (1999), Zabrejko and Nguen (1987) and the references therein. The celebrated Kantorovich theorem is an important tool in numerical analysis, e.g. for providing sufficient criteria for the convergence of Newton's method to zeros of polynomials or of systems of nonlinear equations. This theorem is also important in Nonlinear Functional Analysis, where it is also used as a semilocal result for establishing the existence of a solution of a nonlinear equation in an abstract space.

In the present chapter we are being motivated by the work of Cianciaruso (2007) on approximate zeros for the Newton-Kantorovich method and optimization considerations. We show how to extend the applicability of these results under the same computational cost. In particular, we improve the convergence domain and ratio given in earlier studies by Argyros and Szidarovszky (1993), Argyros and Magreñán (2015), Magreñán and Argyros (2015), Cianciaruso (2007), Smale (1986) and Wang (1999).

The chapter is organized as follows: In Sect. 2 we introduce some definitions and state the earlier results as well as the results of this chapter. The semilocal and local analyses are presented in Sect. 3 and Sect. 4, respectively. Numerical examples are presented in the concluding Sect. 5.

2 Preliminaries

It is well known that if the initial point x_0 is close enough to the solution x^*, then sequence $\{x_n\}$ is ultrafast convergent to x^* (Argyros 2007; Argyros et al. 2012; Kantorovich and Akilov 1982). The ultrafast convergence of sequences $\{x_n\}$ is related to the definition of approximate zero introduced by Smale (1986).

Definition 1 A point x_0 is said to be an approximate-type zero of F if $\{x_n\}$ is well defined and there exist A_0 and A such that $0 < A_0 < 1, 0 < A, 0 < A_0A < 1$ and

$$\|x_{n+1} - x_n\| \leq A_0(A_0A)^{2^{n-1}-1}\|x_1 - x_0\| \text{ for each } n \in \mathbb{N}. \tag{2.1}$$

In the literature they use $A_0 = A_0A = 1/2$ (see e.g. Cianciaruso 2007; Smale 1986). In this chapter A_0 is at least as small as A (see the proof of Theorem 3.2 and Remark 1) which leads to more precise estimates on $\|x_{n+1} - x_n\|$.

Clearly, if x_0 is an approximate-type zero for F, then the sequence $\{x_n\}$ is convergent and its limit point x^* is a zero of F, $F(x^*) = 0$. The corresponding definitions by Smale (1986) and Cianciaruso (2007) are

$$\|x_{n+1} - x_n\| \leq \left(\frac{1}{2}\right)^{2^n-1} \|x_1 - x_0\| \text{ for each } n \in \mathbb{N} \tag{2.2}$$

and

$$\|x_{n+1} - x_n\| \leq A^{2^n-1}\|x_1 - x_0\| \text{ for each } n \in \mathbb{N} \tag{2.3}$$

respectively. Notice that the new error estimates can be smaller than the old ones for sufficiently small A_0 and A.

Let $F : \bar{U}(x_0, R) \to \mathcal{Y}$ be analytic and $F'(x_0)^{-1} \in \L(\mathcal{X}, \mathcal{Y})$. Then, Smale (1986) defined

$$\gamma = \gamma(x_0) = \sup_{n>1}\left\|\frac{F'(x_0)^{-1}F^{(n)}(x_0)}{n!}\right\|^{\frac{1}{n-1}}, \tag{2.4}$$

$$\eta = \eta(x_0) = \|F'(x_0)^{-1}F(x_0)\| \tag{2.5}$$

and

$$\alpha = \gamma\eta, \tag{2.6}$$

where $F^{(n)}$ stands for the n-th Fréchet-derivative of operator F.

Smale proved that if

$$\alpha < 0.130707, \tag{2.7}$$

then x_0 is an approximate zero of F. This result does not hold if F is not analytic on $\mathcal{X}$. Later, Rheinboldt (1988) proved that if $F : d \subseteq \mathcal{X} \to \mathcal{Y}$, where $D \subset \mathcal{X}$ is open and

$$\alpha < 0.11909, \tag{2.8}$$

then, sequence $\{x_n\}$ converges. Smale's proof was based on the Newton-Kantorovich theorem (Kantorovich and Akilov 1982).

Theorem 2.1 *Suppose: $F : \bar{U}(x_0, R) \to \mathcal{Y}$ is Fréchet-differentiable on $U(x_0, R)$ and F' is Lipschitz continuous; $F'(x_0)^{-1} \in Ł(\mathcal{Y}, \mathcal{X})$. Set*

$$l = \sup_{x \neq y} \frac{\|F'(x_0)^{-1}(F'(x) - F'(y))\|}{\|x - y\|},$$

$$h = l\eta, \; t^* = \frac{2\eta}{1 + \sqrt{1 - 2h}},$$

$$t_0 = 0, \; t_1 = \eta,$$

$$t_{n+1} = t_n + \tfrac{l(t_n - t_{n-1})^2}{2(1 - l t_n)} \text{ for each } n \in \mathbb{N}$$

where η is defined in (2.5).

If

$$h \leq \frac{1}{2} \tag{2.9}$$

and

$$t^* \leq R, \tag{2.10}$$

then the sequence $\{x_n\}$ generated by the Newton-Kantorovich method (1.1) *is well defined, remains in $\bar{U}(x_0, t^*)$ and converges to x^*. Moreover, the following error estimates hold*

$$\|x_{n+1} - x_n\| \leq \tfrac{l}{2(1 - l t^*)} \|x_n - x_{n-1}\|^2 \text{ for each } n = 1, 2, \ldots$$

Argyros and Szidarovszky (1993) improved the result of Rheinboldt by proving that if

$$\alpha < 0.134854, \tag{2.11}$$

then x_0 is an approximate zero for F replacing $\frac{1}{2\eta}$ for a constant A such that $0 < A < 1$. This result was shown by using the following generalization of Theorem 2.2 (Argyros 2004).

Theorem 2.2 *Under the hypotheses and notations of Theorem 2.2 excluding* (2.9) *and* (2.10)*, set*

$$l_0 = \sup_{x \in \bar{U}(x_0, R)} \frac{\|F'(x_0)^{-1}(F'(x) - F'(x_0))\|}{\|x - x_0\|},$$

$$s_0 = 0,\ s_1 = \eta,$$

$$s_{n+1} = s_n + \frac{l(s_n - s_{n-1})^2}{2(1 - l_0 s_n)} \textit{ for each } n \in \mathbb{N}.$$

If

$$h_0 = \left(\frac{l + l_0}{2}\right)\eta \leq \frac{1}{2}, \tag{2.12}$$

then, scalar sequence $\{s_n\}$ *converges to its unique least upper bound which we denote by* s^*. *If* $s^* \leq R$, *then the sequence* $\{x_n\}$ *is well defined, remains in* $\bar{U}(x_0, s^*)$ *and converges to* x^*. *Moreover, the following error estimates hold*

$$\|x_{n+1} - x_n\| \leq \frac{l\|x_n - x_{n-1}\|^2}{2(1 - l_0\|x_n - x_0\|)} \leq s_{n+1} - s_n, \textit{ for each } n \in \mathbb{N}.$$

In Cianciaruso (2007), Cianciaruso used Theorem 2.3 (i.e. used (2.12) which is weaker than (2.9) if $l_0 < l$) and showed that if

$$\alpha < 0.1582547, \tag{2.13}$$

then x_0 is an approximate zero for F for some A such that $0 < A < 1$. In Wang (1999), Wang used an approach not based on the Kantorovich theorem and showed that, if

$$\alpha < 0.157670781, \tag{2.14}$$

then x_0 is an approximate zero for F for some A such that $0 < A < 1$. In the present chapter we show that if

$$\alpha < 0.164332458249868\ldots, \tag{2.15}$$

then x_0 is an approximate zero for F for some A_0 and A such that $0 < A_0 < 1, 0 < A$ and $0 < A_0 A < 1$. Notice that (2.15) improves the earlier results. The proof is based on the following refinement of our Theorem 2.3 in Argyros and Hilout (2012).

Theorem 2.3 *Under the hypotheses and notations of Theorem 2.3 excluding* (2.12), *set*

$$r_0 = 0,\ r_1 = \eta, r_2 = r_1 + \frac{l_0(r_1 - r_0)^2}{2(1 - l_0 r_1)},$$

$$r_n = r_{n-1} + \frac{l(r_n - r_{n-1})^2}{2(1 - l_0 r_n)} \textit{ for each } n = 3, 4, \ldots.$$

If

$$h_1 = L_1\eta \leq \frac{1}{2}, \tag{2.16}$$

where

$$L_1 = \frac{1}{8}\left(4l_0 + \sqrt{l_0 l} + \sqrt{l_0 l + 8l_0^2}\right),$$

then the scalar sequence $\{r_n\}$ *converges to its unique least upper bound which we denote by* r^*. *If* $r^* \leq R$, *then the sequence* $\{x_n\}$ *is well defined, remains in* $\bar{U}(x_0, r^*)$ *and converges to* x^*. *Moreover, the following error estimates hold*

$$\|x_{n+1} - x_n\| \leq \frac{l_1 \|x_n - x_{n-1}\|^2}{2(1 - l_0\|x_n - x_0\|)} \leq r_{n+1} - r_n \text{ for each } n = 1, 2, \ldots,$$

where

$$l_1 = \begin{cases} l_0 & \text{if } n = 1 \\ l & \text{if } n > 1. \end{cases}$$

It is worth noticing that $l_0 \leq l$, $\frac{l}{l_0}$ can be arbitrarily large (Argyros 2004, 2007; Argyros and Hilout 2010b; Argyros et al. 2012) and

$$h \leq \frac{1}{2} \Rightarrow h_0 \leq \frac{1}{2} \Rightarrow h_1 \leq \frac{1}{2}$$

but not necessarily vice versa unless $l_0 = l$. We have that

$$\frac{h_0}{h} \to \frac{1}{2}, \quad \frac{h_1}{h} \to 0 \text{ and } \frac{h_1}{h_0} \to 0 \text{ as } \frac{l_0}{l} \to 0.$$

We have that the preceding implications show by how many times (at most) the results of Theorem 2.3 expand the results of Theorem 2.2 which in turn expand the results of the Kantorovich Theorem 2.1. It is expected that since (2.16) is weaker than (2.12) used in Cianciaruso (2007), we can obtain a better result than (2.11). This is the first advantage of our approach.

Moreover, concerning the error estimates sequence $\{r_n\}$ is tighter than $\{s_n\}$ (used in Cianciaruso 2007) which is tighter than $\{t_n\}$ (Argyros 2004, 2007; Argyros and Hilout 2012). Concerning to the local convergence of Newton's method, we need the following definition for approximate zeros.

Definition 2 Let $F : \mathcal{X} \to \mathcal{Y}$ and let x^* be a zero of F. A point x_0 is said to be an approximate-type zero of second kind of F if $\{x_n\}$ is well defined and there exist A_0 and A such that $0 < A_0 < 1$, $0 < A$, $0 < A_0 A < 1$ and

$$\|x_{n+1} - x^*\| \leq A_0 (A_0 A)^{2^{n-1}-1} \|x_0 - x^*\|^2 \text{ for each } n \in \mathbb{N}. \tag{2.17}$$

The corresponding definitions given by Smale and Cianciaruso are

$$\|x_n - x^*\| \leq \left(\frac{1}{2}\right)^{2^n - 1} \|x_0 - x^*\| \text{ for each } n \in \mathbb{N}$$

and

$$\|x_n - x^*\| \leq A^{2^n - 1}\|x_0 - x^*\| \text{ for each } n \in \mathbb{N}, \tag{2.18}$$

respectively.

Notice that the new error estimates can be smaller than the old ones for sufficiently small A_0 and A.

Suppose that $F'(x^*)^{-1} \in Ł(\mathcal{Y}, \mathcal{X})$. Let

$$\gamma^* = \gamma(x^*) = \sup_{n>1}\|\frac{F'(x^*)^{-1}F^{(n)}(x^*)}{n!}\|^{\frac{1}{n-1}}. \tag{2.19}$$

In Smale (1986), Smale showed that if $F : \mathcal{X} \to \mathcal{Y}$ is analytic and

$$\gamma\|x - x^*\| \leq \frac{5 - \sqrt{17}}{4} \approx 0.219224, \tag{2.20}$$

then x_0 is an approximate zero of second kind. Later, Argyros and Szidarovszky (1993) showed that if $F : \bar{U}(x^*, R) \to \mathcal{Y}$ is analytic and

$$\gamma\|x - x^*\| \leq 0.20629947, \tag{2.21}$$

then x_0 is an approximate zero of second kind. Later, In Cianciaruso (2007), Cianciaruso improved Argyros's result. Cianciaruso showed that, if

$$\gamma\|x - x^*\| \leq 0.2390211, \tag{2.22}$$

then x_0 is an approximate zero of second kind. In the present chapter we show that, if

$$\gamma\|x - x^*\| \leq \alpha_0^* = 0.2489069896460221 \ldots \ldots, \tag{2.23}$$

then x_0 is an approximate zero of second kind. Clearly, (2.23) improves the earlier results. It is worth noticing that the ratio of convergence is also improved, if $A_0 < A$.

3 Semilocal Convergence

Let $F : \bar{U}(x_0, R) \to \mathcal{Y}$ be Fréchet differentiable on $U(x_0, R)$. We assume that there exists an increasing function $\varphi : [0, R] \to [0, +\infty)$ and $x_0 \in \mathcal{X}$ such that $F'(x_0)^{-1} \in Ł(\mathcal{Y}, \mathcal{X})$ and

$$\|F'(x_0)^{-1}[F'(x) - F'(y)]\| \leq \varphi(r)\|x - y\| \text{ for each } x, y \in \bar{U}(x_0, r), 0 < r \leq R. \tag{3.1}$$

It follows from (3.1) that there exists an increasing function $\varphi_0 : [0, R] \to [0, +\infty)$ such that

$$\|F'(x_0)^{-1}[F'(x)-F'(x_0)]\| \le \varphi_0(r)\|x-x_0\| \text{ for each } x \in \bar{U}(x_0,r), 0<r\le R. \tag{3.2}$$

Let

$$r_0 := sup\{t \in (0,+\infty) : \varphi_0(t)t < 1\}$$

It also follows from (3.1) that there exists an increasing function $\varphi_1 : [0,R] \to [0,+\infty)$ such that

$$\|F'(x_0)^{-1}[F'(x)-F'(y)]\| \le \varphi_1(r)\|x-y\| \text{ for each } x,y \in \bar{U}(x_0,r)\cap U(x_0,r_0) \tag{3.3}$$

Clearly,

$$\varphi_0(r) \le \varphi(r) \text{ for each } r \in [0,R] \tag{3.4}$$

and

$$\varphi_1(r) \le \varphi(r) \text{ for each } r \in [0,R] \tag{3.5}$$

hold in general and $\frac{\varphi}{\varphi_0}$ can be arbitrarily large (Argyros 2004, 2007; Argyros and Hilout 2010a, b, 2012; Argyros et al. 2012). Notice that the computation of function φ requires the computation of φ_0 or φ_1 as special cases. Hence, (3.2) or (3.3) are not additional hypotheses to (3.1).

We need the following version of a theorem by Zabrejko and Nguen (1987)

Theorem 3.1 *Suppose that* (3.3) *is satisfied. Set*

$$\psi_1(r) = \eta - r + \int_0^r (r-t)\varphi_1(t)dt, \; \eta = \|F'(x_0)^{-1}F(x_0)\|. \tag{3.6}$$

Suppose that ψ_1 *has a minimal zero denoted by* $\bar{r}$ *in* $(0,R]$. *Then, sequence* $\{x_n\}$ *is well defined, remains in* $\bar{U}(x_0,\bar{r})$ *and converges to* x^*.

Next, we show the main semilocal convergence result for Newton's method and approximate zeros.

Theorem 3.2 (Argyros and Magreñán 2015) *Let* $F : \bar{U}(x_0,R) \to \mathcal{Y}$ *be Fréchet-differentiable on* $U(x_0,r)$ *with* F' *satisfying* (3.2) *and* (3.3). *Moreover, suppose that the function* ψ_1 *given in* (3.6) *has a minimal zero denoted by* $\bar{r}$ *in* $(0,R]$. *Furthermore, suppose that*

$$\int_0^{\bar{r}} \varphi_0(t)dt < 1, \tag{3.7}$$

$$\begin{aligned}&\int_0^v (2t-v)\varphi_0(\bar{r}+t)dt \ge 0,\\ &\int_0^v (2t-v)\varphi_1(\bar{r}+t)dt \ge 0, \qquad (3.8)\\ &\varphi_0(v) \le \varphi_1(v), \quad \textit{for } 0 \le v \le \bar{r},\end{aligned}$$

$$A_0 = \frac{\eta \int_0^{\bar{r}} (\bar{r} - t)\varphi_0(\bar{r} + t)dt}{\bar{r}^2 \left(1 - \int_0^{\bar{r}} \varphi_0(t)dt\right)} \tag{3.9}$$

and

$$A = \frac{\eta \int_0^{\bar{r}} (\bar{r} - t)\varphi_1(\bar{r} + t)dt}{\bar{r}^2 \left(1 - \int_0^{\bar{r}} \varphi_0(t)dt\right)}. \tag{3.10}$$

Then x_0 is an approximate-type zero for F. That is, sequence $\{x_n\}$ is well defined. remains in $\bar{U}(x_0, \bar{r})$ and converges to x^*. Moreover, the following error estimates hold

$$\|x_{n+1} - x_n\| \le A_0 (A_0 A)^{2^{n-1}-1} \|x_1 - x_0\| \text{ for each } n \in \mathbb{N}.$$

Proof Simply replace function φ by function φ_1 in the proof in Argyros and George (2015) and notice that the iterates lie in $\bar{U}(x_0, r) \cap U(x_0, r_0)$ which is a more precise information on the location of the iterates, since $\bar{U}(x_0, r) \cap U(x_0, r_0) \subseteq \bar{U}(x_0, r)$. □

Remark 1 (a) If $\varphi_0 = \varphi = \varphi_1$, then Theorem 3.2 reduces to Theorem 3.2 in Argyros and Magreñán (2015). Otherwise, i.e., if $\varphi_0 < \varphi = \varphi_1$ it constitutes an improvement of the work in Argyros and George (2015). Moreover, if $\varphi_1 < \varphi$, then the new result improves both old works in Argyros and George (2015); Argyros and Magreñán (2015).

(b) The results can be improved even further, if we replace $\bar{U}(x_0, r)$ by $\bar{U}(x_1, r - \|x_1 - x_0\|)$ in the hypotheses (3.1)–(3.3). Then, the corresponding functions "φ" will be even tighter, since $\bar{U}(x_1, r - \|x_1 - x_0\|) \subseteq \bar{U}(x_0, r)$.

Next, we apply Theorem 3.2 to an operator F analytic on $U(x_0, R)$, we need the following lemma from Cianciaruso (2007).

Lemma 3.3 *Let $F : U(x_0, R) \to \mathcal{Y}$ be an operator analytic at interior points of $U(x_0, R)$. Suppose $F'(x_0)^{-1} \in Ł(\mathcal{Y}, \mathcal{X})$. Then, F satisfies the two following conditions*

$$\|F'(x_0)^{-1}[F'(x) - F'(y)]\| \le \varphi(r)\|x - y\| \text{ for each } x, y \in \bar{U}(x_0, r) \cap U(x_0, r_0), 0 < r \le R,$$

with

$$\varphi_1(r) = \frac{2\gamma_1}{(1 - \gamma_1 r)^3}; \tag{3.11}$$

$$\|F'(x_0)^{-1}[F'(x) - F'(x_0)]\| \le \varphi_0(r)\|x - x_0\|, \text{ for each } x \in \bar{U}(x_0, r) \cap U(x_0, r_0), 0 < r \le R,$$

with

$$\varphi_0(r) = \frac{\gamma_1(2-\gamma_1 r)}{(1-\gamma_1 r)}. \tag{3.12}$$

Notice that $\gamma_1 \leq \gamma$, *since (3.4) or (3.5) hold, and* $\bar{U}(x_0, r) \cap U(x_0, r_0) \subseteq \bar{U}(x_0, r)$.

To prove the convergence of the method for an operator F analytic at interior points of $U(x_0, R)$ it is sufficient to apply Theorem 3.2. In fact, function ψ_1 becomes

$$\psi_1(r) = \alpha - 2r + \frac{r}{(1-\gamma_1 r)^2}, \left(0 \leq r \leq \frac{1}{\gamma_1}\right).$$

The function ψ_1 admits at least a zero in its domain if, and only if, $\alpha \leq 3 - 2\sqrt{2}$ and

$$\bar{r} = \bar{r}(\alpha) = \frac{1+\alpha-\sqrt{\alpha^2-6\alpha+1}}{4\gamma_1},$$

is the smallest zero of ψ_1 (unique if $\alpha \leq 3-2\sqrt{2}$). Then, if $\alpha \leq 3-2\sqrt{2}$, the Newton–Kantorovich approximations are well defined for all $n \in \mathbb{N}$, converge to a zero x^* of F. Now we can prove our theorem on approximate zeros of operators analytic on $U(x_0, R)$.

Proposition 1 *Let* $F : \bar{U}(x_0, R)$ *by analytic on* $U(x_0, R)$. *Suppose that* $F'(x_0)^{-1} \in$ $Ł(\mathcal{Y}, \mathcal{X})$,

$$\alpha = \eta\gamma_1 < \alpha_0 = 0.164332458249868\ldots$$

and

$$\bar{r} \leq R.$$

Then x_0 *is an approximate-type zero for* F. *That is, sequence* $\{x_n\}$ *is well defined, remains in* $\bar{U}(x_0, R)$ *and converges to* x^*. *Moreover, the following error estimates hold*

$$\|x_{n+1} - x_n\| \leq A_0(A_0 A)^{2^{n-1}-1}\|x_1 - x_0\|,$$

where

$$g(r) = \frac{1+r-\sqrt{r^2-6r+1}}{4},$$

$$A_0(r) = \frac{r\left[2(1-g(r))\log\frac{1-g(r)}{1-2g(r)} - g(r)\right]}{g(r)\left[1-2g(r)+(1-g(r))\ln(1-g(r))\right]},$$

$$A(r) = \frac{r}{(1-g(r))(1-2g(r))\left[1-2g(r)+(1-g(r))\ln(1-g(r))\right]},$$

$$A_0 = A_0(\alpha)$$

and

$$A = A(\alpha).$$

Proof Using weaker hypothesis (2.14) instead of (2.10) used in Cianciaruso (2007) accordingly to our Theorem 2.4 it suffices to show the estimates

$$0 \le A_0(r)A(r) < 1,$$

$$0 \le A_0(r) < 1$$

and

$$0 \le A_0(r) \le A(r)$$

which holds true for each $r \in [0, \alpha_0)$. Indeed, the verification of these estimates can be seen from the Figs. 1 and 2. ■

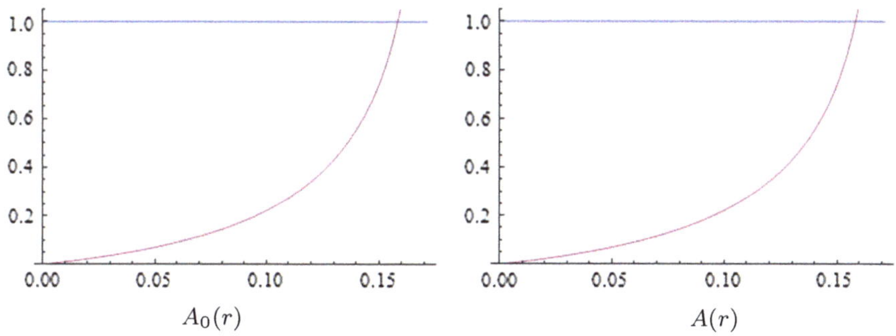

Fig. 1 In the *left hand* it appears the graphic of $A_0(r)$, and in the *right hand* the graphic of $A(r)$, in both cases it is printed the line 1

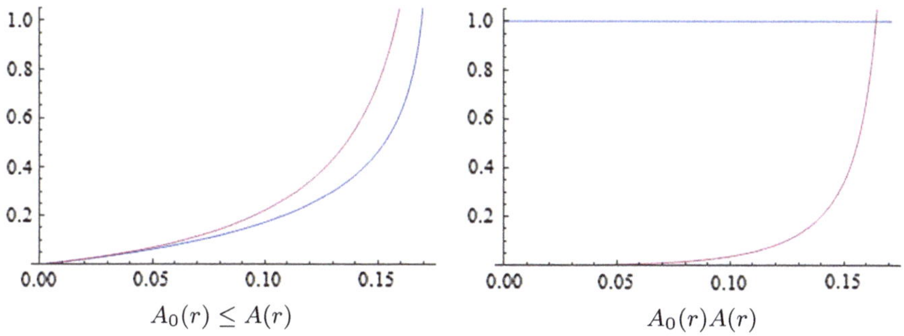

Fig. 2 In the *left hand* it appears the graphic of $A_0(r)$ and $A(r)$ in which we observe that $A_0(r) \le A(r)$. In the *right hand* the graphic of $A_0(r)A(r)$

4 Local Convergence

We suppose that F has a zero x^*. In an analogous way to Sect. 3 we show the local convergence results for Newton's method and approximate zeros.

Theorem 4.1 *Let* $F : \bar{U}(x^*, R) \to \mathcal{Y}$ *be Fréchet-differentiable on* $U(x^*, R)$ *with* $F'(x^*)^{-1} \in Ł(\mathcal{Y}, \mathcal{X})$ *and* $F(x^*) = 0$. *Suppose that there exist increasing functions* $\bar{\varphi}^*, \varphi^*, \bar{\varphi}_1^* : [0, R] \to [0, +\infty)$ *such that* F' *satisfies*

$$\|F'(x^*)^{-1}[F'(x) - F'(y)]\| \le \bar{\varphi}^*(r)\|x - y\| \text{ for each } x, y \in \bar{U}(x^*, r), 0 < r \le R. \tag{4.1}$$

$$\|F'(x^*)^{-1}[F'(x) - F'(x^*)]\| \le \varphi^*(r)\|x - x^*\| \text{ for each } x \in \bar{U}(x^*, r), 0 < r \le R. \tag{4.2}$$

and

$$\|F'(x^*)^{-1}[F'(x) - F'(y)]\| \le \bar{\varphi_1^*}(r)\|x - y\| \text{ for each } x, y \in \bar{U}(x^*, r) \cap U(x^*, r_1), \tag{4.3}$$

where

$$r_1 = sup\{t \in (0, +\infty) : \varphi^*(t)t < 1\}.$$

Moreover suppose that there exists $r \in [0, R]$ *such that*

$$\int_0^r \varphi^*(t)dt < 1, \tag{4.4}$$

$$v^2\varphi^*(v) - 2\int_0^r t\varphi^*(t)dt \ge 0, v^2\bar{\varphi}^*(v) - 2\int_0^r t\bar{\varphi}_1{}^*(t)dt \ge 0 \text{ for each } 0 \le v \le r, \tag{4.5}$$

$$A_0 = \frac{\int_0^r t\varphi^*(t)dt}{r(1 - \int_0^r \varphi^*(t)dt)} < 1 \tag{4.6}$$

and

$$A = \frac{\int_0^r t\bar{\varphi}_1{}^*(t)dt}{r(1 - \int_0^r \varphi^*(t)dt)} < 1. \tag{4.7}$$

Then, x_0 *is an approximate-type zero of second kind. That is, sequence* $\{x_n\}$ *is well defined and converges to* x^*. *Moreover, the following estimates hold*

$$\|x_n - x^*\| \le A_0(A_0A)^{2^{n-1}-1}\|x_0 - x^*\|^2 \text{ for each } n \in \mathbb{N}.$$

Proof It follows from (4.4) that functions $\bar{\rho}^*$ and ρ^* given by

$$\bar{\rho}^*(v) = \frac{\int_0^v t\bar{\varphi}_1{}^*(t)dt}{v^2}$$

and

$$\rho^*(v) = \frac{\int_0^v t\varphi^*(t)dt}{v^2}$$

are increasing. We shall show using induction on n that

$$\|x_n - x^*\| \leq \frac{\int_0^r t\overline{\overline{\varphi}}_1^*(t)dt}{r^2(1 - \int_0^r t\varphi^*(t)dt)}\|x_{n-1} - x^*\|^2 \text{ for each } n \in \mathbb{N}, \tag{4.8}$$

where

$$\bar{\bar{\varphi}} = \begin{cases} \varphi^* & \text{if } n = 1 \\ \bar{\varphi}_1{}^* & \text{if } n > 1. \end{cases}$$

It follows from (4.3) that

$$\int_0^{\|x-x^*\|} \varphi^*(t)dt < 1, \tag{4.9}$$

implies that the operator $F'(x^*)^{-1}F'(x_0)$ is invertible and

$$\|F'(x^*)^{-1}F'(x_0)\| \leq \frac{1}{1 - \int_0^{\|x_0-x^*\|} \varphi^*(t)dt}.$$

Then for $n = 1$, we have

$$\begin{aligned}
\|x_1 - x^*\| &= \|x_0 - x^* - F'(x_0)^{-1}(F(x_0) - F(x^*))\| \\
&\leq \|F'(x_0)^{-1}F(x^*)\|\|F'(x^*)^{-1}[F(x_0) - F(x^*) - F'(x_0)(x_0 - x^*)]\| \\
&\leq \frac{1}{1 - \int_0^{\|x_0-x^*\|} \varphi^*(t)dt} \\
&\quad \times \int_0^1 \|F'(x^*)^{-1}[F'((1-s)x_0 + sx^*) - F'(x_0)]\|ds\|x_0 - x^*\| \\
&\leq \frac{1}{1 - \int_0^{\|x_0-x^*\|} \varphi^*(t)dt} \int_0^1 ds \int_{(1-s)\|x_0-x^*\|}^{\|x_0-x^*\|} \bar{\varphi}_1^* dt\|x_0 - x^*\| \\
&= \frac{1}{1 - \int_0^{\|x_0-x^*\|} \varphi^*(t)dt} \int_0^{\|x_0-x^*\|} t\bar{\varphi}_1^* dt \\
&\leq \frac{\int_0^r t\bar{\varphi}_1^* dt}{r^2(1 - \int_0^r \varphi^*(t)dt)}\|x_0 - x^*\|^2,
\end{aligned}$$

since function ρ^* is increasing. Then $x_2 \in U(x^*, R)$, $F'(x_2)^{-1} \in Ł(\mathcal{Y}, \mathcal{X})$ and

$$\|F'(x^*)^{-1}F'(x_2)\| \le \frac{1}{1-\int_0^{\|x_2-x^*\|}\varphi^*(t)dt}.$$

Moreover, we have

$$\begin{aligned}
\|x_3 - x^*\| &= \|x_2 - x^* - F'(x_2)^{-1}(F(x_2) - F(x^*))\| \\
&= \|F'(x_2)^{-1}[F(x_2) - F(x^*) - F'(x_2)(x_2 - x^*)]\| \\
&\le \|F'(x_2)^{-1}F(x^*)\| \|F'(x^*)^{-1}[F(x_2) - F(x^*) - F'(x_2)(x_2 - x^*)]\| \\
&\le \|F'(x_2)^{-1}F(x^*)\| \\
&\quad \times \int_0^1 \|F'(x^*)^{-1}[F'((1-s)x_2 + sx^*) - F'(x_2)]\|ds\|x_2 - x^*\| \\
&\le \frac{1}{1-\int_0^{\|x_2-x^*\|}\varphi^*(t)dt}\int_0^1 ds\int_{(1-s)\|x_2-x^*\|}^{\|x_2-x^*\|}\bar{\varphi}_1^*(t)dt\|x_2 - x^*\| \\
&\le \frac{\int_0^r t\bar{\varphi}_1^*(t)dt}{r^2(1-\int_0^r\varphi^*(t)dt)}\|x^* - x_2\|^2,
\end{aligned}$$

and $\|x_3 - x^*\| \le A\|x_2 - x^*\|$. Suppose that (4.7) holds for $n = 2, 3, \ldots$. We shall show that it holds for $n + 1$.

$$\begin{aligned}
\|x_{n+1} - x^*\| &= \|x_n - x^* - F'(x_n)^{-1}(F(x_n) - F(x^*))\| \\
&= \|F'(x_n)^{-1}[F(x_n) - F(x^*) - F'(x_n)(x_n - x^*)]\| \\
&\le \|F'(x_n)^{-1}F(x^*)\| \|F'(x^*)^{-1}[F(x_n) - F(x^*) - F'(x_n)(x_n - x^*)]\| \\
&\le \|F'(x_n)^{-1}F(x^*)\| \\
&\quad \times \int_0^1 \|F'(x^*)^{-1}[F'((1-s)x_n + sx^*) - F'(x_n)]\|ds\|x_n - x^*\| \\
&\le \frac{1}{1-\int_0^{\|x_n-x^*\|}\varphi^*(t)dt}\int_0^1 ds\int_{(1-s)\|x_n-x^*\|}^{\|x_n-x^*\|}\bar{\varphi}_1{}^*(t)dt\|x_n - x^*\| \\
&\le \frac{\int_0^r t\bar{\varphi}_1{}^*(t)dt}{r^2(1-\int_0^r\varphi^*(t)dt)}\|x^* - x_n\|^2,
\end{aligned}$$

The induction is complete. Then, using (4.5) we have for $n = 1$,

$$\|x_1 - x^*\| \le A_0^{2^1-1}\|x_0 - x^*\|,$$

then, for $n = 2$, we have that

$$\begin{aligned}\|x_2 - x^*\| &\le \tfrac{A}{\eta}\|x_1 - x^*\|^2\\ &\le A^{2^{2-1}-1}A_0^{2^{2-1}}\|x_0 - x^*\|^2\end{aligned}$$

then, continuing with the process we obtain that

$$\begin{aligned}\|x_n - x^*\| &\le A^{2^{n-1}-1}A_0^{2^{n-1}}\|x_0 - x^*\|^2\\ &\le \tfrac{1}{A}(A_0A)^{2^{n-1}}\|x_0 - x^*\|^2\\ &\le A_0(A_0A)^{2^{n-1}-1}\|x_0 - x^*\|^2.\end{aligned}$$

■

In order to apply Theorem 4.1 to analytic operators, we need the following Lemma given in Cianciaruso (2007).

Lemma 4.2 *Let $F : U(x^*, R) \to \mathcal{Y}$ be an operator analytic on $U(x^*, R)$ with $F(x^*) = 0$, $F'(x^*)^{-1} \in Ł(\mathcal{Y}, \mathcal{X})$ and $\gamma_1^*R < 1$. Then, F' satisfies* (4.1) *and* (4.2) *with*

$$\varphi^*(r) = \frac{\gamma_1^*(2 - \gamma_1^* r)}{(1 - \gamma_1^* r)^2}$$

and

$$\bar{\varphi_1}^*(r) = \frac{2\gamma_1^*}{(1 - \gamma_1^* r)^3}.$$

Notice that $\gamma_1^* \le \gamma^*$.

Then, we can apply Theorem 4.1 for analytic operators to obtain:

Proposition 2 *Let $F : \bar{U}(x^*, R)$ by analytic on $U(x^*, R)$ with $F(x^*) = 0$, $F'(x^*)^{-1} \in Ł(\mathcal{Y}, \mathcal{X})$,*

$$\gamma^* r < \alpha_0^* = 0.2489069896460221\ldots.$$

Then x_0 is an approximate-type zero of second kind for F. That is, sequence $\{x_n\}$ is well defined, and converges to x^*. Moreover, the following error estimates hold

$$\|x_n - x^*\| \leq A_0 (A_0 A)^{2^{n-1}-1} \|x_0 - x^*\|, \textit{ for each } n \in \mathbb{N},$$

where

$$A_0(r) = \frac{r}{1 - 2r + (1 - r)\ln(1 - r)},$$

$$A(r) = \frac{r}{(1 - r)[1 - 2r + (1 - r)\ln(1 - r)]}$$

$$A_0 = A_0(\gamma_1^* r)$$

and

$$A = A(\gamma_1^* r).$$

Proof As in the proof of Theorem 3.3 in Cianciaruso (2007) it suffices to show the estimates

$$0 \leq A_0(r)A(r) < 1,$$

$$0 \leq A_0(r) < 1$$

and

$$0 \leq A_0(r) \leq A(r)$$

which hold true for each $r \in [0, \alpha_0^*)$. Indeed, the verification of the estimates can be seen from the Figs. 3 and 4. ∎

Remark 2 (a) Notice that (4.1) implies (4.2),

$$\varphi^*(r) \leq \bar{\varphi}^*(r) \text{ for each } r \in [0, R]$$

and $\frac{\bar{\varphi}^*}{\varphi^*}$ can be arbitrarily large (Argyros 2004, 2007; Argyros and Hilout 2010b; Argyros et al. 2012).

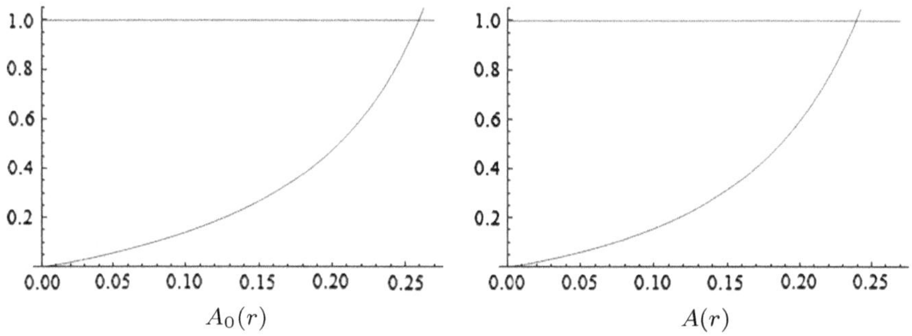

Fig. 3 In the *left hand* it appears the graphic of $A_0(r)$, and in the *right hand* the graphic of $A(r)$, in both cases it is printed the line 1

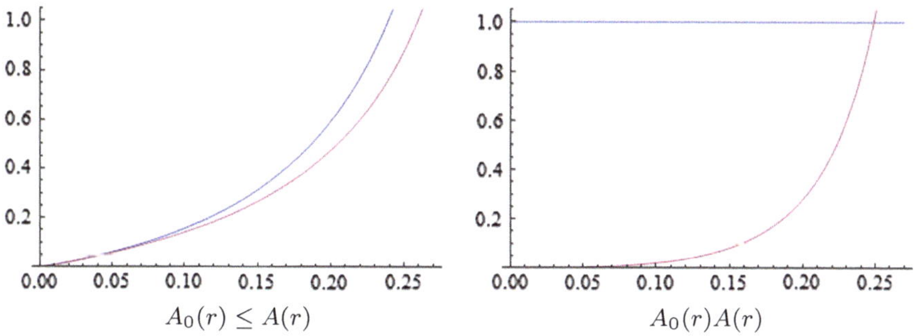

Fig. 4 In the *left hand* it appears the graphic of $A_0(r)$ and $A(r)$ in which we observe that $A_0(r) \leq A(r)$. In the *right hand* the graphic of $A_0(r)A(r)$

(b) If $\varphi^* = \varphi_1^* = \bar{\varphi}_1^*$, then Theorem 4.1 reduces to Theorem 4.2 in Cianciaruso (2007). Otherwise, i.e., if $\varphi^* < \bar{\varphi}^*$ it constitutes an improvement over the results in Argyros and George (2015). If $\varphi_1^* \leq \bar{\varphi}^*$ then new result is better than old one (Argyros and Magreñán 2015).

5 Numerical Examples

We present two examples one for the semilocal case and another for the local case.

Example 1 Let $\mathcal{X} = \mathcal{Y} = \mathbb{R}$ and consider the real function

$$F(x) = x^3 - a$$

with $a \in [0, 1]$ and we are going to apply Newton's method.

First of all, it is easy to see that the derivatives of F are:

$$F'(x) = 3x^2,$$

$$F''(x) = 6x,$$

$$F'''(x) = 6,$$

and the following ones are zero. Moreover we obtain:

$$\varphi(t) = 2(2 - a),$$

$$\varphi_0(t) = 3 - a,$$

$$\varphi_1(t) = 2(1 + \frac{1}{3-a})$$

and

$$r_0 = \frac{1}{3-a}.$$

We choose the starting point $x_0 = 1$ and we consider the domain $\Omega = B(x_0, 1)$. In this case, we obtain

$$\eta = |F'(x_0)^{-1}F(x_0)| = \frac{|1-a|}{3},$$

and considering the following

$$\frac{F'(x_0)^{-1}F''(x_0)}{2} = 1\ldots,$$

$$\frac{F'(x_0)^{-1}F'''(x_0)}{6} = 0.693361\ldots,$$

and the next ones are null, we obtain

$$\gamma = 1,$$

$$\alpha = \eta\gamma = \frac{|1-a|}{3}.$$

We see that with our new conditions we can ensure the convergence for $0.507003 < a < 0.525236$, for which the conditions (2.11), (2.13) and (2.14) are not satisfied but condition (2.15) is satisfied, since $\alpha < 0.164332458$. So we can ensure the convergence to the solution by means of applying our conditions.

Example 2 Let $X = Y = \mathbb{R}^3$, $D = \overline{U}(0, 1)$ and $u^* = (0, 0, 0)^T$. Define function F on D for $w = (x, y, z)^T$ by

$$F(w) = (e^x - 1, \frac{e-1}{2}, y^2 + y, z)^T. \tag{5.1}$$

Then, the Fréchet derivative of F is given by

$$F'(w) = \begin{pmatrix} e^x & 0 & 0 \\ 0 & (e-1)y+1 & 0 \\ 0 & 0 & 1 \end{pmatrix}$$

We choose $w_0 = (0, 0.4, 0)^T$. Then, we obtain that

$$F'(w_0)^{-1}F(w_0) = \begin{pmatrix} 0 \\ 0.318532 \\ 0 \end{pmatrix}$$

$$F''(w_0)^{-1}F(w_0) = \begin{pmatrix} 1\,0\,0\,0 & 0 & 0\,0\,0\,0 \\ 0\,0\,0\,0 & 1.01835 & 0\,0\,0\,0 \\ 0\,0\,0\,0 & 0 & 0\,0\,0\,0 \end{pmatrix}$$

$$F'''(w_0)^{-1}F(w_0) = \begin{pmatrix} 1 & 0 & \cdots & 0 \\ 0 & 0 & \cdots & 0 \\ 0 & 0 & \cdots & 0 \end{pmatrix}$$

Moreover we obtain:

$$\varphi^*(t) = (e-1),$$

$$\varphi_1^*(t) = e^{\left(\frac{1}{e-1}\right)},$$

$$\bar{\varphi}^*(t) = e,$$

$$r_1 = \frac{1}{e-1}$$

and

$$r = 1.$$

And it is easy to see that

$$\eta = |F'(w_0)^{-1}F(w_0)| = 0.318532\ldots,$$

$$\gamma = \frac{|F'(w_0)^{-1}F(w_0)|}{2} = 0.509177\ldots$$

and

$$\alpha = \eta\gamma = 0.162189\ldots$$

Notice that the conditions (2.11), (2.13) and (2.14) are not satisfied but condition (2.15) is satisfied, since $\alpha < 0.164332458$. So we can ensure the convergence to the solution by means of applying our conditions.

Acknowledgements This research was supported by Universidad Internacional de La Rioja (UNIR, http://www.unir.net), under the Plan Propio de Investigación, Desarrollo e Innovación [2015–2017]. Research group: Modelación matemática aplicada a la ingeniería(MOMAIN), by the grant SENECA 19374/PI/14 and by Ministerio de Ciencia y Tecnología MTM2014-52016-C2-{01}-P.

References

Amat, S., Busquier, S., & Negra, M. (2004). Adaptive approximation of nonlinear operators. *Numerical Functional Analysis and Optimization*, *25*, 397–405.

Argyros, I. K., & Szidarovszky, F. (1993). *The theory and applications of iteration methods*, Systems Engineering Series. Boca Ratón, Florida: CRC Press.

Argyros, I. K. (2004). On the Newton-Kantorovich hypothesis for solving equations. *Journal Computational Applied Mathematics*, *169*, 315–332.

Argyros, I. K. (2007). In Chui, C. K. & Wuytack, L. (Eds.), *Computational theory of iterative methods* (Vol. 15), Series: Studies in Computational Mathematics. New York: Elsevier Publishing Company.

Argyros, I. K., & Hilout, S. (2010a). A convergence analysis of Newton-like methods for singular equations unsig recurrent functions. *Numerical Functional Analysis and Optimization*, *31*(2), 112–130.

Argyros, I. K., & Hilout, S. (2010b). Extending the Newton-Kantorovich hypothesis for solving equations. *Journal of Computational Applied Mathematics*, *234*, 2993–3006.

Argyros, I. K., & Hilout, S. (2012). Weaker conditions for the convergence of Newton's method. *Journal of Complexity*, *28*, 364–387.

Argyros, I. K., Cho, Y. J., & Hilout, S. (2012). *Numerical method for equations and its applications*. New York: CRC Press/Taylor and Francis.

Argyros, I. K., & George, S. (2015). Ball convergence for Steffensen-type fourth-order methods. *IJIMAI*, *3*(4), 27–42.

Argyros, I. K., & Magreñán, Á. A. (2015). Extended convergence of Newton-Kantorovich method to an approximate zero. *Journal of Computational and Applied Mathematics*, *286*, 54–67.

Cianciaruso, F. (2007). Convergence of Newton-Kantorovich approximations to an approximate zero. *Numerical Functional Analysis and Optimization*, *28*(5–6), 631–645.

Ezquerro, J. A., Gutiérrez, J. M., Hernández, M. A., Romero, N., & Rubio, M. J. (2010). The Newton method: from Newton to Kantorovich. (Spanish), *Gac. R. Soc. Mat. Esp.*, *13*(1), 53–76.

Kantorovich, L. V., & Akilov, G. P. (1982). *Functional analysis*. Oxford: Pergamon Press.

Magreñán, Á. A., & Argyros, I. K. (2015). An extension of a theorem by Wang for Smale's α-theory and applications. *Numerical Algorithms*, *68*(1), 47–60.

Magreñán, Á. A. (2014a). Different anomalies in a Jarratt family of iterative root-finding methods. *Applied Mathematics and Computation*, *233*, 29–38.

Magreñán, Á. A. (2014b). A new tool to study real dynamics: The convergence plane. *Applied Mathematics and Computation*, *248*, 215–224.

Potra, F. A., & Pták, V. (1984). *Nondiscrete induction and iterative processes* (Vol. 103). Research Notes in Mathematics. Boston, Massachusetts: Pitman (Advanced Publishing Program).

Proinov, P. D. (2010). New general convergence theory for iterative processes and its applications to Newton-Kantorovich type theorems. *Journal of Complexity*, *26*, 3–42.

Rheinboldt, W. C. (1988). On a theorem of S. Smale about Newton's method for analytic mappings. *Applied Mathematics Letter*, *1*, 69–72.

Smale, S. (1986). newton's method estimates from data at one point. In R. Ewing, K. Gross, & C. Martin (Eds.), *The merging of disciplines: new directions in pure* (pp. 185–196). Applied and Computational Mathematics. New York: Springer.

Wang, X. H. (1999). Convergence of Newton's method and inverse function theorem in Banach spaces. *Mathematics of Compattion*, *68*, 169–186.

Zabrejko, P. P., & Nguen, D. F. (1987). The majorant method in the theory of Newton-Kantorovich approximations and the Pták error estimates. *Numerical Functional Analysis Optimization*, *9*, 671–684.

The Non-symmetric L-Nash Bargaining Solution

Ferenc Forgó

Dedicated to Ferenc Szidarovszky for his academic and research achievements in the last 50 years.

Abstract It is demonstrated how the concept of the Limit-Nash bargaining solution as defined in Forgó and Szidarovszky (Eur J Oper Res 147:108–116, 2003) can be carried over to the non-symmetric case. It is studied how externally given weights of the players and the relative magnitude of penalties for not being able to come to an agreement influence the solution.

1 Introduction

The Nash bargaining solution as introduced by Nash (1950) is a fundamental concept in game theory and conflict resolution. In its most simple form it is about two players trying to come to an agreement on choosing an element from a given set of feasible outcomes. The outcomes are evaluated according to the individual utility functions of the players. Normally, this leads to a conflict. To resolve the conflict, mutual concessions have to be made, otherwise a bad outcome (disagreement outcome) will realize where both players are penalized for not having been able to agree. Nash approached the problem from two directions. One is the axiomatic approach, Nash (1950) where reasonable properties (axioms) are required of a solution to hold. Nash showed that his axioms uniquely determine what is now called the Nash bargaining solution. The other, Nash (1953), aims at devising a suitable bargaining process which realizes in subgame perfect equilibrium the same outcome that the axiomatic approach prescribes. This dual approach was later termed the "Nash program" see e.g. Thomson (1994), Serrano (2005).

Since the Nash bargaining solution depends on both the feasible set of outcomes and the disagreement point, it is a valid question to ask how it behaves if either of

F. Forgó (✉)
Corvinus University of Budapest, Budapest, Hungary
e-mail: ferenc.forgo@uni-corvinus.hu

A. Matsumoto (ed.), *Optimization and Dynamics with Their Applications*,
DOI 10.1007/978-981-10-4214-0_2

them changes in some way. Forgó and Szidarovszky (2003) showed that if the disagreement point goes to negative infinity in a given direction, then the Nash bargaining solution converges to a unique outcome which they termed L-Nash bargaining solution (L stands for limit). It was also shown that the L-Nash bargaining solution can also be obtained as a solution of a multiple criteria decision making problem with weights that are the reciprocals of the components of the disagreement direction. The L-Nash solution can also be axiomatized within the context of multiple criteria decision making.

One way of generalizing Nash's bargaining model is to allow assigning weights to the players meant to indicate their "importance" or "power" in the conflict. This is outside information (just as the disagreement point), and critically influences the final outcome. Several axiomatizations of the non-symmetric Nash bargaining solution have emerged throughout the years e.g. Harsanyi and Selten (1972), Kalai (1997), Roth (1979), Anbarci and Sun (2013) as well as non-cooperative bargaining models that implement it, e.g. Kalai (1997), Laruelle and Valenciano (2008), Britz et al. (2010), Anbarci and Sun (2013).

In this paper it is demonstrated how the non-symmetric Nash bargaining solution behaves when the disagreement point goes to negative infinity in a fixed direction. It turns out that in certain cases the two pieces of outside information, the power of the players and the disagreement direction can be treated as one, while in other instances they cannot.

2 Preliminaries

We consider two-person bargaining games. Let $B(F, d, p, q)$ be a two-person non-symmetric bargaining problem with convex, compact feasible set $F \subset \mathbb{R}^2$ which is assumed to have at least one positive element, d is a non-positive disagreement point, and the positive integers p, q represent the "power" of the players. The game is played as follows. If both players agree, then they choose a feasible point $f \in F, f = (f_1, f_2)$ in which the players get the components of f accordingly. If they cannot come to an agreement, then a usually "bad" disagreement point $d = (d_1, d_2)$ realizes.

Consider the following constrained maximization problem

$$P: \quad \max\ (x_1 - d_1)^p\ (x_2 - d_2)^q$$

$$x \in F$$

where p, q are positive and $p + q = 2$.

This problem has a unique solution $\varphi \in F$ which is called the non-symmetric (asymmetric) Nash bargaining solution (*NSNBS*). In the special case $p = q = 1$ we have the classical, symmetric bargaining solution (*NBS*) of Nash (1950) which is uniquely determined by a set of axioms (feasibility, rationality, Pareto-optimality, independence of irrelevant alternatives, scale independence and symmetry). Among the several axiomatizations of the *NSNBS* we only quote Roth's (1979). From among the Nash axioms Roth changed Pareto optimality and symmetry to the following (Roth's axiom).

Consider a bargaining problem $B(G, 0, p, q)$, where the feasible set is defined by

$$G = \{g = (g_1, g_2) \in G, g_1 + g_2 \leq 2\}.$$

It is required by Roth's axiom that the solution of $B(G, 0, p, q)$ be (p, q). Then the unique solution of $B(F, d, p, q)$ is the *NSNBS*.

3 The Main Result

Parametrize $B(F, d, p, q)$ by taking $d = -\alpha r$, where $r > 0$ is the so-called disagreement direction and the positive parameter α represents how far we push the disagreement point in the direction $-r$. Forgó and Szidarovszky (2003) introduced the L-Nash bargaining solution as the limit of the Nash bargaining solution as $\alpha \to \infty$. Several interesting issues were addressed in Forgó and Szidarovszky (2003) concerning the behavior of the L-Nash solution. It is a natural question to ask: what happens if we consider non-symmetric bargaining problems and approach negative infinity with the disagreement point in a given direction?

Let $B(F, -\alpha r, p, q)$ be a two-person bargaining problem with positive parameter α. The parametrized two-person non-symmetric Nash bargaining solution *NSNBS*(α) is the unique solution of the maximization problem

$$P(\alpha): \quad \max\, (x_1 + \alpha r_1)^p\, (x_2 + \alpha r_2)^q$$

$$x \in F.$$

It is not a significant loss of generality if we confine ourselves to rational weights which amounts to allowing p and q to be positive integers.

For any given x and r, the objective function of $P(\alpha)$ is a polynomial of order $p + q$ of the parameter α. The coefficient of the leading term α^{p+q} is $r_1^p r_2^q$, independent of x thus having no role in the maximization of the objective function of $P(\alpha)$. The coefficient $h(x)$ of α^{p+q-1}, however, does depend on x. In particular, as can be shown by the application of the binomial formula

$$h(x) = p r_1^{p-1} r_2^q x_1 + q r_1^p r_2^{q-1} x_2,$$

or equivalently

$$h(x) = r_1^p r_2^q (\frac{p}{r_1} x_1 + \frac{q}{r_2} x_2).$$

If α is large enough, then the linear function $h(x)$ should be as large as possible in order to maximize $P(\alpha)$. If

$$\max h(x) \quad x \in F \tag{1}$$

has a unique solution, then terms with degree less then $p+q-1$ do not count if α is large enough. If the above maximization problem has multiple solutions, then the coefficient $g(x)$ of the term α^{p+q-2} comes into play. In particular, $g(x)$ should be maximized over the optimal set of (1) i.e. the following maximization problem should be solved

$$\begin{aligned} &\max g(x) \\ h(x) = &\max_{x \in F} h(x) \\ &x \in F. \end{aligned} \tag{2}$$

Again, by using the binomial formula, it can easily be seen that

$$g(x) = \frac{p(p-1)}{2} r_1^{p-2} r_2^q x_1^2 + pq r_1^{p-1} r_2^{q-1} + \frac{q(q-1)}{2} r_1^p r_2^{q-2} x_2^2.$$

Define

$$f(x) = r_1^{2p} r_2^{2q} (\frac{p}{r_1^2} x_1^2 + \frac{q}{r_2^2} x_2^2).$$

Then, with simple algebra one can verify that

$$g(x) = \frac{1}{2 r_1^p r_2^q} ((h(x))^2 - f(x)).$$

This means that problem (2) is equivalent to

$$\begin{aligned} &\min f(x) \\ h(x) = &\max_{x \in F} h(x) \\ &x \in F \end{aligned} \tag{3}$$

whose objective function is a strictly convex quadratic function, the feasible set is convex, compact implying that problem (3) has a unique optimal solution x^2.

Define now $x^0 = x^1$ if problem (1) has the unique optimal solution x^1, and $x^0 = x^2$ otherwise.

We can now state the main result.

Theorem 1 *The NSNBS of the two-person nonsymmetric bargaining problem* $B(F, -\alpha r, p, q)$ *converges to* x^0 *if* $\alpha \to \infty$.

Proof Along the lines of Theorem 1 in Forgó and Szidarovszky (2003).

x^0 can rightly be called the non-symmetric limit-Nash bargaining solution, *L-NSNBS*. For polyhedral feasible sets there is no need to go to infinity with α to obtain the *L-NSNBS*.

Theorem 2 *If F is a polytope, then there is an* α_0 *such that for all* $\alpha \geq \alpha_0$, *the two-person NSNBS coincides with the L-NSNBS.*

Proof Along the lines of Theorem 2 in Forgó and Szidarovszky (2003).

4 Example

Let us consider a very simple example of firm-union bargaining over wage and employment as in McDonald and Solow (1981). The firm has a profit function (revenue less labor cost) $R(L) - wL$, where w denotes wage per worker and L denotes the number of employed workers. The union's utility function is given by $L[U(w) - U(w')]$, where w' denotes benefits if worker is unemployed, and U is each union member's utility function. Bargaining takes place in the region constrained by the bounds $0 \leq w \leq W$, $0 \leq L \leq N$. The utility function of the union (total wage) is increasing in both arguments and in order to have a conflict, the profit function of the firm should be decreasing in w and L. For the model to be meaningful we assume that wage is at least as high as the marginal revenue of labor, $w \geq R'(L)$. Bargaining power of the two parties are p and q and we suppose that $p < q$ (union is less powerful than the firm). On the other hand, the firm is more vulnerable to the failure of negotiations, i.e. $r_1 < r_2$.

We will determine the *L-NSNBS* for specific values of the parameters and specific forms of the functions involved. In particular, let

$$\begin{aligned} U(w) - U(w') &= w \\ R(L) &= 320L - 10L^2 \\ W &= 400, N = 200 \\ p &= 2, q = 3 \\ r_1 &= 1, r_2 = 3. \end{aligned}$$

Then, to determine the *L-NSNBS*, problem (1) has first to be solved, which takes now the form

$$\max 2wL + \frac{3}{2}(320L - L^2 - wL)$$
$$320 - 20L \le w \le 400$$
$$0 \le L \le 10.$$

Notice that the objective of the above problem is linear in the utilities of the parties but nonlinear (quadratic) in the original decision variables w, L. The solution is in favor of the union:

$L = 10, w = 400$, full employment and highest possible wages.

5 Discussion

Consider the case when problem (1) has a unique solution. As pointed out and also observed in the context of this paper, in this case the *L-NSNBS* is the solution of a multi-criteria decision problem (*MCDP*) by the method of linear weighting where the weights are represented by the coefficients in the linear objective function of problem (1). In the symmetric case, the additional information is supplied by the relative magnitude of the components in the disagreement direction. The less the first player is hurt relative to the other (r_1 is small) by disagreement getting more costly, the more weight her interest carries through the large coefficient $\frac{1}{r_1}$in the objective function of problem (1). In the non-symmetric case there seem to be two reference points (outside information indicating the weight or importance of the players). One is the direct weights p, q, the other is the relative costs of disagreement r_1, r_2. Our analysis reveals, however, that when combining these together and using only the disagreement directions $\frac{r_1}{p}, \frac{r_2}{q}$ in the symmetric bargaining model, we get the same limiting solution.

This is not the case when problem (1) has multiple optima. Then the coefficients of the quadratic terms in the objective function of problem (3) are $\frac{p}{r_1^2}$ and $\frac{q}{r_2^2}$ while in the corresponding symmetric bargaining model they would be $\frac{p^2}{r_1^2}$ and $\frac{q^2}{r_2^2}$ giving rise to different solutions. It should also be noticed that if F is a polyhedron, then problem (1) is a linear programming problem and multiple optima are unlikely to occur in unstructured problems. Nevertheless, theoretically, *L-NSNBS* is determined by two reference points.

There is, however, a significant difference between the "importance indicators": the direct weights and the disagreement direction. Direct weights do not explicitly require interpersonal comparison of utilities since they are completely external to the model. As we have interpreted the components of the disagreement direction vector as an expression of the relative damage caused by prolonged negotiations, they implicitly mean comparison in utility (damage interpreted as disutility). Comparison of power is less closely related to utilities. It is therefore somewhat of a surprise, that these two things merge together in the *L-NSNBS*.

The whole analysis can be done for an arbitrary number of players and results remain the same if adjustments are made accordingly. We confined ourselves to two players in order to keep technicalities within reasonable bounds and because the two-player case is of interest in its own.

Forgó and Fülöp (2008) showed how the L-Nash solution can be implemented by proper adjustment of Rubinstein's alternating offer bargaining scenario, Rubinstein (1982) either exactly or asymptotically depending on F, r, and exactly by Howard's scheme, Howard (1992) for any F, r. There does not seem any special difficulty to extend their results to the non-symmetric case if the weights p, q are externally given. Howard's implementation makes it possible to internalize not only the penalty parameter α but the weights p, q as well. How this can technically be done in the framework of a bargaining process remains an issue and calls for further research. It is also left for further research how other bargaining processes for *NSNBS* can be adjusted so that they implement *L-NSNBS*.

Acknowledgements The author thanks for the invitation to be a contributor of this book. The research was supported by the grant NKFI K-119930.

References

Anbarci, N., & Sun, C. (2013). Asymmetric Nash bargaining solutions: A simple Nash program. *Economics Letters Elsevier*, *120*, 211–214.
Britz, V., Herings, P. J. J., & Predtetchinski, A. (2010). Non-cooperative support for the asymmetric Nash bargaining solution. *Journal of Economic Theory*, *145*, 1951–1967.
Forgó, F., & Szidarovszky, F. (2003). On the relation between the Nash bargaining solution and the weighting method. *European Journal of Operational Research*, *147*, 108–116.
Forgó, F., & Fülöp, J. (2008). On the implementation of the L-Nash bargaining solution in two-person bargaining games. *Central European Journal of Operations Research*, *16*, 359–378.
Harsanyi, J. C., & Selten, R. (1972). A generalized Nash solution for two-person bargaining games with incomplete information. *Management Science*, *18*, 80–106.
Howard, J. V. (1992). A social choice rule and its implementation in perfect equilibrium. *Journal of Economic Theory*, *56*, 142–159.
Kalai, E. (1977). Non-symmetric Nash solutions and replications of 2-person bargaining. *International Journal of Game Theory*, *6*, 129–133.
Laruelle, A., & Valenciano, F. (2008). Non-cooperative foundations of bargaining power in committees and the Shapley-Shubik index. *Games and Economic Behavior*, *63*, 341–353.
McDonald, I. M., & Solow, R. M. (1981). Wage bargaining and employment. *The American Economic Review*, *71*, 896–908.
Nash, J. F. (1950). The bargaining problem. *Econometrica*, *18*, 155–162.
Nash, J. F. (1953). Two person cooperative games. *Econometrica*, *21*, 128–140.
Roth, A. E. (1979). Axiomatic models of bargaining. Springer Verlag, Berlin.
Rubinstein, A. (1982). Perfect equilibrium in a bargaining model. *Econometrica*, *50*, 97–109.
Serrano, R. (2005). Fifty years of the Nash program 1953–2003. *Invest Economicas*, *29*, 219–258.
Thomson, W. (1994). Cooperative models of bargaining. In R. J. Aumann & S. Hart (Eds.), Handbook of game theory (Vol. II, pp. 1237–1284). Amsterdam: Elsevier Science

Analyzing the Impact of Process Improvement on Lot Sizes in JIT Environment When Capacity Utilization Follows Beta Distribution

József Vörös, Gábor Rappai and Zsuzsanna Hauck

Abstract Even after many years one kind picture still floating in my eyes: I see professor Szidarovszky, facing the blackboard, sponge and chalk at the upheld left and right hands, and writing and cleaning the lines simultaneously he put on the table, to fill up our heads with numerical methods. This style expresses his very dynamic and efficient research work at the same time, and hopefully this paper indicates that his efforts have not been useless as numerical methods are very intensively used in order to characterize the nature of lot sizing problems in JIT environment. In JIT environment the jidoka principle empowers employees to signal quality problems, and these result in frequent stoppages. This way we consider the output of the assembly line random variable that follows Beta distribution, but with low beta values. For specific beta values we derive explicit forms of the expected values of the inventory related and the annual total costs as function of alpha, the other parameter of the Beta distribution. But increasing alpha expresses increasing process quality. We found that increasing process quality decreases the expected annual cost, and the explicit forms give the saved cost volumes. Two simulation analyses are conducted to reveal the development of the variance of annual costs. The estimations of the variance of the minimum total annual costs indicate that with process improvement the variance of the minimum of the annual total costs will decrease.

Keywords JIT · Process quality · Stochastic lot sizing · Backlogs · Beta distribution

---

J. Vörös (✉) · G. Rappai · Z. Hauck
Faculty of Business and Economics, University of Pécs,
Rákóczi u. 80, Pécs 7622, Hungary
e-mail: voros@ktk.pte.hu

G. Rappai
e-mail: rappai@ktk.pte.hu

Z. Hauck
e-mail: hauckzs@ktk.pte.hu

A. Matsumoto (ed.), *Optimization and Dynamics with Their Applications*,
DOI 10.1007/978-981-10-4214-0_3

1 Introduction

Similarly to many categories in business and management, and in economics, sometimes there are more definitions for one concept, like capacity. Meredith and Shafer (2011) say that capacity is generally taken to mean the maximum rate at which a transformation system produces outputs or processes inputs, though the rate may be 'all at once'. Slack et al. (2015) state that capacity is the output that an operation can deliver in a defined unit of time, while Krajewki et al. (2013) define capacity as the maximum rate of output of a process or a system. Seemingly there are conflicts in these definitions, but one is better than the other in certain dimensions. Measuring capacity as the rate of handling inputs is more appropriate when the system produces wide varieties of goods, however when output is homogenous, measuring capacity by outputs provides a better insight. As we are going to analyze JIT systems, we should know that JIT systems are more efficient when JIT principles are applied to standardized products, produced in large volumes (see any operations management book, mentioned above), consequently, we use the volume of outputs to measure capacities. Besides these, the literature usually attaches different additional adjectives to capacity: it talks about designed capacity, effective capacity, realized capacity. Designed capacity is taken as the maximum rate of output during unit time. For example, if we have an assembly line where the designed cycle time is 1 min, then we may say that the design capacity of the assembly line is 480 units of goods per shift when the shift is 8 h long.

However, production problems always occur, independently from its configuration. So it is even in case of a JIT production system. JIT system was configured firstly by Toyota, and the system has achieved tremendous success. Toyota Production System (TPS) has evolved as Toyota's response to the task of 'better cars for more people'. The concept of better cars means flawless quality, and for more people means affordable price with perfect timing. The former CEO of Toyota, Mr. Watanabe, says the Toyota Way rests on two pillars: continuous improvement and respecting people (employees, suppliers and customers) (Watanabe 2007). Behind these pillars there are two important elements of TPS: the principles of jidoka and heijunka (Mishina and Takeda 1992). The task of heijunka is multiple: to connect the total value chain from customers to suppliers, make what customers want and when they want, and smooth the system pulse. The production volume is streamlined as smooth as possible, but product mix is similarly spread out as evenly as possible. The result of this policy is that in each moment the sequence structure of different model types in the assembly line reflects the volume and the structure of the monthly and smoothed daily demand. For example, if the volume ratio of the monthly demand for models A, B, and C is 3:2:1, respectively, then the sequence of cars in the assembly line appears to be as AAABBCAAABBC, and so on. This way, the implementation of the heijunka principle considers the rules of lot sizing as well, which is 3, 2 and 1 for models A, B and C, respectively in this case. To make these lot sizes optimal or less expensive, concerted efforts are required to decrease setup cost to the desired level. Schniederjans and Cao (2000), and later

Cao and Schniederjans (2004) published comparative models for inventory related costs under economic order quantity and JIT policy showing the efficacy of the more cost inclusive models.

From the heijunka principle of TPS it follows that in a JIT environment demand (daily production requirement) may not vary during a short time horizon and this way, demand can be considered constant and stable, even for a month. The demand for the output of a production line is determined by the production volume scheduled by the production plan. This is because there are important pre-assumptions to use lean production system, and level loading plays primary role among the prerequisites (Stevenson 2012). Once established, production schedules are relatively fixed over time and this fact provides certainty to the system. According to Mishina and Takeda (1992) a monthly demand (particularly, of May) is frozen a couple of months before (in February), and the monthly production plan is smoothed as much as possible and it is desirable to produce in small lots and to spread the production of the different products throughout the day to achieve smooth production (Stevenson 2012). But even in the most advanced lean production system production problems and stoppages exist, as to avoid the production of defective items, employees are empowered to signal quality problems. Under the principle of jidoka, an employee must pull the andon cord in case of witnessing quality problems, signaling the need for the help of the team leader. If they are not able to fix the problem, the assembly line stops. By Mishina and Takeda (1992), an employee pulls the andon cord a dozen times during a shift, out of which one results in a stoppage on the average. They mention also, that the run ratio (we believe as it is the actual output over designed output) is around 95%, but sometimes the run ratio was down to a meager 85%. In other words, we would say, the average capacity utilization, which is the ratio of the actual output and the design capacity, is around 95%, but sometimes is down to 85%.

Based upon these observations we consider an actual output of the assembly line operating under JIT principle random variable, whose expected value is close to the design capacity, or saying different way, we consider capacity utilization level random variable, whose expected value is close to one, and taking values around 0.8 is rather rare. But we have to note also that TPS is rather pragmatic. When a stoppage would be very long, the product (car), that causes the stoppage, goes through the assembly line, tagged with a red card, and the car goes to the clinic area. Later it will be fixed here. The number of cars arriving at the clinic area is considered random variable also, thus the output able to satisfy demand is the origin of two random variables.

Vörös and Rappai (2016) has built a model comprehending the circumstances described above and they characterized the optimal lot size and the resulting expected total cost as function of demand for arbitrary distribution functions of the random variables. They found that the expected value of the total set up and inventory cost is increasing function of demand when demand (production schedule) equals design capacity, the expected total cost function may have minimum point (and may have maximum point as well) and an algorithm is given to find the optimal level of contract volume (optimal demand level). Their paper provides an excessive literature overview also, thus the interested readers may turn to this paper to observe the development of the literature and therefore we may omit this literature survey.

Although this paper uses the same base model, the targets are different. First of all, we are interested in the impacts of process improvements, namely, how process improvement impacts on lot sizes and the total setup and inventory costs. We consider a process quality increasing if the expected value of process yield increases while the variance is unchanged, or the variance of process yield decreases while the expected value does not change, or the expected value of the process yield increases and its variance decreases at the same time.

The next significant difference is that we specify the distribution functions. Based upon the observations mentioned above, we believe that the Beta distribution is a good approximation of the real distribution of the utilization level of an assembly line operating in JIT environment. Using Beta distribution, we are able to derive explicit formulas to lot sizing and related cost functions.

The usage of Beta distribution to utilization level gives nice explicit form for the expected values of lot sizes and total costs, but we have rather complicated expressions for the variance of total costs. To get insights into the behavior of the variance of the total cost, we conduct excessive simulation processes, generating thousands of random events to estimate the impact of process improvement on the variance of total costs.

The next section identifies the model to be analyzed, Sect. 3 determines explicit forms for lot sizes and the expected value of total cost, and characterizes the nature of lot sizes and total cost as function of an improving process. Sect. 4 conducts two simulation approaches to get insights into the nature of the variance of the minimum annual total costs, and the last section gives the conclusions.

Among the most important findings it can be mentioned, that in JIT environment improving process quality will result in larger lots under large capacity utilization levels only, which can be considered as rather new contribution. Using explicit formulas we can describe the total cost, and using computer based numerical methods we find that the expected value of the annual total costs will decrease also when process quality improves. The simulation analyses give additional new insights according to which, when the expected value of process output increases, not only the expected value of the minimum total annual cost will decrease, but so does its variance as well, simultaneously.

2 The Model with Beta Distribution

Lot sizing task appears almost everywhere in a production process even in case of a JIT system. The role of Kanban cards clearly indicates the presence of lots in a JIT system: Kanban cards determine the quantity of parts that may be pulled into the system in one lot. The question offers itself: how many parts, products should be produced in one lot? The question is especially interesting taking into the fact, that stoppages may occur, which influences the production rate and lot sizes.

The seminal papers of Harris (1913) and Taft (1918) on how much to order or produce at once have attracted many researchers and many new concepts have been created since. Probably Shih (1980) is among the firsts who mentioned that there may be shortages in the delivery process as not every unit of the product in the accepted lot is of perfect quality. As the pioneer works of Porteus (1985, 1986) point it out, lot sizing and process quality are interrelated. He considers investment into setup cost reduction and process quality improvement possibilities in the EOQ model, and he found that the optimal lot size is strictly increasing with improvement in process quality. Also among the pioneers there is Chand (1989), who discusses the results of worker learning that reduces setup costs and improves process quality, and finally we have small lot sizes, the base of the stockless production philosophy. Our analysis is directly based on the model constructed by Vörös and Rappai (2016), where the authors took into account that stoppages may occur due to quality problems in the assembly line, moreover, jidoka principle may be violated when solving production problems instantaneously would require long time. In this case, cars move to the clinic area, waiting to be fixed. Both two events are considered to be random and in their paper the inventory costs and cycle times are determined, which are also random variables.

We use the following notations (Table 1).

Table 1 Notations used

ξ	The number of units leaving the assembly line in a day (including those entering the clinic area), random variable with probability density function $f(x)$. ξ is the daily production rate
η	The number of units arriving at the clinic/overflow area during a day with a particular known quality problem, random variable with probability density function $g(y)$
D	Daily demand in units, input parameter
m	Maximum number of cars fixed in the clinic/overflow area during a day, input parameter
Q	The lot size in units, decision variable
s	The current setup cost, input parameter
h	Holding cost per unit per day, input parameter
b	Backlogging cost per unit per day, input parameter
z	$=(y/x - m/D)/(D - m)$, assuming $D > m$
K	Design capacity of the assembly line per day, in units (the ratio of the duration of a day over the planned cycle time), input parameter
M	The lowest value of ξ with positive probability ($P(K \geq \xi \geq M) > 0$, and $P(\xi < M) = 0$), input parameter
N	Number of working days in a year
k	At most x/k units enter into the clinic area per day, input parameter
u	The observed utilization level of the assembly line, $u = x/K$
v	The planned capacity utilization level of the assembly line, $v = D/K$
r	The ratio of b and h, i.e. $r = b/h$

To increase the convenience of the reader, we repeat here the base model of Vörös and Rappai (2016) from which we originate our results. Similarly to that paper, let ξ denote the number of cars leaving the assembly line during a day, including tagged cars with known defective problems. We consider ξ a random variable with the probability density function $f(x)$, and we suppose there is zero probability that the assembly line is completely in shutdown state during a day (shift). Let η denote the number of tagged cars arriving at the clinic/overflow area with a particular quality problem during a day. We consider η a random variable with probability density function $g(y)$. We suppose that m cars with the particular quality problem can be fixed during a day. m is a known input parameter.

For simplicity, we assume there is one shift per working day, thus shift or day may be used alternatively, and demand can be expressed as D units per model type per day, which is a known constant input parameter, as a result of applying the heijunka principle. The design capacity of the assembly line can be calculated as the ratio of the duration of a shift over the planned cycle time. Because of stoppages, the real output of the assembly line is lower than the design capacity.

We denote the daily holding and backlogging cost with h and b per unit, respectively, and the set up cost is denoted by s. We seek the optimum lot size, and the lot size is denoted by Q.

Let x denote a realization of ξ, while y a realization of η, i.e. x denotes the observed number of cars leaving the assembly line in a particular day, out of which y are observed as defective with known quality problem. In the followings we summarize all the possible outcomes of these observations in order to calculate the expected cost and cycle length. Case 1 considers events when in a day the number of cars arriving at the clinic area is not larger than the repair capacity of this cell, i.e. we may assume that $y \leq m$. Consequently, clinic area capacity will not contribute to the development of backlogs. So, if there are no serious quality problems with the assembly line, backlogs will not appear at all if the assembly line has the sufficient capacity on the observed day, i.e. $x \geq D$. We identify this situation as Case 1a, and the inventory build-up diagram is represented by Fig. 1.

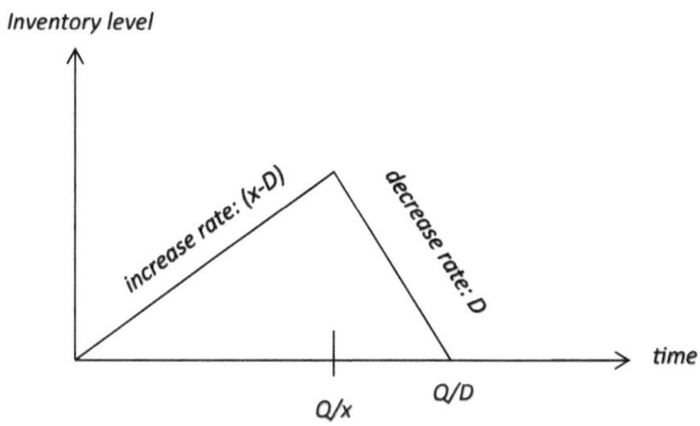

Fig. 1 The inventory build-up diagram for Case 1a

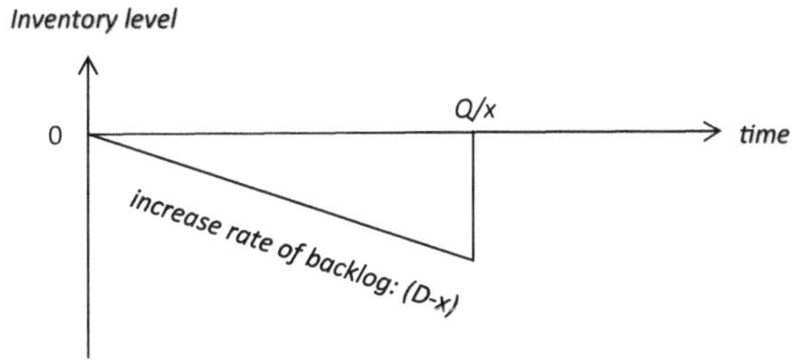

Fig. 2 The inventory build-up diagram for Case 1b

Figure 1 is a well-known diagram (see for example Hax and Candea 1984) and denoting the inventory related costs for Case 1a by H_{1a}, it can be written as

$$H_{1a}(Q,x,y)/h = \left(Q^2/2\right)(1/D - 1/x) \tag{1a}$$

The explanation behind Fig. 1 is that producing Q units in one lot that leaves the assembly line requires Q/x time units and serves demand over Q/D time units. During run time inventory increases at the rate of (x – D) cars per time unit, and when a lot is finished, inventory decreases at the rate of D cars per time unit.

On the other hand, even if $y \leq m$ (the clinic area is not lagging behind), and the assembly line suffers from serious quality problems, backlog may develop, i.e. $x < D$. We identify this situation as Case 1b, and its inventory build-up diagram is represented by Fig. 2.

Figure 2 indicates that there are no cars waiting for customers as in each time unit, the system produces less than the demand. The volume of unsatisfied demand is (D – x) per day, thus until the end of the production run $Q(D - x)/x$ units backlog develops as the length of the production run is Q/x time units. The volume of the accumulated backlog is produced under overtime or weekend shifts and the unit cost of the backlog is denoted by b. The time required to produce the accumulated backlog is considered negligible, for simplicity. In fact, the backlog level is kept till the start of overtime, however the cost of this maybe inserted into b, thus the approach indicated by the figure may be accepted. Denoting the accumulated backlogging cost by H_{1b}, the following can be written:

$$H_{1b}(Q,x,y)/b = \left(Q^2/2\right)(D - x)/x^2 \tag{1b}$$

Now we turn to Case 2, where we assume that the clinic/overflow area does not have the sufficient capacity to fix each car arriving at the clinic area at the same day, i.e. $y > m$. We identify Case 2aa as when although $y > m$, the assembly line does not suffer from serious quality problems and has enough capacity to meet demand, i.e. $x - y + m \geq D$. This form indicates as well that the outcome (the fixed units) of the clinic area is utilized at once to satisfy demand, and defective cars are hold at the same warehouse wearing the same holding cost like non-defective cars, as they are fully assembled. Additionally, we assume that the length of the cycle (the time elapsing between two consecutive points when inventory level is zero—these points are called regeneration points in the dynamic lot sizing literature) is long enough to fix all cars parking at the clinic area. During the production run ($=Q/x$) the number of cars entering the clinic area is yQ/x, and it requires yQ/mx time units to fix them. As we have time to fix all the cars during the cycle, it must be valid that $yQ/mx \leq Q/D$. But in this case no backlog will develop throughout the cycle, thus we have the same inventory build-up diagram like in Case 1a. Denoting the inventory related cost in Case 2aa by $H_{2aa}(Q, x, y)$, i.e. when $y \geq m$, $x - y + m \geq D$ and $y/x \leq m/D$, the following can be written:

$$H_{2aa} = H_{1a}. \tag{2aa}$$

On the other hand, when although backlog does not accumulate during production time, but the length of the cycle is not enough to fix all the cars parking at the clinic/overflow area, backlogs may accumulate after the end of the production runtime.

Let us identify this situation as Case 2ab, i.e. the assumptions are: $y \geq m$, $x - y + m \geq D$ but $y/x > m/D$. Figure 3 represents the inventory build-up diagram where after that production is terminated (at Q/x), backlog starts accumulating at a certain point. Again, the stock level realization diagram indicated by Fig. 3 includes both the flawless and defective items and the fixed units at the clinic area are used to satisfy daily demand.

It is a natural consequence that $y \leq x$, i.e. the number of cars entering the clinic area may not be larger than the number of cars leaving the assembly line. Then from the assumption $y/x > m/D$ follows that $m/D < 1$ as well, so $m < D$ is valid, and Fig. 3 reflects this property. After production terminates non defective cars are sold

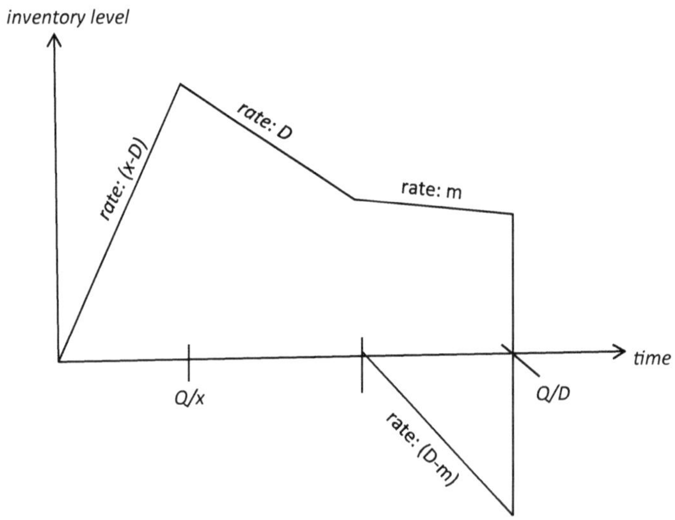

Fig. 3 Inventory build-up diagram for Case 2ab

at the rate of the demand, however from a certain point, when flawless inventory runs out, inventory may deplete at the rate of m instead of D because only the fixed cars can be sold, and non-fixed ones are still waiting at the clinic area.

The number of cars getting into the clinic area during the production run is yQ/x while during the cycle—which is Q/D time units long—only mQ/D units may be fixed. The volume of the accumulated backlog is $Q(y/x - m/D)$, and this volume accumulates during the time length of $Q(y/x - m/D)/(D - m)$ as the rate of backlog accumulation is $(D - m)$. Let us note that this backlog volume is staying at the clinic area in the form of defective cars as well, so positive and negative inventory exist simultaneously. We can say that backlog starts accumulating at time $[Q/D - Q(y/x - m/D)/(D - m)]$. The on hand inventory at this point is: $[Q(y/x - m/D) + mQ(y/x - m/D)/(D - m)]$. The inventory cost during the accumulation of the backlog this way is: $h[[Q(y/x - m/D)/(D - m)][2Q(y/x - m/D) + mQ(y/x - m/D)/(D - m)]]/2 = h[Q(y/x - m/D)]^2[D - m/2]/(D - m)^2$. The occurring backlogging cost during the same interval is: $b[Q(y/x - m/D)]^2$.

By Fig. 3, at point Q/x the inventory level is $(Q/x)(x - D)$, consequently the occurring holding cost from the beginning of the cycle till the point Q/x is: $hQ^2(x - D)/2x^2$ and from the point Q/x till the point when backlog starts developing the occurring holding cost is: $h[Q/D - Q/x - Q((y/x - m/D)/(D - m)][Q(y/x - m/D)(1 + m/(D - m)) + Q(x - D)/x]/2 = hQ^2(Dz + (x - D)/x)(1/D - 1/x - z)/2$, where we used the simplifying notation $z = (y/x - m/D)/(D - m)$. Denoting the total holding and backlogging costs in Case 2ab by H_{2ab}, adding the four types of costs it can be written that

$$\begin{aligned} H_{2ab}(Q,x,y) &= (Q^2/2)\left(h\left[(x - D)/x^2 + (Dz + (x - D)/x)(1/D - 1/x - z) + z^2(2D - m)\right] + b(D - m)z^2\right) \\ &= (Q^2/2)\left[h(1/D - 1/x) + (h + b)(D - m)z^2\right] \end{aligned} \tag{2ab}$$

There is only one subcase waiting for identification inside Case 2: this is when demand cannot be satisfied from the very beginning of the cycle because of the frequent stoppages. Additionally the number of fixed cars is lower than the number of defective cars entering the clinic area, i.e. in Case 2b: $y > m$, and $x - y + m < D$. Figure 4 gives the inventory build-up diagram, where we have the defective items in the system not fixed yet in the clinic area as positive inventory, and simultaneously with the backlogs due to the not sufficient capacity, as negative inventory.

Denoting the occurring holding and backlogging costs in Case 2b by H_{2b}, we can determine this cost like as:

$$H_{2b}(Q,x,y) = (Q^2/2)[h(y - m) + b(D - x + y - m)]/x^2 \tag{2b}$$

Let us note that the length of the cycles (elapsed time between two regenerations points) in Case 1a, Case 2aa, Case 2ab depends on a decision variable (Q), and in Case 1b, Case 2b depends on a decision and a random variable (Q and ξ).

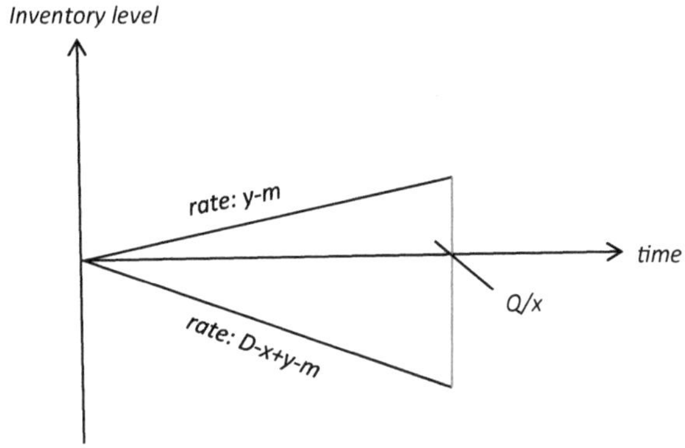

Fig. 4 The inventory build-up diagram for Case 2b

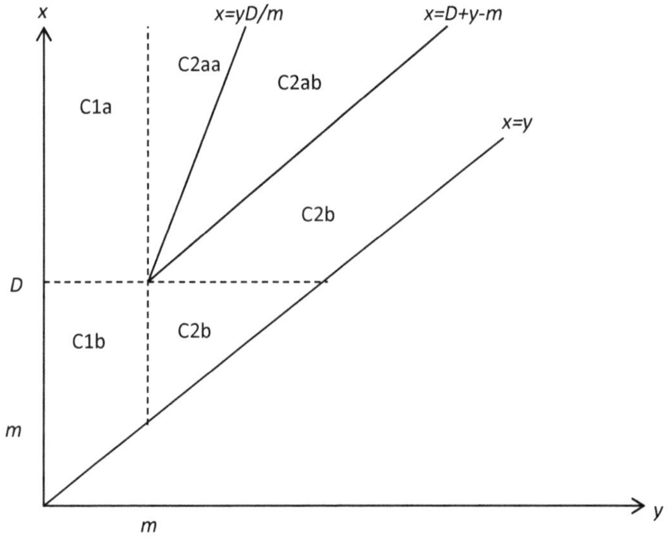

Fig. 5 Probability fields of inventory build-up diagrams with $D > m$ (C refers for Case)

Theoretically there are two cases considering the relationship between D and m, but we visit only one case, namely when $D > m$, i.e. the daily demand is larger than the repairing capacity of the clinic area (the interested reader is directed to the paper of Vörös and Rappai 2016, to see the other outcome). In Fig. 5 the vertical axis represents the observed number of cars leaving the assembly line during a day (denoted by x) and the horizontal one does the observed number of defective cars moved to the clinic area during a day (denoted by y). It is natural that $x \geq y$, as cars entering the clinic area constitute a subset of cars leaving the assembly line. We have to note that there exists a theoretical limit on y (8 by the study of Mishina and Takeda 1992), however the limit is rarely kept. Anyway, the analysis is richer when we consider wider sets of possible outcomes, and we analyze the full upper half of the first quarter.

The vertical axis is divided into two segments by the horizontal line expressing the daily demand level, D. The horizontal axis is also divided into two sections, namely by the repairing capacity m represented by a vertical dashed line. The symbols among lines indicate the valid inventory models elaborated above.

Denoting the expected value of inventory related costs per cycle by EH_C, assuming that the two random variables are independent, this expected value is determined by the following expression (Vörös and Rappai 2016):

$$\begin{aligned} EH_C(Q) = & \int_0^m \int_0^x H_{1b}(Q,x,y)g(y)dyf(x)dx + \int_m^D \int_0^m H_{1b}(Q,x,y)g(y)dyf(x)dx \\ & + \int_m^D \int_m^x H_{2b}(Q,x,y)g(y)dyf(x)dx \\ & + \int_D^K \int_0^m H_{1a}(Q,x,y)g(y)dyf(x)dx + \int_D^K \int_m^{xm/D} H_{2aa}(Q,x,y)g(y)dyf(x)dx \\ & + \int_D^K \int_{xm/D}^{x-D+m} H_{2ab}(Q,x,y)g(y)dyf(x)dx + \int_D^K \int_{x-D+m}^{x} H_{1b}(Q,x,y)g(y)dyf(x)dx \end{aligned} \quad (3)$$

In (3) K denotes the design capacity of the assembly line per day. Using the detailed formulas for the inventory costs, it can be rewritten that

$$EH_C(Q) = (Q^2/2)H_C, \quad (4a)$$

where

$$\begin{aligned} H_C = b & \int_0^m \int_0^x ((D-x)/x^2)g(y)dyf(x)dx + b\int_m^D \int_0^m ((D-x)/x^2)g(y)dyf(x)dx \\ & + \int_m^D \int_m^x [[h(y-m)+b(D-x+y-m)]/x^2]g(y)dyf(x)dx \\ & + h\int_D^K \int_0^{xm/D} \left(\frac{1}{D}-\frac{1}{x}\right)g(y)dyf(x)dx \\ & + \int_D^K \int_{xm/D}^{x-D+m} [h(1/D-1/x)+(h+b)(D-m)z^2]g(y)dyf(x)dx \\ & + \int_D^K \int_{x-D+m}^{x} [[h(y-m)+b(D-x+y-m)]/x^2]g(y)dyf(x)dx, \end{aligned} \quad (4b)$$

Now, we describe the expected value of the length of the inventory cycles. As we noted, in Case 1a, Case 2aa and Case 2ab it depends on a decision variable (on Q/D), and in Case 1b, Case 2b depends on a decision and a random variable (on Q/x). Using the probability fields and the corresponding inventory build-up diagrams, denoting the expected value of the length of the inventory cycles by EL_c, it can be written that

$$EL_c(Q) = QL_C, \tag{5a}$$

where

$$\begin{aligned} L_C = & \int_0^m \int_0^x (1/x)g(y)dyf(x)dx + \int_m^D \int_0^m (1/x)g(y)dyf(x)dx \\ & + \int_m^D \int_m^x (1/x)g(y)dyf(x)dx \\ & + \int_D^K \int_0^m (1/D)g(y)dyf(x)dx + \int_D^K \int_m^{xm/D} (1/D)g(y)dyf(x)dx \\ & + \int_D^K \int_{xm/D}^{x-D+m} (1/D)g(y)dyf(x)dx + \int_D^K \int_{x-D+m}^{x} (1/x)g(y)dyf(x)dx, \end{aligned} \tag{5b}$$

and L_C is positive constant as we suppose that there is zero probability that the assembly line is completely in a shutdown state throughout a shift.

Denoting the expected cycle costs by EC_c, it can be written that

$$EC_c(Q) = s + EH_c(Q) = s + \left(Q^2/2\right)H_C,$$

and we expect that $N/EL_c(Q) = N/(QL_c)$ cycles will appear in a year where N denotes the number of working days in a year. Similarly to the implemented idea in Vörös (2013) how to calculate the total cost, we can calculate the expected annual total cost (denoting it by ETC) as

$$ETC(Q)/N = [1/(QL_c)]\left[s + \left(Q^2/2\right)H_C\right] = (s/L_c)/Q + (Q/2)(H_c/L_c) \tag{6}$$

Vörös and Rappai (2016) pointed out that the optimal lot size can be determined as:

$$Q_{opt} = \sqrt{2s}\sqrt{1/H_c} = \sqrt{2sD/h}\sqrt{h/DH_c}, \tag{7}$$

and the minimum expected total annual cost can be described as:

$$ETC\left(Q_{opt}\right) = \frac{N}{L_c}\sqrt{2sH_c} = \frac{\sqrt{H_c}}{L_c}(constant). \tag{8}$$

In order to gain deeper insights into these two crucial expressions, we specify the distribution functions of the random variables in this paper. First of all we assume that y may not have larger values than the kth fraction of x, i.e. we assume that $y \leq x/k$, $k > 1$. Moreover, we assume that η has uniform distribution in the $0 \leq y \leq x/k$ interval. Thus its density can be written as:

$$g(y) = \left\{ \begin{matrix} \frac{k}{x} & \text{if } 0 \le y \le \frac{x}{k} \\ 0 & \text{otherwise} \end{matrix} \right\} \tag{9}$$

On the hand, we assume that the utilization level of the assembly line follows Beta distribution, and we define the probability density function as

$$f(x) = \left\{ \begin{matrix} \frac{1}{B(\alpha,\beta)} \left(\frac{\mathrm{X}}{\mathrm{K}}\right)^{\alpha-1} (1 - \frac{x}{K})^{\beta-1} & \text{if } 0 \le x \le K \\ 0 & \text{otherwise} \end{matrix} \right\} \tag{10}$$

We recall that K is the design capacity, while k is used to limit the number of cars moving into the clinic area. Depending on the relationship between k and D/m we have to distinct cases again. One of the outcomes of the relationship is when $k < D/m$. As we restrict the number of cars getting into the clinic area into the interval $0 \le y \le \frac{x}{k}$, the probability fields represented by Fig. 5 are modified, as from this assumption it follows that $yk \le x$. Inserting the $x = ky$ line into Fig. 5, we gain Fig. 6, which represents the new probability fields.

The other outcome is when $k \ge D/m$, and inserting the line $x = ky$ again into Fig. 5, we will have Fig. 7, indicating the possible outcomes of inventory cases. In the followings we stick up for this case as the previous one requires significantly more efforts and the estimated results would be developed in another new article.

The particular reason we use Beta distribution for random variable ξ is the flexibility of the probability density function. JIT systems, especially the well-functioning ones like TPS, exhibit very high utilization levels with low variations. If we start from the observation that the average utilization level of the

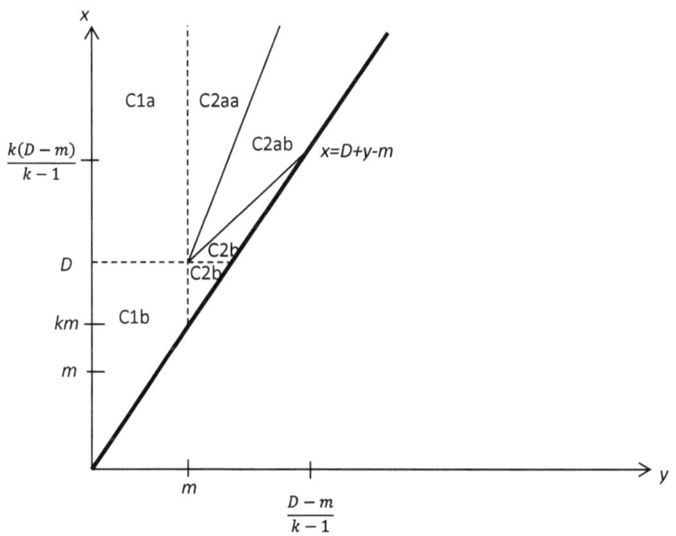

Fig. 6 Probability fields of inventory build-up diagrams with $k < D/m; D > m$ (C refers for Case)

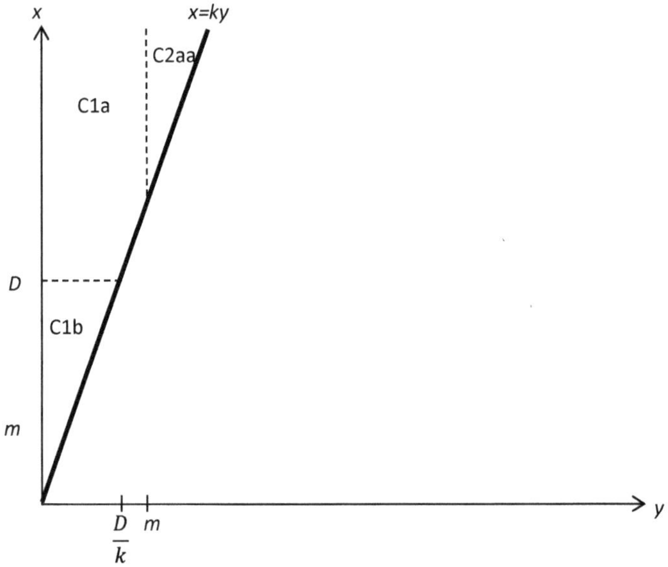

Fig. 7 Probability fields of inventory build-up diagrams with $k \geq D/m; D > m$ (C refers for Case)

assembly line is around 95, and 100% actually never exists, this fact indicates the usage of low $\beta \geq 1$ values. On the other hand, if we wish to have a given expected value, let it be denoted by $\bar{u}$, if we fix β and $\bar{x}$, the other parameter of the Beta distribution, α, is determined also, as

$$\bar{u} = \alpha/(\alpha + \beta) \tag{11a}$$

Figure 8 depicts the form of two density functions of Beta distribution, in both cases the expected utilization level is set on $\bar{u} = 0.95$ ($u = x/K$), while β takes the values of 2 and 3. The figure indicates that for $\beta = 3$ the variance is lower than in case $\beta = 2$, and for $\beta = 2$ capacity utilization levels around 80% occur with larger probability. Although the mathematical technique we are going to use can be applied to betas with larger integer values as well, it seems satisfactory to analyze the problem for betas taking values 2 or 3, additionally we can derive closed formulas providing good insights into the nature of the problem. An intensive use of the Beta distribution in economics can be seen also in Müller-Bungart (2007) as well.

Property 1 *In a JIT system if the utilization level follows Beta distribution, process quality improvement means larger utilization levels with lower variance.*

JIT systems are developed for standard products, produced in large volume, in an efficient way, and operated under high average capacity utilization level in order the dedicated, highly expensive capacities be paid back. Considering the structure of the expected level of capacity utilization given in (11a), when utilization level is high, the α/β ratio must be also high. Writing (11a) in different way, we have: $\alpha/$

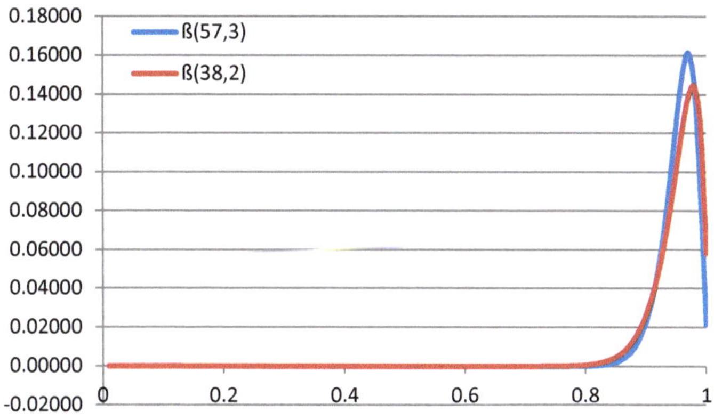

Fig. 8 The probability density function of Beta distribution for beta values 2 and 3 for 95% expected value (alphas are 38 and 75, respectively)

$\beta = -1 + 1/(1 - \bar{u})$ and as $\bar{u} \to 1$, $\alpha/\beta \to \infty$. Thus we may assume, that $\alpha > \beta \geq 1$. Fixing β, larger utilization levels mean larger α values. Now we show, that under the assumption that $\alpha > \beta \geq 1$, larger α involves lower variance.

The variance of the Beta distribution (denoted by σ^2):

$$\sigma^2 = \frac{\alpha\beta}{(\alpha+\beta)^2(\alpha+\beta+1)} \tag{11b}$$

Taking the derivative of (11b) with respect to α, we can write the followings:

$$\frac{d\sigma^2}{d\alpha} = \frac{\beta\left[(\alpha+\beta)^3 + (\alpha+\beta)^2\right] - \alpha\beta[3(\alpha+\beta)^2 + 2(\alpha+\beta)]}{\left[(\alpha+\beta)^3 + (\alpha+\beta)^2\right]^2}. \tag{12}$$

Focusing on the numerator, we show that it is negative for $\alpha > \beta \geq 1$. Let us suppose in contrary that

$$\beta\left[(\alpha+\beta)^3 + (\alpha+\beta)^2\right] - \alpha\beta[3(\alpha+\beta)^2 + 2(\alpha+\beta)] \geq 0,$$

then

$$\left[(\alpha+\beta)^3 + (\alpha+\beta)^2\right] - \alpha[3(\alpha+\beta)^2 + 2(\alpha+\beta)] \geq 0$$

as well. From this

$$-2\alpha^3 - \alpha^2(3\beta + 1) + \beta^3 + \beta^2 \geq 0$$

follows. As $\alpha > \beta \geq 1$, we can write that

$$-2\alpha^3 - \alpha^2(3\beta + 1) + \beta^3 + \beta^2 < -2\beta^3 - \beta^2(3\beta + 1) + \beta^3 + \beta^2 = -4\beta^3,$$

which is contradiction. Consequently, the variance is decreasing in α.

3 The Impact of Process Improvement on Lot Sizes and Total Costs When $x \geq ky$

Based on Fig. 7, we can identify that when $x/k \geq y$, i.e. when the number of cars entering the clinic area never exceeds the limit x/k, then inventory case C1b, C1a, and C2aa occur. The relevant inventory cost expressions in these cases do not contain the variable y, thus we may omit the expressions connected with y. This way, based on (4b), the expected inventory related costs can be determined as

$$\begin{aligned} H_c^{k \geq D/m} &= \int_0^D b \frac{D-x}{x^2} \frac{1}{B(\alpha,\beta)} \left(\frac{x}{K}\right)^{\alpha-1} \left(1 - \frac{x}{K}\right)^{\beta-1} dx \\ &+ \int_D^K h \left(\frac{1}{D} - \frac{1}{x}\right) \frac{1}{B(\alpha,\beta)} \left(\frac{x}{K}\right)^{\alpha-1} \left(1 - \frac{x}{K}\right)^{\beta-1} dx \\ &= \frac{b}{B(\alpha,\beta)} \int_0^D \frac{D-x}{x^2} \left(\frac{x}{K}\right)^{\alpha-1} \left(1 - \frac{x}{K}\right)^{\beta-1} dx \\ &+ \frac{h}{B(\alpha,\beta)} \int_D^K \left(\frac{1}{D} - \frac{1}{x}\right) \left(\frac{x}{K}\right)^{\alpha-1} \left(1 - \frac{x}{K}\right)^{\beta-1} dx. \end{aligned} \tag{13a}$$

Similarly, the expected cycle time can be determined as:

$$\begin{aligned} L_c^{k \geq D/m} &= \int_0^D \frac{1}{x} f(x) dx + \int_D^K \frac{1}{D} f(x) dx = \frac{1}{B(\alpha,\beta)} \int_0^D \frac{1}{x} \left(\frac{x}{K}\right)^{\alpha-1} \left(1 - \frac{x}{K}\right)^{\beta-1} dx \\ &+ \frac{1}{B(\alpha,\beta)} \frac{1}{D} \int_D^K \left(\frac{x}{K}\right)^{\alpha-1} \left(1 - \frac{x}{K}\right)^{\beta-1} dx \end{aligned} \tag{13b}$$

For example, for $\beta = 2$, we gain the following explicit expressions:

$$\begin{aligned}
H_{c,\beta=2}^{k\geq D/m} &= \frac{b}{B(\alpha,2)}\int_0^D \frac{D-x}{x^2}[\left(\frac{x}{K}\right)^{\alpha-1} - \left(\frac{x}{K}\right)^{\alpha}]dx + \frac{h}{B(\alpha,2)}\int_D^K \left(\frac{1}{D}-\frac{1}{x}\right)[\left(\frac{x}{K}\right)^{\alpha-1} - \left(\frac{x}{K}\right)^{\alpha}]dx \\
&= \frac{b}{B(\alpha,2)}\int_0^D [\frac{D}{K^{\alpha-1}}x^{\alpha-3} - \frac{1}{K^{\alpha-1}}x^{\alpha-2} - \frac{D}{K^{\alpha}}x^{\alpha-2} + \frac{1}{K^{\alpha}}x^{\alpha-1}]dx \\
&+ \frac{h}{B(\alpha,2)}\int_D^K [\frac{1}{DK^{\alpha-1}}x^{\alpha-1} - \frac{1}{K^{\alpha-1}}x^{\alpha-2} - \frac{1}{DK^{\alpha}}x^{\alpha-1} + \frac{1}{K^{\alpha}}x^{\alpha-1}]dx \\
&= \frac{1}{B(\alpha,2)}[\frac{D^{\alpha}}{K^{\alpha}}(\frac{b}{\alpha} - \frac{b}{\alpha-1} + \frac{h}{\alpha+1} - \frac{h}{\alpha}) + \frac{D^{\alpha-1}}{K^{\alpha-1}}(\frac{b}{\alpha-2} - \frac{b}{\alpha-1} + \frac{h}{\alpha-1} - \frac{h}{\alpha}) \\
&+ \left(\frac{h}{\alpha} - \frac{h}{\alpha-1}\right) + \frac{K}{D}(\frac{h}{\alpha} - \frac{h}{\alpha+1})]
\end{aligned} \tag{14a}$$

and

$$B(\alpha,2) = \int_0^1 [\left(\frac{x}{K}\right)^{\alpha-1} - \left(\frac{x}{K}\right)^{\alpha}]dx = \frac{K(\alpha+1)-\alpha}{\alpha(\alpha+1)K^{\alpha}} \tag{14b}$$

In these expressions D/K can be considered as planned utilization level as demand D is fixed well before the production time and K is the designed capacity. D/K should be at the $D/K \leq 1$ range as it has no sense to promise more than what can be satisfied. It is also reasonable to set $K = 1$, i.e. taking design capacity as one unit and D would mean the planned capacity utilization level. Let v denote the planned capacity utilization level, i.e. $v = D/K$, and let K be set at one. Then $B(\alpha,2) = 1/\alpha(\alpha+1)$, and substituting this value in (14a), we have the explicit form of the expected value of the inventory related costs:

$$\begin{aligned}
H_{c,\beta=2}^{k\geq D/m} &= \alpha(\alpha+1)[v^{\alpha}(\frac{b}{\alpha} - \frac{b}{\alpha-1} + \frac{h}{\alpha+1} - \frac{h}{\alpha}) \\
&+ v^{\alpha-1}(\frac{b}{\alpha-2} - \frac{b}{\alpha-1} + \frac{h}{\alpha-1} - \frac{h}{\alpha}) + \left(\frac{h}{\alpha} - \frac{h}{\alpha-1}\right) + \frac{1}{v}(\frac{h}{\alpha} - \frac{h}{\alpha+1})] \\
&= v^{\alpha}(-b\frac{\alpha+1}{\alpha-1} - h) + v^{\alpha-1}(b\frac{\alpha(\alpha+1)}{(\alpha-1)(\alpha-2)} + h\frac{\alpha+1}{\alpha-1}) - h\frac{\alpha+1}{\alpha-1} + h\frac{1}{v}
\end{aligned} \tag{15a}$$

This expression will not change its shape when we divide both sides with a positive constant, practically let this constant be h. Let the ratio of b and h be denoted by r, i.e. $r = b/h$. Let us note it is generally accepted in economics that $r > 1$. Then (15a) takes the form:

$$\frac{1}{h}H_{c,\beta=2}^{k\geq D/m} = v^{\alpha}(-\mathrm{r}\frac{\alpha+1}{\alpha-1} - 1) + v^{\alpha-1}(r\frac{\alpha(\alpha+1)}{(\alpha-1)(\alpha-2)} + \frac{\alpha+1}{\alpha-1}) - \frac{\alpha+1}{\alpha-1} + \frac{1}{v} \tag{15b}$$

Property 2 *When we plan full capacity utilization, for $\beta = 2$, the expected value of inventory related costs decreases as process quality improves.*

When we plan full capacity utilization, then $v = 1$, and (15b) can be written as

$$\frac{1}{h}H_{c,\beta=2}^{k\geq D/m} = \left(-\mathrm{r}\frac{\alpha+1}{\alpha-1} - 1\right) + \left(r\frac{\alpha(\alpha+1)}{(\alpha-1)(\alpha-2)} + \frac{\alpha+1}{\alpha-1}\right) - \frac{\alpha+1}{\alpha-1} + 1 = \mathrm{r}\frac{\alpha+1}{\alpha-1}\frac{2}{\alpha-2} \tag{16}$$

Then

$$\frac{d\frac{1}{hr}H_{c,\beta=2}^{k\geq D/m}}{d\alpha} = \frac{-4}{(\alpha-1)^2(\alpha-2)} - \frac{2(\alpha+1)}{(\alpha-2)^2(\alpha-1)},$$

which is negative expression as $\alpha > \beta = 2$. Consequently, the expected inventory related costs decrease when process quality increases.

Property 3 *Under high capacity utilization, for $\beta = 2$, the expected value of inventory related costs decreases as process quality improves.*

Let us note that the statement in Property 3 is more general, than in Property 2. In Property 2 the planned capacity utilization is set to one, now we allow it to vary in a range economically reasonable. In part, Property 3 is valid by continuity, from Property 2, but we expanded our research for a wide variety of r and v values as well. As we have very complicated formulas, we used a machine supported numerical approximation method (Mathematica software). In this method we focused on around a 95% expected value of the random variable x/K, and then for $\beta = 2$ it follows that it has sense when we let α at the interval $36 \leq \alpha \leq 42$ (at $\alpha = 38$ the expected value is 0.95). We let v to change at the interval $0.8 \leq v \leq 1$. Figure 9 shows the form of the expected values of the total costs when $r = 2$.

As it can be seen on Fig. 9, for example when $v = 0.8$, function H increases in α, i.e. it seems that the expected inventory related cost function increases in case of low planned capacity utilization while process improves. To check this, we made cuttings at $v = 0.9$ and at $v = 0.95$, and Figs. 10 and 11 indicate that function H increases when $v = 0.9$ and decreases when $v = 0.95$, respectively. Let us note that the first cutting is before while the second one is after the minimum point of the function H as function of v.

Corollary 1 *Under large planned capacity utilization, when process quality improves, the optimal lot size increases.*

The explanation behind this statement is the expected value H can be found at the denominator of the optimal lot sizing formula (7), thus when H decreases, the optimal lot size will increase. This findings support the idea of Porteus (1985, 1986). Let us note that it has no sense to deal with the low planned capacity utilization segment, i.e. when v < *minimum point of* H, as at this segment costs are at the same level, while output is lower.

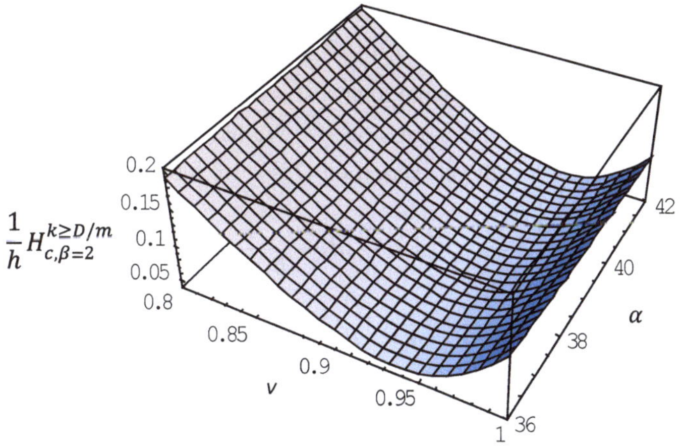

Fig. 9 The expected value of inventory related costs as function of process quality (α) and the planned capacity utilization (v) when Beta = 2, and $r = 2$

Fig. 10 Function H, when $v = 0.9$, Beta = 2, and $r = 2$

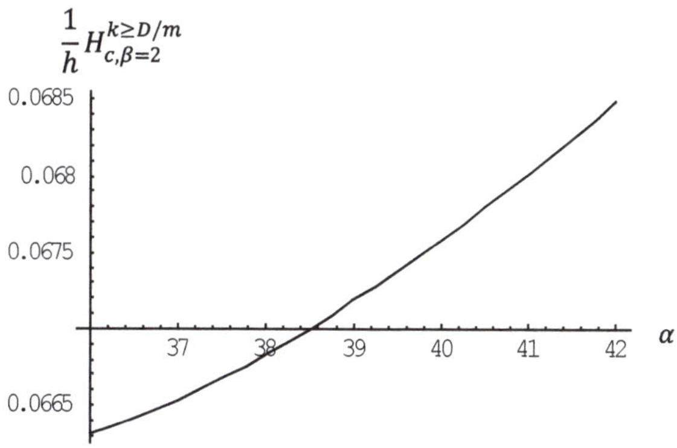

Fig. 11 Function H, when $v = 0.95$, Beta = 2, and $r = 2$

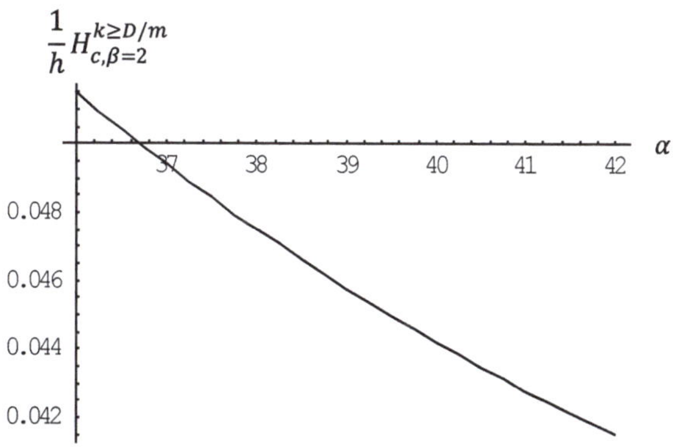

Property 4 *When we plan full capacity utilization, for* $\beta = 2$*, the expected annual total cost function decreases when process quality improves.*

The expected value of the annual total cost is determined in (8) and the shape of this function is formed by the expression $\frac{\sqrt{H_c}}{L_c}$. We have not determined the form of L, the expected cycle length yet. Based on (13b), we have

$$L_{c,\beta=2}^{k \geq D/m} = \frac{1}{B(\alpha,2)} \int_0^D \frac{1}{x}\left(\frac{x}{K}\right)^{\alpha-1}\left(1-\frac{x}{K}\right)dx + \frac{1}{B(\alpha,2)}\frac{1}{D}\int_D^K \left(\frac{x}{K}\right)^{\alpha-1}\left(1-\frac{x}{K}\right)dx$$
$$= \frac{1}{B(\alpha,2)}[(\frac{D}{K})^{\alpha}(\frac{1}{\alpha+1}-\frac{1}{\alpha})+(\frac{D}{K})^{\alpha-1}\left(\frac{1}{\alpha-1}-\frac{1}{\alpha}\right)+\frac{K}{D}\left(-\frac{1}{\alpha+1}+\frac{1}{\alpha}\right)]. \tag{17a}$$

As $B(\alpha,2) = \frac{K(\alpha+1)-\alpha}{\alpha(\alpha+1)K^{\alpha}}$, $v = D/K$, letting $K = 1$, this expression can be written as:

$$L_{c,\beta=2}^{k \geq D/m} = -v^{\alpha} + v^{\alpha-1}\frac{\alpha+1}{\alpha-1} + \frac{1}{v} \tag{17b}$$

Considering full planned capacity utilization, i.e. $v = 1$, (17a) and (17b) results in $\frac{\alpha+1}{\alpha-1}$, and utilizing (16), we can write that

$$\frac{\sqrt{H_c}}{L_c} = \frac{\sqrt{b\frac{\alpha+1}{\alpha-1}\frac{2}{\alpha-2}}}{\frac{\alpha+1}{\alpha-1}} = \sqrt{b\frac{\alpha-1}{\alpha+1}\frac{2}{\alpha-2}} \tag{18}$$

Taking the derivative of $\frac{\alpha-1}{\alpha+1}\frac{2}{\alpha-2}$ (b is constant) with respect to α, we have:

$$\left(\frac{\alpha-1}{\alpha+1}\frac{2}{\alpha-2}\right)' = \frac{2}{(\alpha+1)^2}\frac{2}{\alpha-2} + \frac{-2}{(\alpha-2)^2}\frac{\alpha-1}{\alpha+1} = 2\frac{-\alpha^2+2\alpha-3}{(\alpha+1)^2(\alpha-2)^2}.$$

The expression $-\alpha^2+2\alpha-3$ at the numerator is a concave parabola, with negative discriminant, thus always taking negative values. Consequently, as the derivative is negative, the minimum expected annual total cost in case of full planned capacity utilization is decreasing when process quality improves.

Property 5 *In case of high capacity utilization, for* $\beta = 2$ *the annual expected total cost function decreases when process quality increases.*

Similarly to the case in Property 3, we used machine supported numerical approximation method (Mathematica software). Figure 12 gives a typical result for $b = 2$, $h = 1$ ($r = 2$) when the planned capacity utilization varies between 80–100% and we investigate the nature of the minimum expected annual total cost

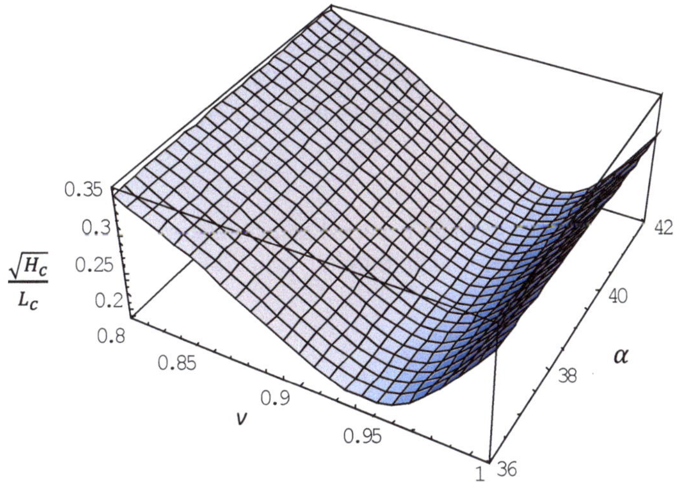

Fig. 12 The expected annual total cost function for $b = 2$, $h = 1$, $\beta = 2$

function around 95% process quality level (as $\beta = 2$, for $\alpha = 38$, the expected value is 0.95).

Now, we turn to the case, when the value of Beta is three. In order to keep the length of this paper reasonable, we give the main results only. For $\beta = 3$, the expected inventory related costs can be calculated as

$$\begin{aligned} H_{c,\beta=3}^{k\geq D/m} &= \frac{b}{B(\alpha,3)}\int_0^D \frac{D-x}{x^2}\left(\frac{x}{K}\right)^{\alpha-1}\left(1-\frac{x}{K}\right)^2 dx + \frac{h}{B(\alpha,3)}\int_D^K \left(\frac{1}{D}-\frac{1}{x}\right)\left(\frac{x}{K}\right)^{\alpha-1}\left(1-\frac{x}{K}\right)^2 dx \\ &= \frac{1}{B(\alpha,3)}[v^{\alpha+1}(\frac{b}{\alpha}-\frac{b}{\alpha+1}+\frac{h}{\alpha+1}-\frac{h}{\alpha+2}) + 2v^{\alpha}(\frac{b}{\alpha}-\frac{b}{\alpha-1}+\frac{h}{\alpha+1}-\frac{h}{\alpha}) \\ &\quad + v^{\alpha-1}(\frac{b}{\alpha-2}-\frac{b}{\alpha-1}+\frac{h}{\alpha-1}-\frac{h}{\alpha})\left(\frac{2h}{\alpha}-\frac{h}{\alpha-1}-\frac{h}{\alpha+1}\right) + \frac{1}{v}(\frac{h}{\alpha}-\frac{2h}{\alpha+1}+\frac{h}{\alpha+2})], \end{aligned} \tag{19a}$$

where $v = D/K$, and

$$B(\alpha,3) = \int_0^1 \left(\frac{x}{K}\right)^{\alpha-1}\left(1-\frac{x}{K}\right)^2 dx = \frac{1}{\alpha K^{\alpha-1}} - \frac{2}{(\alpha+1)K^{\alpha}} + \frac{1}{(\alpha+2)K^{\alpha+1}}, \tag{19b}$$

or writing this in different way:

$$\frac{1}{B(\alpha,3)} = \frac{\alpha(\alpha+1)(\alpha+2)}{2},$$

and applying this expression in (19a), and assuming unit design capacity, i.e. $K = 1$, we have the closed form for *H:*

$$\frac{2}{h}H_{c,\beta=3}^{k\geq D/m} = v^{\alpha+1}(r(\alpha+2)+\alpha) - v^{\alpha}\left(\frac{2r}{\alpha-1}(\alpha+2)(\alpha+1)+2(\alpha+2)\right) + v^{\alpha-1}\left(\frac{r}{(\alpha-1)(\alpha-2)}\alpha(\alpha+1)(\alpha+2)+\frac{(\alpha+1)(\alpha+2)}{\alpha-1}\right) - 2\frac{\alpha+2}{(\alpha-1)} + \frac{2}{v} \tag{20}$$

In order to have the minimum expected annual total cost function, we need the expected value of the cycle length. Similarly to (17a),

$$\begin{aligned} L_{c,\beta=3}^{k\geq D/m} &= \frac{1}{B(\alpha,3)}\int_0^D \frac{1}{x}\left(\frac{x}{K}\right)^{\alpha-1}\left(1-\frac{x}{K}\right)^2 dx + \frac{1}{B(\alpha,3)}\frac{1}{D}\int_D^K \left(\frac{x}{K}\right)^{\alpha-1}\left(1-\frac{x}{K}\right)^2 dx \\ &= \frac{1}{B(\alpha,3)}\left[\left(\frac{D}{K}\right)^{\alpha+1}\left(\frac{1}{\alpha+1}-\frac{1}{\alpha+2}\right) + 2\left(\frac{D}{K}\right)^{\alpha}\left(\frac{1}{\alpha+1}-\frac{1}{\alpha}\right)\right. \\ &\left. + \left(\frac{D}{K}\right)^{\alpha-1}\left(\frac{1}{\alpha-1}-\frac{1}{\alpha}\right) + \frac{K}{D}\left(-\frac{2}{\alpha+1}+\frac{1}{\alpha}+\frac{1}{\alpha+2}\right)\right] \end{aligned} \tag{21a}$$

Then substituting (19b) in (21a), we have with $v = D/K$, and assuming $K = 1$,

$$2L_{c,\beta=3}^{k\geq D/m} = v^{\alpha+1}\alpha - 2v^{\alpha}(\alpha+2) + \frac{v^{\alpha-1}(\alpha+1)(\alpha+2)}{\alpha-1} + \frac{2}{v} \tag{21b}$$

Now, we are able to give the functional form of $\frac{\sqrt{H_c}}{L_c}$ as each of the expressions required is known by (20) and (21b), respectively. As we have gained rather similar results like in case of $\beta = 2$, we modified the ranges of both v and α. We extended the searching area for v from very low planned utilization levels like 0.7 (sometimes 0.5), obviously till 1, while we let α to take lower values. This is because of the interest of mathematical nature, economically has no large relevance to investigate business situations when the process quality is around 50%, i.e. the assembly line is down during the half shift on average because of quality problems. Figure 13 represents the expected value of the minimum total annual cost, when we took $h = 1$, $b = 5$, i.e. $r = 5$, which means that backlogging costs are rather high compared to previous cases where r was only 2. On Fig. 13 the planned utilization level (v) may change at the interval $0.7 \leq v \leq 1$, while we allow α to be at the interval $30 \leq \alpha \leq 70$, which means that process quality (defined by the expected value of the output of the process) improves from 91 to 96%. Interestingly, under low planned capacity utilization the minimum total annual cost grows as process quality increases. In case of high capacity utilization the function decreases.

This nature can be spotted in Fig. 14 as well where the function is the same, but the ratio of the backlogging and holding cost is lower, we reduce it to 2 ($r = 2$). We restricted the planned utilization level into the interval $0.9 \leq v \leq 1$ to enlarge this property.

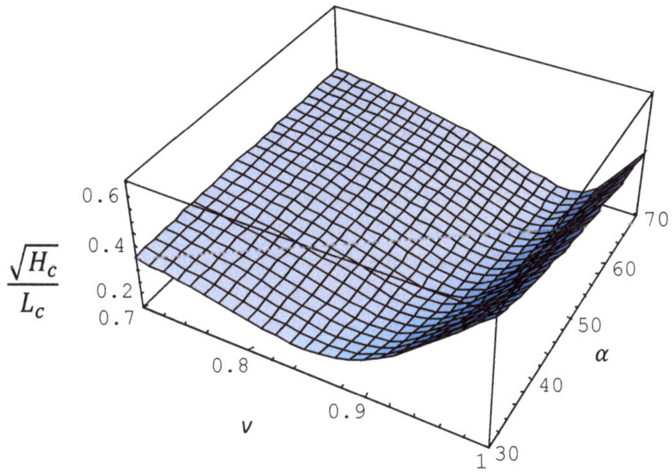

Fig. 13 The expected annual total cost function for $\beta = 3$, $h = 1$, $b = 2$

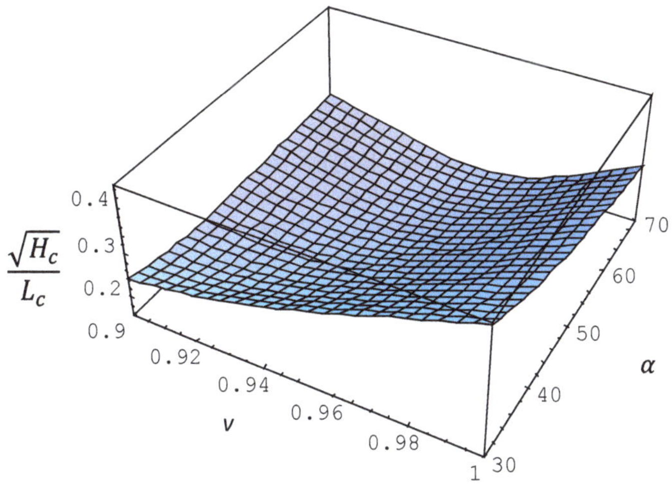

Fig. 14 The expected annual total cost function for $\beta = 3$, $h = 1$, $b = 5$

Based on these analyses we state the following property.

Property 6 *In case of high planned capacity utilization, for* $\beta = 3$, *the minimum annual expected total cost function decreases when process quality increases (as function of* α*).*

4 Estimating the Impact of Process Quality Improvement on the Variance of the Minimum Total Annual Cost

Up to this point we know that when a well operating assembly line follows Beta distribution, an improving process quality exhibits larger expected value of the output of the assembly line, which involves lower variance of the output at the same time. As function of parameter alfa, for fixed beta, we derived explicit expressions for two components of lot sizing rules: the expected value of the inventory related costs (H) and the expected length of the lot size cycles (L). Then we were able to characterize the expected value of the annual total cost, whose key component is $\frac{\sqrt{H_c}}{L_c}$, and suggested a lot size, which is affected by H. So, the decision making process goes like this:

Step 1 Determine contract volume, so we have the value of D
Step 2 Estimate probability density functions $f(x)$ and $g(x)$
Step 3 Determine expected values H and L
Step 4 Determine optimal lot size
Step 5 Determine the expected value of annual total cost as function of alpha for a given beta
Step 6 Improve process if it is economical
Step 7 Modify contract if it is economical, and go to Step 1, otherwise Stop.

As it can be seen in the previous steps, the procedure is based on expected values. When in Step 4 the optimal lot size is determined and this suggestion is implemented, the random events follow this decision, so actually the minimum annual total cost is random variable, too. Now, we provide two approaches to estimate the variance of the minimum annual total cost, as variance is an important information about the risk.

Because we are not able to derive explicit form of the variance of the minimum annual total cost, in this section we carry out two analyses to estimate the development of the variance of the minimum annual total cost through a simulation process. Doing so, the procedure is that we generate 1000-1000 random events under Beta distribution first for $\beta = 2$ and a series of alphas. Once a random event is generated under a given β and α pair, we calculate the inventory cost and cycle length. Let $H_c(\xi)$ denote the inventory cost, and $L(\xi)$ the corresponding cycle length and we have a 1000-1000 $H_c(\xi)$ and $L(\xi)$ values for each alpha-beta pair. The first simple procedure to estimate the variance of the minimum annual total cost is that we calculate the variance of the thousand $\frac{\sqrt{H_c(\xi)}}{L_c(\xi)}$ values, based on the structure of expression (8), which determine the expected minimum value of the annual total costs. So, we will have 1000 values for $\frac{\sqrt{H_c(\xi)}}{L_c(\xi)}$, and we take the average and the variance of these values. We carry out this simulation for a series of alphas and then we observe how the variance takes shape.

Table 2 The simulated average and variance of $\frac{\sqrt{H_c(\xi)}}{L_c(\xi)}$ when $\beta = 2$, $v = 0.95$, and $r = 2$

α values	Average process quality	Average annual total cost	Standard deviation of annual total cost	α values	Average process quality	Average annual total cost	Standard deviation of annual total cost
30	0.9375	0.217502	0.123061	41	0.9535	0.17814	0.089211
31	0.9394	0.209786	0.119538	42	0.9545	0.175805	0.081522
32	0.9412	0.205334	0.115072	43	0.9556	0.177543	0.08611
33	0.9429	0.206677	0.11678	44	0.9565	0.178147	0.087045
34	0.9444	0.198891	0.108373	45	0.9574	0.172653	0.078223
35	0.9459	0.195322	0.102818	46	0.9583	0.17267	0.080557
36	0.9474	0.190353	0.10097	47	0.9592	0.170681	0.076624
37	0.9487	0.187617	0.097878	48	0.9600	0.172856	0.075077
38	0.9500	0.189389	0.099052	49	0.9608	0.169486	0.069354
39	0.9512	0.18479	0.089917	50	0.9615	0.167133	0.067432
40	0.9524	0.184283	0.08773				

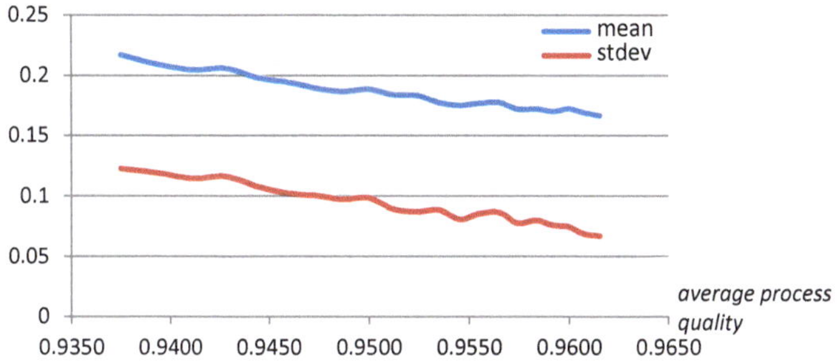

Fig. 15 The simulated mean and standard deviation of $\frac{\sqrt{H_c(\xi)}}{L_c(\xi)}$ for $\beta = 2$, $v = 0.95$, and $r = 2$

Below we see Table 2, giving the results of this simulation process. The inputs of the simulation process are: the planned capacity utilization level $v = 0.95$, inventory cost ratio $r = 2(b = 2, h = 1)$, $\beta = 2$, and α varies at the range of [30, 50].

Figure 15 summarizes the main results. Based on these we can state the following property.

Property 7 *The variance of the minimum annual total cost is decreasing when process quality improves.*

We conducted similar simulation process for $\beta = 3$, $v = 0.95$, and $r = 2$, but we let α at the range of [40, 80]. The results are summarized by Fig. 16, and the trends of the lines support the statement of Property 7.

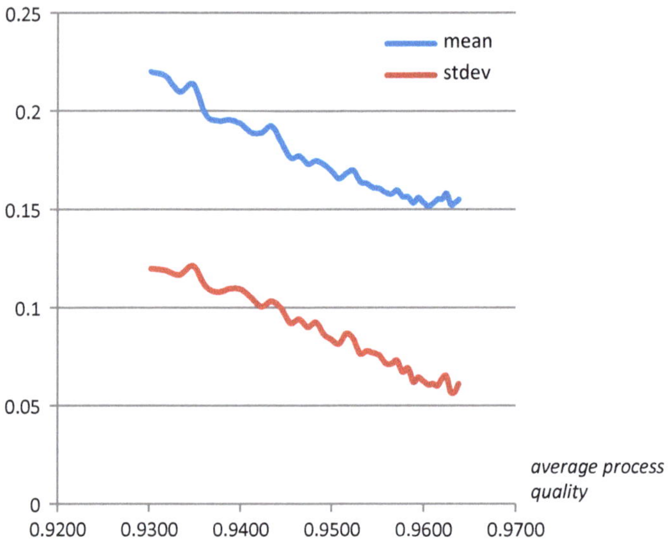

Fig. 16 The simulated mean and standard deviation of $\frac{\sqrt{H_c(\xi)}}{L_c(\xi)}$ for $\beta = 3$, $v = 0.95$, and $r = 2$

In the course of the next variance estimation process we start from the observation that based on (4a) and (5a), the annual total cost can be obtained in the following way: in a cycle the inventory related costs are determined by the expression $Q^2H(\xi)/2$, so the total set up and inventory cost in a cycle is: $s + Q^2H(\xi)/2$. The cycle length can be expressed as $QL(\xi)$, where ξ is the output of the assembly line, a random variable. Thus in a year we have $1/QL(\xi)$ cycles, and the annual cost is

$$TC(Q,\xi) = (1/QL(\xi))\left(s + Q^2H(\xi)/2\right) \tag{22}$$

To have the suggested lot size, in the literature usually the expected value of $H(\xi)$ and the expected value of $L(\xi)$ is taken, and then the optimal lot size is given, which is (by (8) as well)

$$Q_0 = \sqrt{2s}\sqrt{1/H_c}, \tag{23}$$

where H_c is given by (4b), the expected value of inventory related costs.

Substituting (23) in (22), we have

$$TC(\xi) = \frac{\sqrt{s/2}}{L(\xi)}\left(\sqrt{H_c} + \frac{H(\xi)}{\sqrt{H_c}}\right) \tag{24}$$

Table 3 The simulated average and variance of $\frac{1}{L(\xi)\sqrt{2}}(\sqrt{H_c} + \frac{H(\xi)}{\sqrt{H_c}})$ when $\beta = 2$, $v = 0.95$, and $r = 2$

α values	Average process quality	Average annual total cost	Standard deviation of annual total cost	α values	Average process quality	Average annual total cost	Standard deviation of annual total cost
30	0.938	0.247	0.127	41	0.953	0.208	0.111
31	0.939	0.241	0.135	42	0.955	0.207	0.117
32	0.941	0.243	0.134	43	0.956	0.198	0.098
33	0.943	0.237	0.130	44	0.957	0.196	0.103
34	0.944	0.231	0.129	45	0.957	0.192	0.087
35	0.946	0.226	0.125	46	0.958	0.191	0.092
36	0.947	0.224	0.126	47	0.959	0.192	0.089
37	0.949	0.217	0.117	48	0.960	0.186	0.083
38	0.950	0.211	0.113	49	0.961	0.185	0.087
39	0.951	0.215	0.118	50	0.962	0.185	0.076
40	0.952	0.207	0.112				

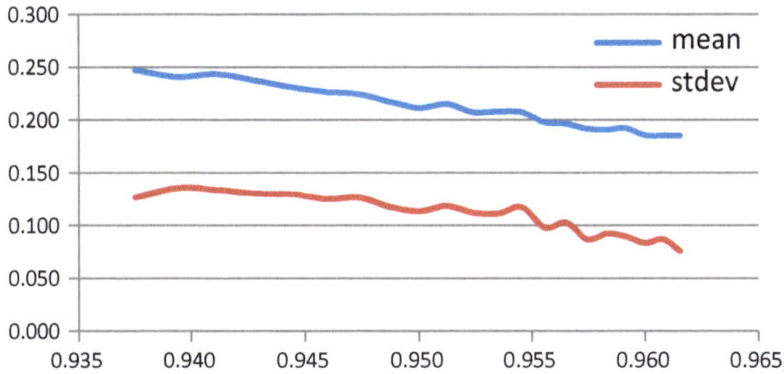

Fig. 17 The simulated mean and standard deviation of $\frac{1}{L(\xi)\sqrt{2}}(\sqrt{H_c} + \frac{H(\xi)}{\sqrt{H_c}})$ for $\beta = 2$, $v = 0.95$, and $r = 2$

Then we generated thousand random events for ξ with Beta = 2, and then Beta = 3 for a range of alphas, and the variance was calculated. Table 3 gives the numerical result for Beta = 2, and Fig. 17 gives the graphical representation.

Similarly to the previous case, we conducted simulation process for $\beta = 3$, $v = 0.95$, and $r = 2$, but we let α at the range of [40, 80]. The results are illustrated by Fig. 18.

All these findings support Property 7.

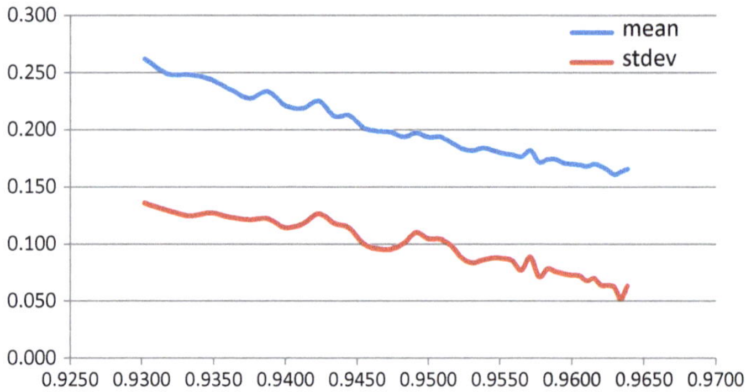

Fig. 18 The simulated mean and standard deviation of $\frac{1}{L(\xi)\sqrt{2}}(\sqrt{H_c} + \frac{H(\xi)}{\sqrt{H_c}})$ for $\beta = 3$, $v = 0.95$, and $r = 2$

5 Conclusions

This paper analyzes one of the oldest topics of management science literature, the lot sizing problem. Lots exist in any operations process, and we analyze lot sizing problem in JIT environment. Two key components of JIT production system are emphasized: the heijunka principle spread workloads as evenly as possible, and daily demands are fixed well in advance when production is planned. This fact suggests to take demand (orders) constant, but on the other hand, production problems may occur in each production system. Implementing the jidoka principle of JIT is about to solve these production problems. Employees are empowered to signal quality problems occurring in the production process and these signals frequently result in stoppages of the production process. This way we considered the output of the assembly line random variable following Beta distribution. The fact that well operating assembly lines in JIT system exhibit large capacity utilization levels suggests the usage of low Beta values and for $\beta = 2$ and $\beta = 3$ we derived explicit forms for the inventory related costs and the minimum expected annual costs as function of α. Let us note that the procedure we implemented can be used to any integer β with larger values as well.

For fixed β, when we increase α values, this expresses the improvement of the quality of the production process as well, as the expected value of the random output increases, additionally we found that the variance of the output decreases simultaneously. Using the explicit functions we are able to measure the impact and cost saving as a result of process improvement. We could point out that the expected value of the annual total cost would decrease when we can increase the process quality. We could point out also, that inventory related (holding and backlogging) costs and annual costs will decrease through process quality improvement, but under high capacity utilization. We produced counter example, when the planned capacity utilization level is low, but higher utilization level means higher output at the same cost, thus in case of economically reasonable planned capacity utilization levels we can state that process quality improvements results in lower expected minimum annual total costs and larger runs.

Because we think that closed forms may not be revealed for the variance of the annual total cost, we conducted two simulation analyses, generating thousand random events for a particular alpha-beta value. Betas were fixed at two and three, and we allow alpha to move in a wide range to see the behavior of the variance of the annual total cost. The results indicate that with process quality improvement the variance decreases.

Altogether, the results indicate that process improvement decreases costs, however process improvement is not free, and cost and benefits must be compared. At the same time, the analysis open windows for new research tracks as our work has been restricted for the case when the fraction of defective products, that go through the assembly line, but having obvious quality problems (getting to the clinic area this way), is relatively low (more definitely, when $k \geq D/m$). Extending the research for the complementary case probably requires the intensive usage of numerical and simulation techniques considering the complicated and hard to tackle formulas.

References

Cao, Q., & Schniederjans, M. J. (2004). A revised EMQ/JIT production-run model: An examination of inventory and productions costs. *International Journal of Production Economics, 87*(1), 83–95.

Chand, S. (1989). Lot sizes and setup frequency with learning in setups and process quality. *European Journal of Operational Research, 42,* 190–202.

Haris, F. W. (1913). How many parts to make at once. *The Magazine of Management, 10*(2), 135–136, 152.

Hax, A. C., & Candea, D. (1984). *Production and inventory management.* NJ: Prentice Hall.

Krajewski, L. J., Ritzman, L. P., & Malhotra, M. N. (2013). *Operations management.* Pearson.

Meredith, J. R., & Shafer, S. M. (2011). *Operations management.* Wiley.

Mishina, K., & Takeda, K. (1992). *Toyota motor manufacturing, U.S.A., Inc.* Harvard Business School 1–693-019.

Müller-Bungart, M. (2007). Revenue management with flexible products. In *Lecture notes in economics and mathematical systems* (Vol. 596). Springer.

Porteus, E. (1985). Investing in reduced setups in the EOQ model. *Management Science, 31,* 998–1010.

Porteus, E. (1986). Optimal lot sizing, process quality improvement and setup cost reduction. *Operations Research, 34*(1), 137–144.

Schniederjans, M. J., & Cao, Q. (2000). A note on JIT purchasing vs EOQ with a price discount: An expansion of inventory costs. *International Journal of Production Economics, 65,* 289–294.

Shih, W. (1980). Optimal inventory policies when stockouts result from defective products. *International Journal of Production Research, 18*(6), 677–686.

Slack, N., Brandon-Jones, A., Johnston, R., & Betts, A. (2015). *Operations and production management.* Pearson.

Stevenson, W. J. (2012). *Operations management.* McGraw-Hill/Irwin.

Taft, E. W. (1918). The most economical production lot. *The Iron Age, 101,* 1410–1412.

Vörös, J. (2013). Economic order and production quantity models without constraint on the percentage of defective items. *Central European Journal of Operations Research, 21*(4), 867–885.

Vörös, J., & Rappai, G. (2016). Process quality adjusted lot sizing and marketing interface in JIT environment. *Applied Mathematical Modelling, 40*(13–14), 6708–6724.

Watanabe, K. (2007). Lessons from Toyota's long drive. *Harvard Business Review*, 74–84.

Exploring Efficient Reward Strategies to Encourage Large-Scale Cooperation Among Boundedly Rational Players with the Risk and Impact of the Public Good

Yi Luo

Abstract In a public goods game, while cooperators need to make contributions, defectors can take a free ride after the realization of the public good. The Nash equilibrium in this game is simply zero contribution from all the players. A conventional approach to encourage cooperation and achieve the public good is using rewards to compensate the difference between the cooperators' and the defectors' payoffs. However, the public good may not be realized due to the uncertainty in the game, and the conventional way could underestimate the required rewards to achieve the public good. On the other hand, public good did realize in human history when people cannot survive from a natural disaster, such as a big flood, without cooperating to build a solid embankment, and most of them are willing to choose cooperation without rewards. The realization of the public good leads to the reduction of the defection cost and the contribution, which has a potential to encourage the players' cooperation. Then the conventional method may overestimate the necessary rewards to realize the public good. In this paper, a public goods game is employed to model the interaction among boundedly rational players with the rewards for large-scale cooperation, and a behavioral game-theoretic approach is developed to describe their decision making processes with the consideration of the risk and impact of public good in the game. It turns out that the conventional rewards to achieve the large-scale cooperation can be reduced for a favorable group of the players.

Y. Luo (✉)
Center for Sustainable Systems, School of Natural Resources and Environment, The University of Michigan, Ann Arbor, MI, USA
e-mail: Luo1@email.arizona.edu

A. Matsumoto (ed.), *Optimization and Dynamics with Their Applications*,
DOI 10.1007/978-981-10-4214-0_4

1 Introduction

Public goods are special class of goods holding the criterions of non-excludability and non-rivalrous consumption, and they had been intensively studied in recent years due to the rising of large-scale cooperation issues in social, environmental fields. In the large-scale cooperation, the interaction among the cooperators and the defectors with a potential public good can be modeled as a public goods game, and a contribution or a defection cost occurs when players choose cooperation or defection in the game. As the number of the cooperators reaches certain level, a public good can be realized, where the cooperators' contribution can be reduced and the defectors are able to take free rides. If the public good doesn't realize, the contribution of the cooperators will be in vain. When the public good is not important to the players or its importance has not been realized by them, the payoff of choosing cooperation is generally less than that of selecting defection with and without the public good in the game, since it takes more risks to choose cooperation. Then zero contribution is the Nash equilibrium of the public goods game, rewards are necessary to compensate the risk of realizing the public good and the difference between the contribution and the defection cost.

When the realization of the public good is critical to the players, for example, survival from natural or manmade disasters, the spread of new breakthrough technology in a region, etc., or the number of the cooperators is close to the realization of the public good, the payoff of choosing cooperation could be greater than that of selecting defection, some players would like to choose cooperation without rewards or with few rewards. Thus, potential public good has an impact to encourage the players' cooperation in these cases. Overall, in order to develop efficient reward strategies to encourage individual players' cooperation, the risk and impact of the public good should be evaluated in their decision making processes. However, classic game theory and equilibrium approach have limited capability to capture them.

The uncertainty in the public goods game is two-fold, although the number of cooperators can be observed, each defector has limited knowledge of other defectors' decisions, and the realization of the public good is also unknown. Due to the uncertainty, a player incurs the risk of obtaining his/her expected gain regardless of choosing cooperation or defection. It is assumed that the players are boundedly rational, they consider both the expected payoff of choosing cooperation or defection and its associated risk and balance them based on their own risk taking attitudes; the individual players' risk taking attitudes are different, which depend on their personal characteristics such as age, experience, background, education, etc., and they choose cooperation if the payoffs of being cooperators are greater than those of being defectors. Therefore, a behavioral game-theoretic approach is developed to model the interaction among the boundedly rational players with the risk and impact of the public good and their decision making processes with the consideration of their risk taking attitudes.

At the initial stage of cooperation, it is risky to choose cooperation. The minimal reward for the cooperation is to make sure that the payoff of being a cooperator is greater than that of being a defector after the realization of the public good. As the number of the cooperators increases, the impact of the public good shows up and it can encourage more players' participation in cooperation. It turns out that the conventional reward to encourage cooperation can be reduced by taking advantage of the impact of the public good, the sequence of the players' choosing cooperation depends on their risk taking attitudes, and the process of realizing the public good with the rewards can be simulated as discrete events. Based on the behavioral game-theoretic approach, efficient reward strategies to achieve large-scale cooperation can be obtained from individual players' decision making processes along the process of realizing the public good.

The organization of the rest of the paper can be described as follows. Section 2 introduces the players' payoffs of choosing cooperation or defection before and after the realization of the public good in a public goods game; Sect. 3 proposes a behavioral game-theoretic decision model to describe individual players' decision making processes in the game; Sect. 4 evaluates efficient reward strategies for large-scale cooperation based on the behavioral game-theoretic decision model; conclusions are given in Sect. 5.

2 The Payoffs of Being a Cooperator and a Defector with Rewards in a Public Goods Game

Let T be the total number of the players needs to cooperate for the realization of a public good. Suppose all of T players' risk taking attitudes have been evaluated, and then these players can be ranked by a sequence of numbers 1, 2, 3, …, T based on the values of their risk taking attitudes starting from the most risk tolerant player to the most risk averter. Let I be the set of T players, i be the index of the players' choosing cooperation in the public goods game, $i \in I$. Let C or D be the cooperator's economic contribution or the defector's cost before the realization of the public good $(C > 0, D > 0)$, and the former and the latter are reduced to C' and D' respectively when the public good is realized $(C' > 0,\ D' > 0)$. In general, the defection cost is less than the contribution in the public goods game irrespective of whether the public good is realized $\left(C > D,\ C' > D'\right)$. Let R_i be the minimal reward to encourage a player to be the ith cooperator in the public goods game $(R_i > 0)$, $\bar{R}$ be the average reward to achieve the public good, which can be obtained from

$$\bar{R} = \frac{\sum_{i=1}^{T} R_i}{T}. \tag{1}$$

If the public good cannot be realized, the reward may not lead to its intended result, and it can be considered as another kind of contribution. The *direct* payoffs of the cooperator or the defector with and without the public good can be described as $\bar{R}-C'$ or $-D'$ and $\bar{R}-C$ or $-D$ respectively. However, it is not trifle to obtain the values of C' and D' in reality. Let θ or γ be the reduction of the contribution or the defection cost before and after the realization of the public good in terms of a unit of initial investments, we have

$$\theta = \frac{C - C'}{C + \bar{R}} \tag{2}$$

and

$$\gamma = \frac{D - D'}{C + \bar{R}}, \tag{3}$$

where their values can be obtained from similar situations in other industries through data analysis. Then the *direct* payoffs of being a cooperator or a defector after realizing the public good can be updated as $R_i - C + \theta(C + \bar{R})$ or $-D + \gamma(C + \bar{R})$ based on Eqs. (2) and (3).

When the players' cooperation reaches certain level, a public good can be realized, and it benefits all the players irrespective of whether they choose cooperation or defection. Let E be economic impact factor to describe its comprehensive effect in terms of one percent increment of the cooperators and one unit of initial contribution, economic literature shows that it has been used to analyze the benefits of the public goods in biomass production (Perez-Verdin et al. 2008), food chain (Sonntag 2008), public markets (Econsult Corporation 2007) and a variety of other sectors on the local economy. Since the benefit from the realization of the public good depends on the threshold of the public good, the value of E, and the initial contributions including C and $\bar{R}$, a player's *additional* payoff after realizing the public good can be estimated as $T(C + \bar{R})E$ irrespective of he/she chooses to be ith cooperator or not. Then the payoffs of choosing cooperation or defection with and without the public good can be summarized in Table 1.

If the useful patterns related to parameters D, E, θ, and γ in the players' payoffs exist in similar industries, their values can be obtained from advanced statistical and machine learning such as logistic regression, Bayesian network analysis; otherwise, their values can be approximated using a scientific consensus methodology such as expert elicitation. Expert elicitation is the synthesis of authorities of a subject where there is uncertainty due to insufficient data or when such data is unattainable because of physical constraints or lack of resources (Apostolakis 1990).

Table 1 The payoffs of a player being ith cooperator or not with and without the public good

Has the public good been realized?	Payoff of being a cooperator	Payoff of being a defector
No	$R_i - C$	$-D$
Yes	$R_i - C + \theta(C + \bar{R}) + T(C + \bar{R})E$	$-D + \gamma(C + \bar{R}) + T(C + \bar{R})E$

As stated previously, the Nash equilibrium of the public goods game without reward is zero cooperation. Conventional reward $C - D$ to encourage individual players' cooperation and push the Nash equilibrium to realize the public good can be obtained from the player's payoffs without the realization of the public good as shown in Table 1. However, in addition to considering the payoff of being a cooperator, each boundedly rational player also concerns about his/her contribution could be in vain once the public good cannot be realized. As the number of cooperators increases, the chance of realizing the public good becomes large, which can encourage more players' participation in cooperation. It turns out that the conventional reward cannot efficiently encourage cooperation to realize the public good due to the uncertainty in the public goods game. Therefore, a behavioral game-theoretic decision model is proposed in the next section to consider the risk and impact of the public good in individual players' decision making processes.

3 Large-Scale Cooperation with the Risk and Impact of the Public Good

3.1 Behavioral Game-Theoretic Decision Model

Let $\tilde{X}$ be a player's potential payoff from being a cooperator or a defector in a public goods game, and it is a random variable due to the uncertainty in the game. In the behavioral game-theoretic decision model, the random variable is approximated by the player's expected payoff and the risk of obtaining it based on his/her risk taking attitude. Let $\bar{X}$ be the estimation of $\tilde{X}$, and its value is described by the following equation

$$\bar{X} = E(\tilde{X}) - \lambda\delta(\tilde{X}), \tag{4}$$

where $E(\tilde{X})$ represents the player's expected payoff of choosing cooperation or defection, standard deviation $\delta(\tilde{X})$ indicates the risk of obtaining it, and coefficient λ denotes his/her risk taking attitude to the risk. If the player has risk tolerant attitude, the value of his/her risk taking attitude is not greater than zero ($\lambda \leq 0$); if he/she is a risk averter, the value of his/her risk taking attitude is greater than zero ($\lambda > 0$).

Individuals' risk attitudes have been studied by many game theory scholars in recent years (Egbue and Long 2012; Eckel and Grossman 2008; Fullenkamp et al. 2003; Pennings 2002; Wang et al. 2009). In this paper, it is assumed that there are both risk tolerant players and risk averters in the game, and the change of exogenous factors, such as the rewards and the number of the cooperators in the game, can only affect the players' expected payoff and its associated risk rather than his/her risk taking attitude. The values of this parameter can be evaluated from large representative surveys combined with complementary field experiments (Dohmen et al. 2011). Finally, the players' risk taking attitudes λ can be estimated by normalizing the evaluated values into an interval between lower bound $-W$ and upper bound V $(W, V > 0)$.

In the public goods game, each player's payoff depends on other players' decisions and the realization of the public good. The progress of realizing the public good indicates the outcome of their interaction with the rewards, which can be described by the number of the cooperators in the public goods game. Let P_i be the probability of realizing a public good with i cooperators $(0 \leq P_i \leq 1, i \in I)$, and it can be evaluated from similar industries using data analysis. In general, the more cooperators lead to the larger probability of reaching the public good. In the meantime, the probability of un-realizing the public good can be denoted as $1 - P_i$. If no defector chooses to be the ith cooperator, the probability of realizing a public good becomes P_{i-1} because of $i-1$ cooperators. Due to the relationship between the players' payoffs and the realization of the public good, probabilities $1 - P_i$ or $1 - P_{i-1}$ and P_i or P_{i-1} also indicate the chances that a player obtains his/her payoffs from choosing cooperation or defection before and after the realization of the public good as shown in Table 1 respectively. Thus, the individual players' payoff in the public goods game can be estimated from the progress of cooperation and their risk taking attitudes based on Eq. (4).

Let $\tilde{G}_i$ be the ith cooperator's payoff with reward R_i. If the probability of realizing the public good is P_i, the player's expected payoff $E(\tilde{G}_i)$ of choosing cooperation can be calculated from the following equation based on Table 1

$$\begin{aligned} E(\tilde{G}_i) &= (R_i - C)(1 - P_i) + [R_i - C + (\theta + TE)(C + \bar{R})]P_i \\ &= R_i - C + (\theta + TE)(C + \bar{R})P_i, \quad i \in I \end{aligned} \tag{5}$$

and its value keeps rising as the number of cooperators or/and the amount of rewards increases. In the meantime, the risk $\delta(\tilde{G}_i)$ of choosing cooperation and obtaining the expected payoff can be estimated from its standard deviation

$$\begin{aligned} \delta(\tilde{G}_i) &= \sqrt{E[(\tilde{G}_i - E(\tilde{G}_i))^2]} = \sqrt{E(\tilde{G}_i^2) - (E(\tilde{G}_i))^2} \\ &= (\theta + TE)(C + \bar{R})\sqrt{P_i(1 - P_i)}. \quad i \in I. \end{aligned} \tag{6}$$

The value of $\delta(\tilde{G}_i)$ increases at the initial stage of cooperation, and then decreases after the value of P_i is greater than 0.5 due to the properties of item $\sqrt{P_i(1 - P_i)}$, which captures the risk of realizing the public good during the process of cooperation. Let λ_i be the ith cooperator's risk taking attitude to balance the expected payoff and its associated risk, and the approximation $\bar{G}_i$ of cooperator's payoff $\tilde{G}_i$ can be obtained as follows from Eq. (1)

$$\bar{G}_i = E(\tilde{G}_i) - \lambda_i \delta(\tilde{G}_i) = R_i - C + (\theta + TE)(C + \bar{R})[P_i - \lambda_i \sqrt{P_i(1 - P_i)}]. \quad i \in I. \tag{7}$$

If the player is not willing to be the ith cooperator, his/her payoff of being a defector is denoted as $\tilde{H}_i$. The probability of realizing the public good becomes P_{i-1}, and the estimation $\bar{H}_i$ of value of $\tilde{H}_i$ can be described as

$$\begin{aligned}\bar{H}_i &= E(\tilde{H}_i) - \lambda_i \delta(\tilde{H}_i) \\ &= -D + (\gamma + TE)(C + \bar{R})[P_{i-1} - \lambda_i \sqrt{P_{i-1}(1 - P_{i-1})}], \quad i \in I\end{aligned} \tag{8}$$

where its derivative is similar to that of Eq. (7). It is assumed that the impact of an individual farmer's choice on probability P_i is small and can be ignored for a large-scale cooperation. Then probability P_{i-1} in Eq. (8) can be replaced by P_i for the sake of mathematical simplicity. Since a player's gain of being a cooperator is the difference of his/her payoffs between choosing cooperation and defection, its value can be computed from

$$\begin{aligned}\bar{G}_i - \bar{H}_i &= E(\tilde{G}_i) - E(\tilde{H}_i) - \lambda_i[\delta(\tilde{G}_i) - \delta(\tilde{H}_i)] \\ &= R_i - C + D + (\theta - \gamma)(C + \bar{R})[P_i - \lambda_i \sqrt{P_i(1 - P_i)}], \quad i \in I\end{aligned} \tag{9}$$

where the player chooses to cooperate if its right-hand-side value is greater than zero. Equation (9) indicates individual players' decision making process in the public goods game based on the behavioral game-theoretic decision approach. The minimal value ($-W$) or the maximal value (V) of the players' risk taking attitudes as mentioned earlier in this section can be evaluated from Eq. (9). For example, the player with the most risk aversion attitude may choose to defection even if his/her reward is $C - D$ and the public good is closed to be realized.

3.2 The Realization of the Public Good with the Rewards

Using the progress of cooperation to represent the outcome of the public goods game, the behavioral game-theoretic decision model explores the players' interaction with the rewards, and describes their individual decision making processes based on their risk taking attitudes. In a public goods game with T players, suppose a player chooses cooperation and becomes the ith cooperator, and the probability of realizing the public good increases to P_i. All the players can observe that, the rest $T - i$ defectors will decide who will be the next cooperator. Each of them can update his/her gain of choosing cooperation, but has no idea of the others' decisions and the realization of the public good due to limited information in the game. If a defector's gain of choosing cooperation with the current cooperation level is greater than zero, he/she will be the $i + 1$th cooperator; otherwise, the reward needs to be increased until one of them is willing to choose cooperation based on his/her gain. Therefore, the conventional reward can be reduced along the process of realizing the public good due to the impact of the public good.

At certain point of the cooperating process, the gains of choosing cooperation for different players depend on their risk taking attitudes. If $\theta \geq \gamma$, the risk of choosing cooperation is greater than that of selecting defection based on Eqs. (6) and (8). It is necessary to use rewards to promote cooperation and realize the public good. With certain reward, the most risk tolerant player is more likely to choose cooperation based on Eq. (9), and the index to indicate the sequence of the cooperator's appearance can be described as $i = 1, 2 \ldots T$. If $\theta < \gamma$, the risk of choosing defection is greater than that of selecting cooperation, and the impact of the public good can greatly promote the player's cooperation and some players are willing to avoid the risk by choosing cooperation even without the reward. Since the most risk aversion player is more likely to choose cooperation in this case from Eq. (9), the index to represent the order of the players' choosing cooperation can be denoted as $i = T, T - 1, T - 2 \ldots 1$. So, the sequence of the players to choose cooperation is determined by their risk taking attitudes. Due to the players' different risk taking attitudes as mentioned before, the large-scale cooperation in the public goods game with the reward can be simulated as discrete events.

4 Individual Reward Strategies Based on Behavioral Game-Theoretic Decision Model

For certain level of the reward to compensate the difference between the contribution and defection cost, whether a player chooses to be the *i*th cooperator or not depends on how he/she treats the potential public good based on his/her risk taking attitude λ_i. If the value of $(\theta - \gamma)(C + \bar{R})[P_i - \lambda_i \sqrt{P_i(1 - P_i)}]$ in Eq. (9) is less than zero, it represents the risk of the public good, and it is necessary to use a reward more than $C - D$ to compensate the risk and make this player choose cooperation; otherwise, it denotes the impact of the public good, and he/she would like to be a cooperator with a reward less than $C - D$. The risk or impact of the public good depends on the number of the cooperators in the game, and then efficient individual reward strategies can be obtained along the process of large-scale cooperation.

From Eq. (9), the value of reward R_i to encourage a player to be the *i*th cooperator can be obtained from the following equation

$$R_i = \begin{cases} C - D - (\theta - \gamma)(C + \bar{R})[P_i - \lambda_i \sqrt{P_i(1 - P_i)})], & \text{if } \theta \geq \gamma \quad i = 1, 2, 3 \ldots T \\ C - D - (\gamma - \theta)(C + \bar{R})[\lambda_i \sqrt{P_i(1 - P_i)} - P_i], & \text{if } \gamma > \theta \quad i = T, T - 1, T - 2 \ldots 1 \end{cases} \tag{10}$$

where the value of average individual-based reward $\bar{R}$ can be derived as follows based on Eqs. (1) and (10)

$$\bar{R} = \frac{(2C-D)T}{T + \sum_{i=1}^{T} \{(\theta-\gamma)[P_i - \lambda_i\sqrt{P_i(1-P_i)}]\}} - C. \tag{11}$$

If the players have risk tolerant attitudes with $\theta \geq \gamma$ and risk aversion attitudes with $\gamma > \theta$, conventional reward $C-D$ to encourage their cooperation can be reduced based on Eq. (10) due to the impact of the public good; otherwise, besides the conventional reward, additional reward is required to promote the players' cooperation by compensating the risk of the public good. Whether the total amount of the conventional rewards can be saved depends on average reward $\bar{R}$ in Eq. (11). Let $\sum_{i=1}^{T}\{(\theta-\gamma)[P_i - \lambda_i\sqrt{P_i(1-P_i)}]\}$ be the accumulate risk/impact of the public good for all T players in terms of one unit of initial investment, and the value of $\bar{R}$ can be less than $C-D$ when the accumulated risk/impact is greater than zero.

5 Conclusions

A public goods game is employed to describe the interaction among players for large-scale cooperation, and a novel behavioral game-theoretic approach is developed to model individual players' choosing cooperation or defection in the public goods game. The players are assumed to make the best decisions by balancing the expected gain and its associated risk with their own risk taking attitudes. It turns out the behavioral game-theoretic model is able to capture the individual players' decision making processes in the game and explore the necessary conditions to develop efficient reward strategies for the realization of the public good.

While the public good's achievement depends on a large variety factors in the public goods game, the approach to describe the individual players' decision making processes reveals their basic relationships, which can be considered as an important component to simulate large-scale cooperation. In this paper, the impact of the public good on the public goods game is analyzed from the viewpoint of individual players' decision-makings. However, as the number of players increases, the increased interaction between them makes their decisions more rational. Then future research will explore the impact of the public good from the viewpoint of group-based decision-making and model the whole process of large-scale cooperation by integrating both of them.

References

Apostolakis, G. (1990). The concept of probability in safety assessments of technological systems. *Science, 250*(4986), 1359–1364.

Dohmen, T., Falk, A., Huffman, D., Sunde, U., Schupp, J., & Wagner, G. G. (2011). Individual risk attitudes: Measurement, determinants, and behavioral consequences. *Journal of the European Economic Association, 9*(3), 522–550.

Eckel, C. C., & Grossman, P. J. (2008). Forecasting risk attitudes: An experimental study using actual and forecast gamble choices. *Journal of Economic Behavior & Organization, 68,* 1–17.

Econsult Corporation. (2007). Estimating the economic impact of public markets. http://www.pps.org/pdf/mpps_public_markets_eis.pdf.

Egbue, O., & Long, S. (2012). Barriers to widespread adoption of electric vehicles: An analysis of consumer attitudes and perception. *Energy Policy, 48,* 717–729.

Fullenkamp, C., Tenorio, R., & Battalio, R. (2003). Assessing individual risk attitudes using field data from lottery games. *Review of Economics and Statistics, 85*(1), 218–226.

Pennings, M. E. J. (2002). Pulling the trigger or not: Factors affecting behavior of initiating a position in derivatives markets. *Journal of Economic Psychology, 23,* 263–278.

Perez-Verdin, G., Grebner, D. L., Munn, I. A., Sun, C., & Grado, S. C. (2008). Economic impacts of woody biomass utilization for bioenergy in Mississippi. *Forest Products Journal, 58*(11), 75–83.

Sonntag, V. (2008). *Why local linkages matter? Findings from the local food economy study.* Seattle, WA: Sustainable Seattle. http://www.usask.ca/agriculture/plantsci/hort2020/local_linkages.pdf.

Wang, J., Fu, F., Wu, T., & Wang, L. (2009). Emergence of social cooperation in threshold public goods games with collective risk. *Physical Review E, 80,* 016101.

Part II
Dynamics

Periodicity Induced by Production Constraints in Cournot Duopoly Models with Unimodal Reaction Curves

Gian-Italo Bischi, Laura Gardini and Iryna Sushko

Abstract In the Cournot duopoly game with unimodal piecewise-linear reaction functions (tent maps) proposed by Rand (J Math Econ, 5:173–184, 1978) to show the occurrence of robust chaotic dynamics, a maximum production constraint is imposed in order to explore its effects on the long run dynamics. The presence of such constraint causes the replacement of chaotic dynamics with asymptotic periodic behaviour, characterized by fast convergence to superstable cycles. The creation of new periodic patters, as well as the possible coexistence of several stable cycles, each with its own basin of attraction, are described in terms of border collision bifurcations, a kind of global bifurcation recently introduced in the literature on non-smooth dynamical systems. These bifurcations, caused by the presence of maximum production constraint, give rise to quite particular bifurcation structures. Hence the duopoly model with constraints proposed in this paper can be seen as a simple exemplary case for the exploration of the properties of piecewise smooth dynamical systems.

Keywords Oligopoly games · Dynamical systems · Constraints · Border collision bifurcations

JEL classification C61 · C73 · L13

G.-I. Bischi (✉) · L. Gardini
DESP-Department of Economics, Society, Politics, University of Urbino, Urbino, Italy
e-mail: gian.bischi@uniurb.it

L. Gardini
e-mail: laura.gardini@uniurb.it

I. Sushko
Institute of Mathematics NASU, 3, Tereshchenkivska st., Kiev 01601, Ukraine
e-mail: sushko@imath.kiev.ua

A. Matsumoto (ed.), *Optimization and Dynamics with Their Applications*,
DOI 10.1007/978-981-10-4214-0_5

1 Introduction

After the duopoly model with linear demand and cost functions proposed by Cournot (1838), where a Cournot-Nash equilibrium is achieved in the long run as the game is repeated by two players endowed with naïve expectations, oligopoly models have been extended by many authors in several directions. A stream of literature studies the stability of oligopolistic markets as the number of competing firms increases (see e.g. Teocharis 1960; Hahn 1962; Okuguchi 1964; Okuguchi and Szidarovszky 1999) or different kinds of expectations are considered (see e.g. Szidarovszky and Okuguchi 1997; Szidarovszky 1999; Bischi and Kopel 2001) or different levels of market knowledge (see e.g. Bischi et al. 2010 and references therein). Another stream considers duopoly models with nonlinear demand and/or cost functions, from which several kinds of reaction functions can be obtained, even non monotonic ones, which may lead to periodic or quasi-periodic or chaotic behaviors (Rand 1978; Dana and Montrucchio 1986; Puu 1991; Kopel 1996; Bischi et al. 2000, 2010). In particular, David Rand, in a seminal paper published in 1978, proposed unimodal reaction functions represented by symmetric piecewise linear tent maps, and by using a formal approach based on symbolic dynamics showed that with such reaction functions a Cournot tâtonnement can be chaotic, i.e. erratic bounded oscillations arise with sensitive dependence on initial conditions. Moreover, such behaviour persists if the shape of the reaction functions is slightly changed, i.e. the chaotic behaviour is structurally stable (also denoted as "robust chaos").

Economic motivations for unimodal reaction functions have been given in Huyck et al. (1984) and Witteloostuijn and Lier (1990) in terms of goods that are strategic substitutes and complements in the sense of Bulow et al. (1985), whereas (Dana and Montrucchio 1986) proved that any kind of reaction function can be obtained from a sound economically microfounded problem with suitable demand and cost functions. In Puu (1991), Puu shows how an hill-shaped reaction function can be obtained by using linear costs and a hyperbolic demand function, i.e. an isoelastic demand, and that complex behavior emerges provided that agents are sufficiently heterogeneous; in Kopel (1996) and Bischi and Lamantia (2002) unimodal reaction curves are obtained starting from a linear demand function and a nonlinear cost function with positive cost externalities. In all these papers complex (periodic or chaotic) dynamics arise through the well-known period doubling route to chaos, typical of nonlinear smooth discrete dynamical systems. Moreover, global dynamical properties have been studied in Bischi et al. (2000); Bischi and Kopel (2001); Bischi and Lamantia (2002); Agliari et al. (2002) where the method of critical curves for continuously differentiable maps is used to bound chaotic attractors and to characterize global bifurcations that cause qualitative modifications of the basins of attraction.

All these oligopoly models are based on the implicit assumption that firms can adjust outputs to their desired levels, without constraints on minimum and maximum production. Only a few works on the subject relax this assumption (see for instance Puu and Norin 2003; Tramontana et al. 2011; Bischi et al. 2010; Bischi and Lamantia 2012). Of course, the presence of an upper limit of production capacity constitutes a quite realistic assumption, that may be related to exogenous factors (e.g. maximum production rules imposed by authorities of scarcity of limiting input factors) or endogenous limitations that have not been considered in the optimization problems leading to the best reply decision expressed by the reaction curve. As a matter of fact, a maximum production constraint introduces an upper cut in the shape of the reaction functions, which consequently become *piecewise smooth* maps characterized by a horizontal portion (*flat-top reaction functions*). The state space of the corresponding discrete dynamical system, obtained by introducing the usual Cournot tâtonnement with naïve expectations, can be partitioned into regions where the functional form of the map changes (see Mosekilde and Zhusubaliyev 2003 and Bernardo et al. 2008). This implies that interesting dynamic scenarios, typical of piecewise differentiable maps, can be observed and explained as consequences of the presence of *borders* in the phase space (or *switching manifolds*) where the functional form defining the map changes, and consequently the Jacobian matrix of the dynamical system is discontinuous along such borders. The collision of an invariant set of the piecewise smooth map with such a border may lead to a bifurcation often followed by drastic changes in the dynamic scenarios. Such contacts are called *Border Collision Bifurcations*, a term introduced in Nusse and Yorke (1992) (see also Nusse and Yorke 1995), and then adopted by many authors. The simplest case occurs when a fixed point (or a periodic point) crosses a border of non differentiability in a piecewise smooth map. In Banerjee et al. (2000a, b), it is shown that such a contact may produce any kind of effect (transition to another cycle of any period or to chaos), depending on the eigenvalues of the two Jacobian matrices involved on the two opposite sides of the border.

The effects of these bifurcations in oligopoly models with constraints have already been considered in Bischi et al. (2010); Bischi and Lamantia (2012). In particular, the latter examines a classical linear Cournot model and shows that the introduction of capacity constraints leads to complex time patters, i.e. the creation of both periodic and chaotic attractors in a model whose dynamics without constraints only exhibits convergence to the unique Cournot-Nash equilibrium. Similar results are shown in Bischi et al. (2010), where smooth nonlinear models with constraints are considered as well, and standard bifurcations typical of smooth dynamical systems are combined with border collision bifurcations when constraints are imposed.

In this paper we consider a different dynamic experiment: Starting from the chaotic piecewise linear model of Rand we introduce the constraints that transform the tent maps into flat-top reaction functions, and we show that this transforms chaotic motion into periodic behaviour, so that predictability is enhanced due to the presence of maximum production constraints. Moreover, the convergence to these cycles is very fast, as they are superstable due to the periodic points inside the flat branches of the modified reaction function. In particular, we investigate the existence of stable cycles and prove that they are created through border collision bifurcations. Cases of coexistence of stable cycles, each with its own basin of attraction, are also discussed.

The duopoly model with constraints proposed in this paper can also be seen as a simple exemplary case for the exploration of the properties of piecewise smooth dynamical systems. Indeed, we believe that it is nowadays interesting to relate such phenomena to the rich literature on piecewise smooth dynamical systems arising in relevant applications in electrical engineering (see Bernardo et al. 1999; Banerjee and Grebogi 1999; Banerjee et al. 2000a, b; Avrutin and Schanz 2006; Avrutin et al. 2006; Tramontana and Gardini 2011) or physics (see e.g. Zhusubaliyev et al. 2006, 2007), and even to the works of some mathematical precursors of Nusse and Yorke that already studied the particular bifurcations associated with piecewise smooth maps, such as Leonov (1959, 1962); Mira (1978, 1987); Maistrenko et al. (1993, 1995, 1998).

The paper is organized as follows. In Sect. 2 the setup of the classical Cournot duopoly model and some basic properties are recalled, in particular the possibility to understand the dynamic behaviour of the duopoly model through the study of a proper one-dimensional map defined as the composition of the two reaction functions. In Sect. 3 the properties of the one-dimensional map are studied in detail, and in Sect. 4 the corresponding dynamics of the two-dimensional Cournot model are considered, together with the effects of the presence of an adaptive adjustment with inertia, or anchoring, where any firm computes the next period production as a convex combination between the current output and the one computed according to the reaction function. Section 5 concludes and gives suggestions on further researches about the proposed model.

2 The Constrained Cournot-Rand Model with Piecewise Linear Reaction Functions

The classical Cournot duopoly game is obtained by considering a market composed of two firms producing homogeneous goods. At each discrete time period $t = 0, 1, ...,$ the two firms decide their outputs, let's say x_t and y_t respectively, by solving a profit maximization problem

$$x_{t+1} = \arg\max \Pi_x(x, y^{(e)}_{t+1}); \quad y_{t+1} = \arg\max \Pi_y(x^{(e)}_{t+1}, y)$$

where $x^{(e)}_{t+1}$ and $y^{(e)}_{t+1}$ represent the expectations of each producer about the production decision of the other one. Under the assumptions that each maximization problem has a unique solution and each firm has naïve expectations, i.e. $x^{(e)}_{t+1} = x_t$ and $y^{(e)}_{t+1} = y_t$ as in the original Cournot paper (Cournot 1838), the classical Cournot tâtonnement is obtained, given by $(x_{t+1}, y_{t+1}) = T(x_t, y_t) = (R_x(y_t), R_y(x_t))$ where $R_x(y_t)$ and $R_y(x_t)$ are called *reaction functions*.

We recall (see e.g. Bischi et al. 2000; Agliari et al. 2002) that the second iterate of the map T is a decoupled map:

$T^2(x, y) = T(R_x(y), R_y(x)) = (R_x(R_y(x)), R_y(R_x(y))) = (F(x), G(y))$.

The two one-dimensional maps F and G are defined as the compositions of reaction functions:

$$F(x) = R_x \circ R_y(x) \;,\; x \in X, \quad and \quad G(y) = R_y \circ R_x(y) \;,\; y \in Y \tag{1}$$

where the strategy sets X and Y are assumed to be such that the maps F and G are well defined. Hence, all the properties of the classical Cournot map can be deduced from the study of these two (conjugate) one-dimensional maps, see again (Bischi et al. 2000) where it is shown, among other properties, that a point (x_i, y_i) is a periodic point of period n for T if and only if $x = x_i$, and $y = y_i$ are periodic points of F and G of period n (if n is odd) or a divisor of n (if n is even). Moreover, any cycle C of the two-dimensional map T is associated with one or two cycles of F, say C_1 and C_2 ($C_2 = C_1$ or $C_2 \neq C_1$), with the periodic points of C belonging to the cartesian product $(C_1 \cup C_2) \times (R_y(C_1 \cup C_2))$, where C is attracting for T if and only C_1 and C_2 are attracting for F. A similar result holds for cyclic chaotic intervals of the map F, giving rise to cyclic chaotic rectangles of T according to the corresponding cartesian products. In case of coexistence of attractors, also the basins of attraction have the form of a rectangular shaped grid, according the cartesian products of the corresponding basins of the maps F and G.

In the original work of Cournot, as well as in many textbooks, the reaction functions are decreasing and intersect in a unique point of the positive quadrant, which is also the unique equilibrium point, now denoted as Cournot-Nash equilibrium. In this case the trajectories of the discrete dynamical system can either converge to the fixed point, or to a cycle of period 2 or diverge. Instead, if more general reaction functions are considered, then the Cournot tâtonnement may display more complex behaviors. In the pioneering paper (Rand 1978) it is shown that quite complex dynamics can emerge when unimodal reaction functions are considered, in particular the occurrence of robust chaos is proved by considering a tent map, defined by the following piecewise linear function:

$$R(z) = \begin{cases} az & if \;\; 0 \le z \le \frac{1}{2} \\ a(1-z) & if \;\; \frac{1}{2} \le z \le 1 \end{cases} \tag{2}$$

In this paper, following (Bischi et al. 2010), we introduce an upper capacity limit L_z, with $z = x, y$, so that each firm can produce an output z bounded inside the interval $[0, L_z]$. Moreover, we introduce the possibility of inertia (or anchoring attitude) given by a convex combination between the current production and the one computed according to the reaction function, namely

$$T: \begin{cases} x_{t+1} = (1-\alpha_x)x_t + \alpha_x R_x(y_t) \\ y_{t+1} = (1-\alpha_y)y_t + \alpha_y R_y(x_t) \end{cases} \tag{3}$$

where $\alpha_z \in [0, 1]$, $z = x, y$, expresses the attitude to stick at the current production as the parameter α_z decreases. Of course, for $\alpha_x = \alpha_y = 1$, the usual Cournot tâtonnement without inertia is obtained, whose dynamics is reduced to the study of the one-dimensional map F.

2.1 The Piecewise Linear One-Dimensional Map for the Case Without Inertia

Let us consider the following reaction function, given by a piecewise linear continuous maps depending on two parameters, the slope $a_x > 1$ and the upper production constraint (the "roof") $L_x > \frac{1}{2}$:

(I) if $L_x \geq \frac{a_x}{2}$, then the reaction function is the standard tent map (2), i.e.:

$$R_{x(I)}(y) = \begin{cases} a_x y & \text{if } 0 \leq y \leq \frac{1}{2} \\ a_x(1-y) & \text{if } \frac{1}{2} \leq y \leq 1 \end{cases} \tag{4}$$

(II) If $\frac{1}{2} < L_x < \frac{a_x}{2}$ then the reaction function is a tent map with a flat top branch:

$$R_{x(II)}(y) = \begin{cases} a_x y & \text{if } 0 \leq y \leq \frac{L_x}{a_x} \\ L_x & \text{if } \frac{L_x}{a_x} \leq y \leq 1 - \frac{L_x}{a_x} \\ a_x(1-y) & \text{if } 1 - \frac{L_x}{a_x} \leq y \leq 1 \end{cases} \tag{5}$$

whose shape in the two different cases is qualitatively shown in Fig. 1. An analogous definition holds for $R_y(x)$ with slope a_y and upper production L_y.

Let us first consider the Cournot duopoly model (3) without inertia, i.e. $\alpha_x = 1$ and $\alpha_y = 1$, so that map T becomes a two-dimensional map having the second iterate with separate variables:

$$T^2 : \begin{cases} x'' = F(x) := R_x \circ R_y(x) \\ y'' = G(y) := R_y \circ R_x(y) \end{cases} \tag{6}$$

whose properties can be studied by use of the composite function $F(x) = R_x \circ R_y(x)$ (or the conjugate one $G(y)$) as described in Bischi et al. (2000). Thus, let us focus on the possible shapes of the map related to $F(x)$. Depending on the values of the four constants of the map, a_x, L_x, a_y and L_y, we can distinguish four cases:

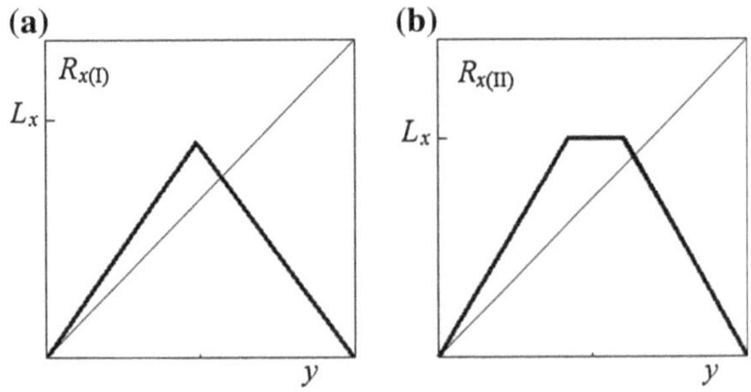

Fig. 1 Qualitative picture of the reaction function $R_x(y)$ in the two possible configurations: $R_{x(I)}(y)$ in (**a**), and $R_{x(II)}(y)$ in (**b**)

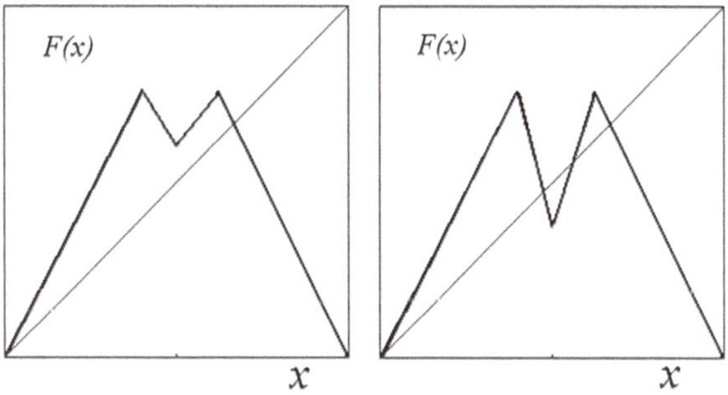

Fig. 2 Qualitative shape of $F(x)$ in the case N1

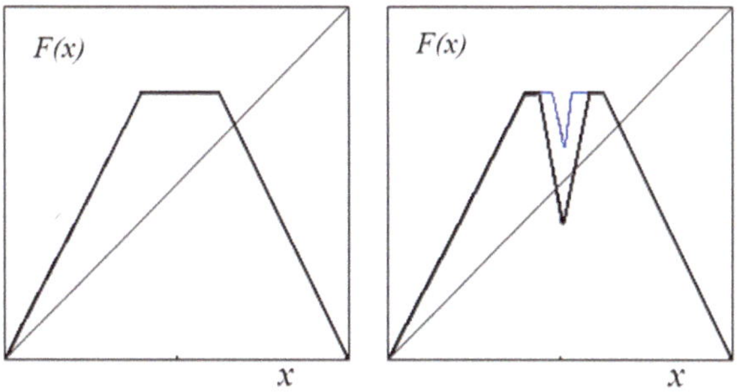

Fig. 3 Qualitative shape of $F(x)$ in the case N2

N1 (I-I) when $L_y \geq \frac{a_y}{2}$ and $L_x \geq \frac{a_x}{2}$, then the function $F(x)$ is defined as follows (the qualitative shape of the map is shown in Fig. 2):

$$F(x) = \begin{cases} a_x a_y x & \textit{if } 0 \leq x \leq \frac{1}{2a_y} \\ -a_x a_y x + a_x & \textit{if } \frac{1}{2a_y} \leq x \leq \frac{1}{2} \\ -a_x(-a_y x + a_y) + a_x & \textit{if } \frac{1}{2} \leq x \leq 1 - \frac{1}{2a_y} \\ a_x(-a_y x + a_y) & \textit{if } 1 - \frac{1}{2a_y} \leq x \leq 1 \end{cases} \tag{7}$$

N2 (I-II) when $L_y \geq \frac{a_y}{2}$ and $L_x < \frac{a_x}{2}$. Then depending on the parameters' values we can have the following two cases (see Fig. 3):

if $\frac{a_y}{2} \leq (1 - \frac{L_x}{a_x})$ then

$$F(x) = \begin{cases} a_x a_y x & \textit{if } 0 \leq x \leq \frac{L_x}{a_x a_y} \\ L_x & \textit{if } \frac{L_x}{a_x a_y} \leq x \leq 1 - \frac{L_x}{a_x a_y} \\ -a_x(-a_y x + a_y) + a_x & \textit{if } 1 - \frac{L_x}{a_x a_y} \leq x \leq 1 \end{cases} \tag{8}$$

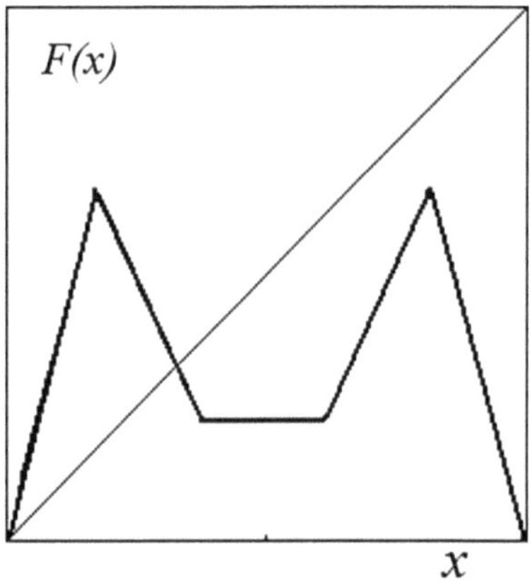

Fig. 4 Qualitative shape of $F(x)$ in the case N3

if $\frac{a_y}{2} > (1 - \frac{L_x}{a_x})$ then

$$F(x) = \begin{cases} a_x a_y x & \text{if } 0 \le x \le \frac{L_x}{a_x a_y} \\ L_x & \text{if } \frac{L_x}{a_x a_y} \le x \le \frac{1}{a_y} - \frac{L_x}{a_x a_y} \\ -a_x a_y x + a_x & \text{if } \frac{1}{a_y} - \frac{L_x}{a_x a_y} \le x \le \frac{1}{2} \\ -a_x(-a_y x + a_y) + a_x & \text{if } \frac{1}{2} \le x \le 1 - \frac{1}{a_y} + \frac{L_x}{a_x a_y} \\ L_x & \text{if } 1 - \frac{1}{a_y} + \frac{L_x}{a_x a_y} \le x \le 1 - \frac{L_x}{a_x a_y} \\ a_x(-a_y x + a_y) & \text{if } 1 - \frac{L_x}{a_x a_y} \le x \le 1 \end{cases} \tag{9}$$

N3 (II-I) when $L_y < \frac{a_y}{2}$ and $L_x \ge \frac{a_x}{2}$ then the function $F(x)$ is defined as follows (the qualitative shape of the map is shown in Fig. 4):

$$F(x) = \begin{cases} a_x a_y x & \text{if } 0 \le x \le \frac{1}{2a_y} \\ -a_x a_y x + a_x & \text{if } \frac{1}{2a_y} \le x \le \frac{L_y}{a_y} \\ -a_x L_y + a_x & \text{if } \frac{L_y}{a_y} \le x \le 1 - \frac{L_y}{a_y} \\ -a_x(-a_y x + a_y) + a_x & \text{if } 1 - \frac{L_y}{a_y} \le x \le 1 - \frac{1}{2a_y} \\ a_x(-a_y x + a_y) & \text{if } 1 - \frac{1}{2a_y} \le x \le 1 \end{cases} \tag{10}$$

N4 (II-II) when $L_y < \frac{a_y}{2}$ and $L_x < \frac{a_x}{2}$. Then depending on values of the parameters we can have the following two cases:

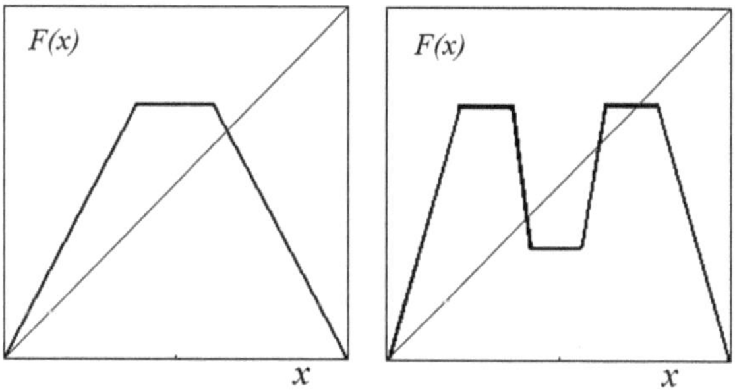

Fig. 5 Qualitative shape of $F(x)$ in the case N4

if $\frac{a_y}{2} \leq (1 - \frac{L_x}{a_x})$ then

$$F(x) = \begin{cases} a_x a_y x & \textit{if } 0 \leq x \leq \frac{L_x}{a_x a_y} \\ L_x & \textit{if } \frac{L_x}{a_x a_y} \leq x \leq 1 - \frac{L_x}{a_x a_y} \\ a_x(-a_y x + a_y) & \textit{if } 1 - \frac{L_x}{a_x a_y} \leq x \leq 1 \end{cases} \tag{11}$$

if $\frac{a_y}{2} > (1 - \frac{L_x}{a_x})$ then

$$F(x) = \begin{cases} a_x a_y x & \textit{if } 0 \leq x \leq \frac{L_x}{a_x a_y} \\ L_x & \textit{if } \frac{L_x}{a_x a_y} \leq x \leq \frac{1}{a_y} - \frac{L_x}{a_x a_y} \\ -a_x a_y x + a_x & \textit{if } \frac{1}{a_y} - \frac{L_x}{a_x a_y} \leq x \leq \frac{L_y}{a_y} \\ -a_x L_y + a_x & \textit{if } \frac{L_y}{a_y} \leq x \leq 1 - \frac{L_y}{a_y} \\ -a_x(-a_y x + a_y) + a_x & \textit{if } 1 - \frac{L_y}{a_y} \leq x \leq 1 - \frac{1}{a_y} + \frac{L_x}{a_x a_y} \\ L_x & \textit{if } 1 - \frac{1}{a_y} + \frac{L_x}{a_x a_y} \leq x \leq 1 - \frac{L_x}{a_x a_y} \\ a_x(-a_y x + a_y) & \textit{if } 1 - \frac{L_x}{a_x a_y} \leq x \leq 1 \end{cases} \tag{12}$$

(the qualitative shape of the map is shown in Fig. 5).

Before investigating the properties of the one-dimensional map $F(x)$ in the different cases, let us recall those related to the tent map given in (4) and to the tent map with a flat top defined in (5). Due to the constraints assumed for the slopes in our system (always larger than 1), we have that the tent map cannot have any attracting cycle, thus only one chaotic interval or cyclic chaotic intervals can exist as attracting sets (we refer to the survey Sushko et al. 2015 for further details). Differently, the occurrence of a flat top introduces the existence of attracting cycles (which are also superstable, having the eigenvalue equal to zero), which occurs whenever a cycle has a periodic point in the flat branch, and it attracts almost all the points of the interval (except the repelling cycles and related preimages of any order, which may belong to a Cantor set, also called a chaotic repellor), or of an unstable cycle which is a Milnor attractor (see Milnor 1985), attracting almost all the points of the interval. The cycles existing in the flat top case are not different from those existing in unimodal maps, as already described in Metropolis et al. (1973), but their appearance/disappearance is related to the crossing of a point of the cycle through one border of the interval in which the map has a flat branch, leading to a so-called border collision bifurcations (BCB for short) in the terminology commonly used since its introduction by Nusse and Yorke (1992, 1995) (see also Sushko et al. 2016).

3 Dynamics of the Equivalent One-Dimensional Map

The dynamics of the tent map is clearly strictly related to that of the one-dimensional map $F(x)$ in the case N1 described above, $F(x)$ is a bimodal map with slopes everywhere larger than 1, from which it follows that also now an attracting cycle cannot exist. However, the existence of two critical values (a local minimum and a maximum) may lead to coexistence of two chaotic attracting sets. An example is shown in Fig. 6: the local minimum and its image, as well as the maximum and its image, bound two intervals in which the dynamics are chaotic. The two basins of attraction are separated by the repelling fixed point x^* shown in Fig. 6a, and all its preimages of any rank.

More frequently, when the Cournot game is described by map $F(x)$ in the case N1, the attracting set is unique, and given by a chaotic interval or cyclical chaotic intervals. Differently, the cases considered below, involving a flat top, lead to attracting cycles.

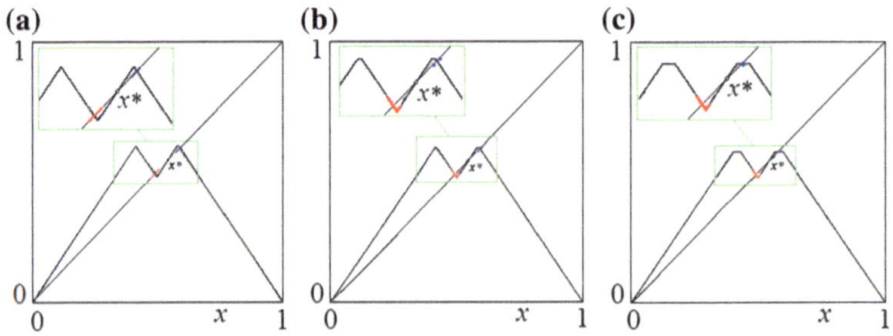

Fig. 6 Map $F(x)$ at $a_y = 1.2$, $L_y = 0.7$, $a_x = 1.2$. In **a** Two coexisting chaotic intervals as attracting sets in case N1 at $L_x = 0.61$. In **b** at $L_x = 0.595$ case N2 with two attracting sets: a chaotic interval and a superstable 2-cycle. In **c** at $L_x = 0.59$ case N2 with two attracting sets: a chaotic interval and a superstable fixed point

3.1 Case N2

Let us consider the dynamics of map $F(x)$ when the parameters satisfy the conditions of case N2. Although the generic case consists of a unique superstable cycle, attracting almost all the points of the interval, except those of repelling cycles and related preimages, which may belong to a chaotic repellor, we can have situations leading to coexistence. In fact, considering the example of case N1 shown above, in Fig. 6a, by decreasing the slope a_x we enter in case N2. In Fig. 6b, c we show the coexistence of the attracting chaotic interval, related to the local minimum and its image, with a superstable cycle related to the maximum. In Fig. 6b the maximum is a periodic point of a superstable 2-cycle, while in Fig. 6c it is a superstable fixed point. In both cases the two basins of attraction are separated as before by the repelling fixed point x^* shown in Fig. 6b, c and all its preimages of any rank.

In order to show the bifurcations related to the superstable cycles let us consider a two-dimensional bifurcation diagram in the parameter plane (a_x, L_x), as shown in Fig. 7a, at fixed $a_y = 1.6$ and $L_y = 0.81$. It can be seen that the diagonal (the straight line $L_x = a_x/2$) is the bifurcation curve which separates two different regimes: above it case N1 occurs (white points denote chaotic dynamics) while below it we have case N2, and the colored regions indicate periodicity regions associated with superstable cycles for which the maximum value is a periodic point (different colors correspond to different periods of the cycles).

Whenever the periodic point belonging to a flat branch collides with one border of the interval on which the map is flat, the cycle undergoes a border collision. For example, considering a parameter point belonging to the yellow region shown in Fig. 7a, map F has a unique attractor, a superstable fixed point x_s^* on the rightmost flat branch (and in value it is $x_s^* = L_x$), as shown in Fig. 7b. Decreasing the parameter L_x the fixed point x_s^* undergoes border collision when it collides with the lower border of the flat interval, i.e. when it is $L_x = 1 - \frac{1}{a_y} + \frac{L_x}{a_x a_y}$ leading to the BCB curve Bl of equation (which also corresponds to $L_x = x^*$, or $x_s^* = x^*$)

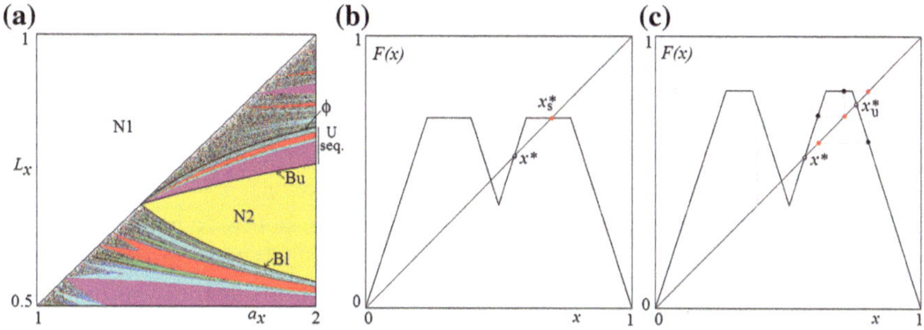

Fig. 7 In **a** two-dimensional bifurcation diagram in the parameter plane (a_x, L_x); In **b** map $F(x)$ in case N2 at $a_x = 1.9$ and $L_x = 0.7$ with a superstable fixed point; In **c** map $F(x)$ in case N2 at $a_x = 1.9$ and $L_x = 0.8$ with a superstable 3-cycle. Here $a_y = 1.6$ and $L_y = 0.81$

$$Bl : L_x = \frac{a_x a_y - a_x}{a_x a_y - 1} \tag{13}$$

while increasing the parameter L_x the fixed point x_s^* undergoes border collision when it collides with the upper border of the flat interval, i.e. when it is $L_x = 1 - \frac{L_x}{a_x a_y}$ leading to the BCB curve Bu of equation

$$Bu : L_x = \frac{a_x a_y}{a_x a_y + 1} \tag{14}$$

The two BCB curves Bl and Bu are intersecting in the point $(a_x, L_x) = (2 - \frac{1}{a_y}, 1 - \frac{1}{2a_y})$ and thus it belongs to the diagonal, the line of equation $L_x = a_x/2$, as it is clearly visible in Fig. 7a. The BCB occurring when the upper boundary Bu is crossed is a flip BCB: the superstable fixed point becomes unstable (moving to the steep decreasing rightmost branch) and leading to the appearance of a superstable 2-cycle. Notice that increasing L_y further the dynamics can be studied by use of the tent map with a flat top given in (5), up to the homoclinic bifurcation of the repelling fixed point $x^* = \frac{a_x a_y - a_x}{a_x a_y - 1}$ occurring when the image of the maximum is mapped into it, that is, when $F(L_y) = x^*$ leading to the (homoclinic) bifurcation curve ϕ of equation

$$\phi : L_x = 1 - \frac{a_y - 1}{a_y(a_x a_y - 1)} \tag{15}$$

which is also shown in Fig. 7a.

In fact, in this interval of values of L_x, the absorbing interval is given by $[F(L_x), L_x]$ and map F in this interval has the same shape of map given in (5), thus the whole sequence of cycles occurring in the U-sequence occurs also here, leading to infinitely many periodicity regions related to attracting cycles. The borders of the periodicity regions are related to flip BCB curves accumulating on homoclinic bifurcation curves which, on their turn, are related to parameter values in which the image of some finite rank $F^k(L_x)$ of the maximum L_x merges with a periodic point of some repelling cycle. When a parameter point belongs to such particular homoclinic curves, as for the bifurcation curve ϕ given above, the repelling cycle which includes the periodic point $F^k(L_x)$ attracts the intervals in which the map is flat, as well as all the related preimages, thus the repelling cycle becomes a Milnor attractor (see Sushko et al. 2014 for further examples and discussion).

The values of the parameters used in the case shown in Fig. 7c belong to the red region inside the portion where the U-sequence commented above occurs, and is associated with an attracting 3-cycle with periodic points inside the absorbing interval $[F(L_x), L_x]$. Notice that in Fig. 7a we can observe other regions associated with superstable 3-cycles, they differ in the position of the periodic points with respect to the branches of map F. For example in Fig. 8a we can see an enlarged portion of the two-dimensional bifurcation diagram and on the line $a_x = 1.3$ two more red regions

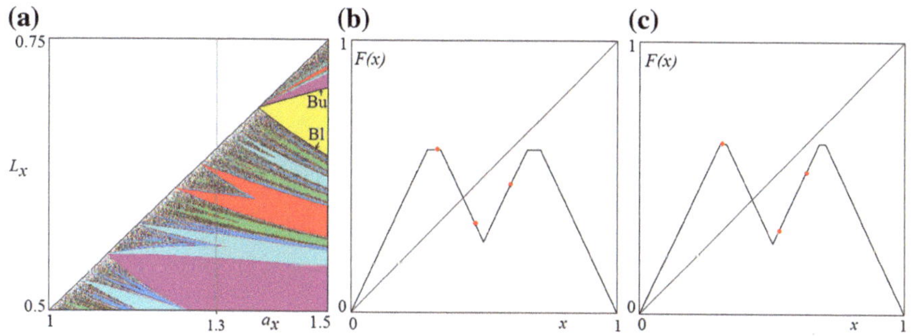

Fig. 8 In **a** enlargement of a portion of Fig. 7a; In **b** map $F(x)$ in case N2 at $a_x = 1.3$ and $L_x = 0.6$; In **c** map $F(x)$ in case N2 at $a_x = 1.3$ and $L_x = 0.626$. Superstable 3-cycles with periodic points belonging to different branches of $F(x)$ and different from the 3-cycle of Fig. 7c. Here $a_y = 1.6$ and $L_y = 0.81$

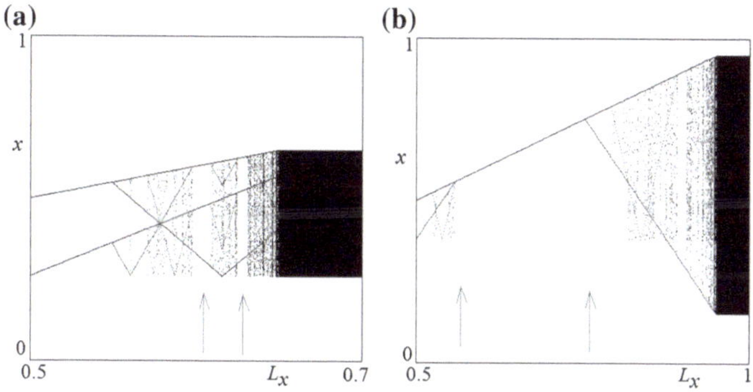

Fig. 9 One-dimensional bifurcation diagrams as a function of L_x at fixed $a_y = 1.6$ and $L_y = 0.81$. In **a** $a_x = 1.3$; In **b** $a_x = 1.9$

are crossed, and the related 3-cycles are shown in Fig. 8b, c. This also explains the particular swallow tail shape of the regions below the BCB curve *Bl*.

The two arrows in the one-dimensional bifurcation diagram shown in Fig. 9a at fixed $a_x = 1.3$ and increasing L_x are evidencing the intervals related to the superstable 3-cycles shown in Fig. 8. While in Fig. 9b we show the one-dimensional bifurcation diagram at fixed $a_x = 1.9$ and increasing L_x and the two arrows are evidencing the interval related to the superstable fixed point bounded by the BCBs *Bl* and *Bu* discussed above and shown in Fig. 7a. In both cases of Fig. 9a, b, as the parameters cross the diagonal entering the region related to case N1 the dynamics become chaotic in a unique interval.

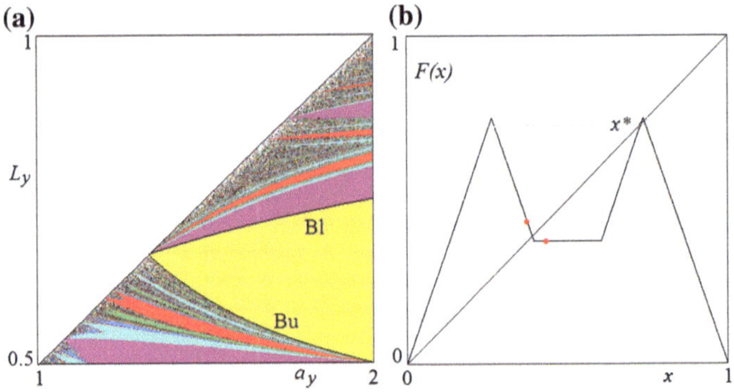

Fig. 10 In **a** two-dimensional bifurcation diagram in the parameter plane (a_y, L_y); In **b** map $F(x)$ in case N3 at $a_y = 1.9$ and $L_y = 0.75$ with a superstable 2-cycle having a periodic point in the local minimum. Here $a_x = 1.5$ and $L_x = 0.8$

3.2 *Case N3*

The dynamics of map $F(x)$ when the parameters satisfy the conditions of case N3 are similar to those of case N2 considered above, and can be studied via the conjugation property. In fact, let us assume that the parameters are those of case N3, then instead of map $F(x)$ we can consider map $G(y)$, for which the conditions are those of "case N2" considered above. Then via conjugation as described in Bischi et al. (2000) we can obtain the properties of map $F(x)$.

As an example, consider the two-dimensional bifurcation diagram in the parameter plane (a_y, L_y) in Fig. 10a at fixed $a_x = 1.5$ and $L_x = 0.8$ It can be seen that the diagonal (the straight line $L_y = a_y/2$) is the bifurcation curve which separates two different regimes: above it case N1 occurs (white points denote chaotic dynamics) while below the diagonal we have case N3, and the colored regions indicate periodicity regions associated with superstable cycles for which the local minimum value $a_x(1 - L_y)$ is a periodic point.

It can be seen that the bifurcation structure is similar to the one shown in the previous case N2, the BCB curves bounding the region related to the superstable fixed point can be easily obtained from those already detected in the previous case exchanging $a_x \Leftrightarrow a_y$ and $L_x \Leftrightarrow L_y$ leading to

$$Bu : L_y = \frac{a_x a_y}{a_x a_y + 1} \tag{16}$$

$$Bl : L_y = \frac{a_x a_y - a_y}{a_x a_y - 1} \tag{17}$$

which are also shown in Fig. 10a. At fixed value of a_y, increasing L_y the flip BCB occurring crossing Bl leads to a repelling fixed point and to a superstable 2-cycle (see an example in Fig. 10b), and as long as the local minimum is mapped by F below the repelling fixed point x^* shown in Fig. 10b, (i.e. $F(a_x(1 - L_y)) < x^*$), we have that the existing cycles are those related to the U-sequence.

3.3 *Transition N2–N4 (N3–N4)*

From the comments given so far it is enough to consider only one transition, for example we are interested in describing what happens when map $F(x)$ changes from case N2 to case N4, as similar behavior occurs also in the transition from case N3 to case N4. The map in case N4 is characterized by both extrema, maximum and local minimum, which occur in a flat branch. Thus the main difference with respect to the cases commented above, in which the superstable cycle is necessarily unique, is that the bistability of two superstable cycles may now occur, one with a periodic point in the maximum L_x and one with a periodic point in the local minimum $a_x(1 - L_y)$.

We can have a global view of the many superstable cycles that occur in this regime, considering the two-dimensional bifurcation diagram in the parameter plane (a_y, L_y) in Fig. 11a at fixed $a_x = 1.5$ and $L_x = 0.7$. It can be seen that the diagonal (the straight line $L_y = a_y/2$) is the bifurcation curve which separates two different regimes: above it case N2 occurs, and the vertical strips are related to the existing superstable cycles having a periodic point in the local maximum L_x. The vertical strips may continue also in the region below the diagonal, as in fact the related cycles (superstable cycles with L_x as periodic point) may continue to exist, but now there are also different regions having a horizontal shape, issuing from the diagonal, which are periodicity regions associated with superstable cycles having a periodic point in the local minimum $a_x(1 - L_y)$. For example, the vertical lines issuing from the points $a_y(Bu)$ and $a_y(Bl)$, where

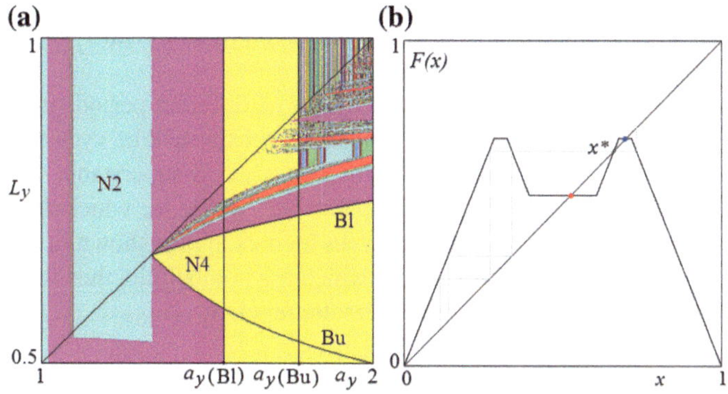

Fig. 11 In **a** two-dimensional bifurcation diagram in the parameter plane (a_y, L_y); In **b** map $F(x)$ in case N4 at $a_y = 1.65$ and $L_y = 0.65$ with two coexisting superstable fixed points. Here $a_x = 1.5$ and $L_x = 0.7$

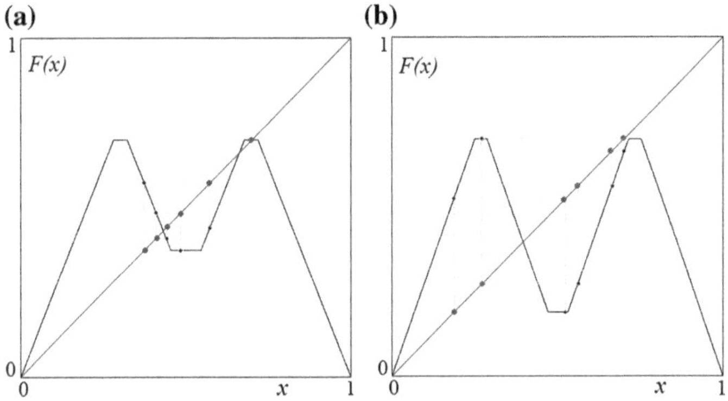

Fig. 12 Two examples in case N4 at fixed $a_x = 1.5$ and $L_x = 0.7$. In **a** $a_y = 1.65$ and $L_y = 0.75$, a superstable fixed point coexists with a superstable 5-cycle. In **b** $a_y = 1.86$ and $L_y = 0.875$, a superstable 4-cycle coexists with a superstable 2-cycle

$$a_y(Bu) = \frac{L_x}{a_x(1 - L_x)}, \quad a_y(Bl) = \frac{a_x - L_x}{a_x(1 - L_x)} \tag{18}$$

have been obtained from the expressions given in (14) and (13), respectively, denote a vertical strip in which the superstable fixed point $x_s^* = L_x$ exists. The periodicity region bounded by the BCB curve Bl and Bu given in (17) and (16), respectively, denote that also the fixed point $x^* = a_x(1 - L_y)$ related to the local minimum exists and is superstable. Thus in the intersection of the two regions we have that the two superstable fixed points coexist. An example of this situation is shown in Fig. 11b, and the two basins of attraction are separated by the unstable fixed point x^* and its preimages of any rank, leading to alternating intervals which are accumulating to $x = 0$ and $x = 1$.

From the parameters used in Fig. 11b, increasing L_y the bifurcation curve Bl is crossed and leading to an unstable fixed point and a superstable 2-cycle. Further increasing of L_y leads to cycles of different period coexisting with $x_s^* = L_x$, an example is shown in Fig. 12a, with a 5-cycle.

Clearly when vertical strips related to cycles of different periods intersect periodicity regions issuing from the diagonal, then coexistence of cycles of different periods occur. An example is shown in Fig. 12b where a superstable 4-cycle (with a periodic point in L_x) coexists with a superstable 2-cycle for which the local minimum value $a_x(1 - L_y)$ is a periodic point. In both examples shown in Fig. 12 there are also fixed points and repelling cycles which are homoclinic, thus the two basins have a fractal boundary, as it includes a chaotic repeller.

4 Rand Duopoly with Inertia

The dynamics of the two-dimensional Cournot map T given in (3) in the case $\alpha_x = 1$ and $\alpha_y = 1$ are related to those of the composite function $F(x)$ defined in (1). As already mentioned, these are studied in Bischi et al. (2000) where it is shown that the relevant property of map T is multistability (whenever map F has attracting cycles of period larger than 2). In the case considered above, except for case N1, we have abundance of regions related to superstable cycles of F and thus we have multistability of superstable cycles of map T, with basins that are separated by vertical and horizontal straight lines.

As an example, let us consider the parameter values leading to the attracting 3-cycle of $F(x)$ shown in Fig. 7c. We can see in that figure that the fixed points x^* and x_u^* are both homoclinic, thus a chaotic repellor exists in this case. We know that for the two-dimensional map T (Cournot-Rand modified map) there are 9 periodic points of the attracting sets belonging to two coexisting cycles, one of period 3 and one of period 6. Their basins are constituted by rectangles with a fractal structure, as infinitely many repelling cycles belong to a Cantor set which is included in the frontier between the two basins of attraction, as shown in Fig. 13a.

Similarly, considering the example of case N4 shown in Fig. 12a, in which map $F(x)$ has two attractors, a fixed point and a 5-cycle, and the fixed point close to the 5-cycle is homoclinic, thus a chaotic repellor exists. We know that for T the periodic points of the attracting sets are 36 and belong to five coexisting cycles, one fixed point, a 5-cycle, and three different 10-cycles, whose basins are constituted by rectangles with a fractal structure, as shown in Fig. 13b.

The introduction of the adaptive mechanism (with $\alpha_x < 1$ and $\alpha_y < 1$) leads to the disappearance of many cycles, and the map (no longer with separate variables in

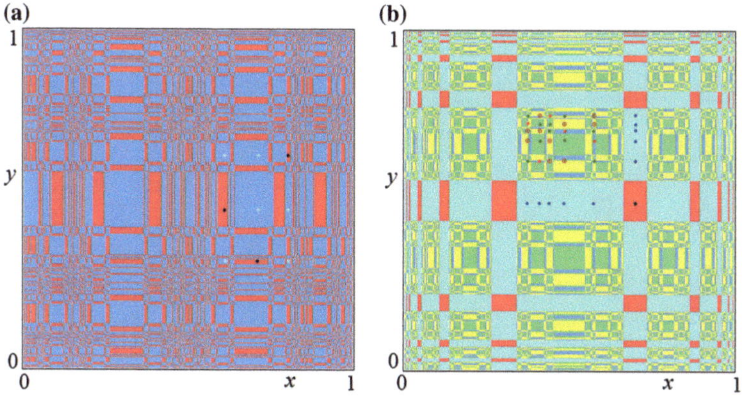

Fig. 13 Two examples of the two-dimensional map T in the case $\alpha_x = 1$ and $\alpha_y = 1$. In **a** corresponding to the example shown in Fig. 7c ($a_y = 1.6$, $L_y = 0.81$, $a_x = 1.9$ and $L_x = 0.8$) map T has two attractors, a 3-cycle and a 6-cycle. In **b** corresponding to the example shown in Fig. 12a ($a_y = 1.65$, $L_y = 0.75$, $a_x = 1.5$ and $L_x = 0.7$), map T has 5 coexisting attracting cycles, as commented in the text

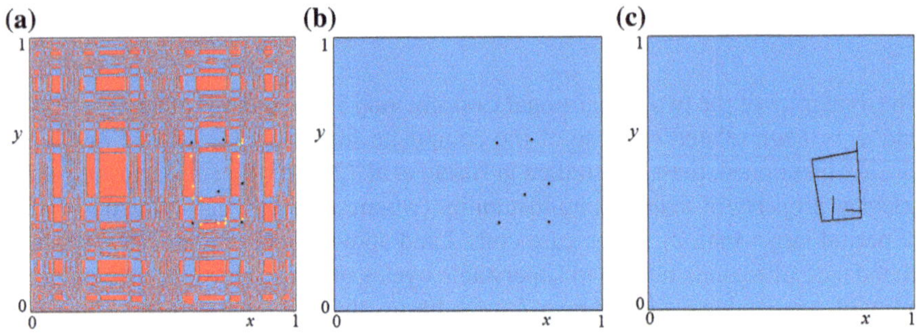

Fig. 14 Three examples of the two-dimensional map T in the case with inertia, corresponding to changes in the example shown in Fig. 13a ($a_y = 1.6$, $L_y = 0.81$, $a_x = 1.9$ and $L_x = 0.8$). In **a** $\alpha_x = 0.995$ and $\alpha_y = 0.99$, the map T has two coexisting attracting cycles. In **b** $\alpha_x = 0.989$ and $\alpha_y = 0.99$, the map T has a unique attractor, a 6-cycle. In **c** $\alpha_x = 0.9$ and $\alpha_y = 0.95$, the map T has a unique chaotic attractor

the second iterate) also has basins which are no longer bounded by straight lines. In Fig. 14 we slightly decrease the values of α_x and α_y. In Fig. 14a we can see that the attractors are still two, but now a 6-cycle coexists with a 13-cycle, and the basins also are slightly modified in shape (they are no longer bounded by vertical and horizontal straight lines). With a further decrease of α_x we see the disappearance of the 13-cycle and the 6-cycle is left as unique attractor (see Fig. 14b), which attracts almost all the points (as repelling cycles and a chaotic repellor most likely still exist), and in Fig. 14c a chaotic attractor is observed as unique attracting set.

Similarly, considering the case shown in Fig. 13b and slightly decreasing the values of α_x and α_y, we observe the disappearance of some attractors. The coexistence of three attractors is evidenced in Fig. 15a (a fixed point, a 4-cycle, and a 20-cycle),

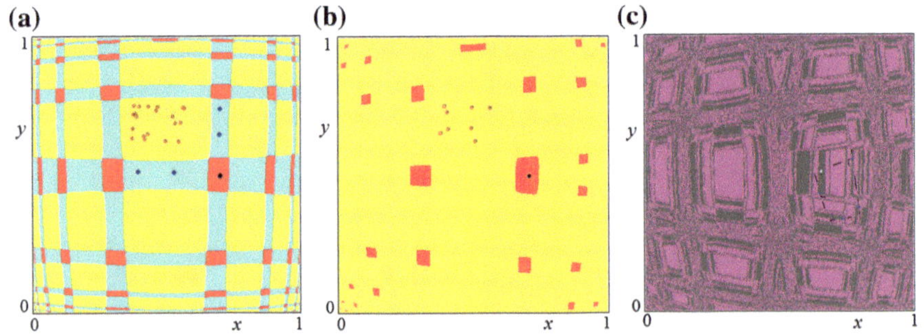

Fig. 15 Three examples of the two-dimensional map T in the case with inertia, corresponding to changes in the example shown in Fig. 13b ($a_y = 1.65$, $L_y = 0.75$, $a_x = 1.5$ and $L_x = 0.7$). In **a** at $\alpha_x = 0.99$ and $\alpha_y = 0.95$ map T has three coexisting attracting cycles. In **b** $\alpha_y = 0.9$ and $\alpha_x = 0.95$, the map T has two coexisting attracting cycles. In **c** $\alpha_x = 0.8$ and $\alpha_y = 0.95$, the map T has a chaotic attractor coexisting with an attracting 4-cycle

while decreasing further α_x two attractors are left (a fixed point and an 8-cycle are visible in Fig. 15b). Here also a transition to chaos is observed, in Fig. 15c a chaotic set coexists with an attracting 4-cycle. The shape of the basin shows that a chaotic repellor still exists.

The two-dimensional map T results now in a continuous piecewise smooth map, thus contact bifurcations and other global bifurcations can be studied by using the usual tools of noninvertible maps (see Mira et al. 1996), coupled with the properties and BCBs related to piecewise smooth maps. This field of research in two-dimensional maps still has many open problems, and we leave this as further work for the future.

5 Conclusions and Outline of Further Investigations

In this paper we considered the Cournot duopoly model proposed by Rand (1978) characterized by unimodal piecewise linear reaction functions in the form of tent maps, and we added maximum production constraints, thus giving flat-top shaped reaction functions. We have shown that the robust chaotic motion proved by Rand is replaced by superstable periodic dynamics whenever production constraints are imposed. By taking the limiting values of upper production constraints as bifurcation parameters, we have analyzed the appearance of bifurcation structures which are specific to border collision bifurcations, global (or contact) bifurcation that have recently become a focus topic in the literature on applied non smooth dynamical systems. The analysis is performed by studying the equivalent one-dimensional map obtained from the composition of the two piecewise linear top-flat reaction functions, a general property specific to two-dimensional mappings with decoupled second iterate.

Then inertia is added to the model, expressed by assuming a convex combination between the current production and the one computed according to the reaction functions. After this modification the property of decoupled second iterate no longer holds, and a true two-dimensional dynamical system must be considered. However, the results obtained without inertia can provide an useful benchmark case that helps one to approach the more complete study of the adaptive dynamics with inertia. Just a few numerical examples are given in this paper about the model with inertia, and much more can be done in the future to understand the effects of inertia and of heterogeneity between firms related to different degrees of inertia as well as different reaction functions.

Acknowledgements This work is developed in the framework of the research project on "Dynamic Models for behavioural economics" financed by DESP-University of Urbino.

References

Agliari, A., Bischi, G.I., & Gardini L. (2002). Some methods for the global analysis of dynamic games represented by noninvertible maps. In T. Puu & I. Sushko (Eds.), *Oligopoly dynamics: models and tools*. Springer Verlag.

Avrutin, V., & Schanz, M. (2006). Multi-parametric bifurcations in a scalar piecewise-linear map. *Nonlinearity*, *19*, 531–552.

Avrutin, V., Schanz, M., & Banerjee, S. (2006). Multi-parametric bifurcations in a piecewise-linear discontinuous map. *Nonlinearity*, *19*, 1875–1906.

Banerjee, S., & Grebogi, C. (1999). Border-collision bifurcations in two-dimensional piecewise smooth maps. *Physical Review E*, *59*(4), 4052–4061.

Banerjee, S., Karthik, M. S., Yuan, G., & Yorke, J. A. (2000a). Bifurcations in one-dimensional piecewise smooth maps—theory and applications in switching circuits. *IEEE Transactions on Circuits and System I: Fundamental Theory and Applications*, *47*(3), 389–394.

Banerjee, S., Ranjan, P., & Grebogi, C. (2000b). Bifurcations in two-dimensional piecewise smooth maps—theory and applications in switching circuits. *IEEE Transactions on Circuits and System I: Fundamental Theory and Applications*, *47*(5), 633–643.

Bischi, G. I., & Lamantia, F. (2002). Nonlinear duopoly games with positive cost externalities due to spillover effects. *Chaos, Solitons & Fractals*, *13*, 805–822.

Bischi, G. I., Chiarella, C., Kopel, M., & Szidarovszky, F. (2010). *Nonlinear oligopolies: Stability and bifurcations*. Springer-Verlag.

Bischi, G. I., & Kopel, M. (2001). Equilibrium selection in a nonlinear duopoly game with adaptive expectations. *Journal of Economic Behavior and Organization*, *46*(1), 73–100.

Bischi, G. I., & Lamantia, F. (2012). Routes to complexity induced by constraints in Cournot oligopoly games with linear reaction functions. *Studies in Nonlinear Dynamics & Econometrics*, *16*(2), 1–30.

Bischi, G. I., Mammana, C., & Gardini, L. (2000). Multistability and cyclic attractors in duopoly games. *Chaos, Solitons & Fractals*, *11*, 543–564.

Bulow, J., Geanokoplos, J., & Klemperer, P. (1985). Multimarket oligopoly: Strategic substitutes and complements. *Journal of Political Economy*, *93*, 488–511.

Cournot, A. (1838). *Recherches sur les principes matematiques de la theorie de la richesse*. Paris: Hachette.

Dana, R. A., & Montrucchio, L. (1986). Dynamic complexity in duopoly games. *Journal of Economic Theory*, *40*, 40–56.

Di Bernardo, M., Budd, C. J., Champneys, A. R., & Kowalczyk, P. (2008). *Piecewise-smooth dynamical systems*. London: Springer Verlag.

Di Bernardo, M., Feigen, M. I., Hogan, S. J., & Homer, M. E. (1999). Local analysis of C-bifurcations in n-dimensional piecewise smooth dynamical systems. *Chaos, Solitons & Fractals*, *10*(11), 1881–1908.

Hahn, F. (1962). The stability of the Cournot solution. *Journal of Economic Studies*, *29*, 329–331.

Kopel, M. (1996). Simple and complex adjustment dynamics in Cournot duopoly models. *Chaos, Solitons & Fractals*, *7*(12), 2031–2048.

Leonov, N. N. (1959). Map of the line onto itself. *Radiofisica*, *3*(3), 942–956.

Leonov, N. N. (1962). Discontinuous map of the straight line. *Dokl. Acad. Nauk. SSSR.*, *143*(5), 1038–1041.

Metropolis, N., Stein, M. L., & Stein, P. R. (1973). On finite limit sets for transformations on the unit interval. *Journal of Combinatorial Theory*, *15*, 25–44.

Maistrenko, Y. L., Maistrenko, V. L., & Chua, L. O. (1993). Cycles of chaotic intervals in a time-delayed Chua's circuit. *International Journal Bifurcation and Chaos*, *3*(6), 1557–1572.

Maistrenko, Y. L., Maistrenko, V. L., Vikul, S. I., & Chua, L. O. (1995). Bifurcations of attracting cycles from time-delayed Chua's circuit. *International Journal Bifurcation and Chaos*, *5*(3), 653–671.

Maistrenko, Y. L., Maistrenko, V. L., & Vikul, S. I. (1998). On period-adding sequences of attracting cycles in piecewise linear maps. *Chaos, Solitons & Fractals, 9*(1), 67–75.

Milnor, J. (1985). On the concept of attractor. *Communications in Mathematical Physics, 99*, 177–195.

Mira, C. (1978). Sur les structure des bifurcations des diffeomorphisme du cercle. *C.R.Acad. Sc. Paris 287 Series A*, 883–886.

Mira, C. (1987). *Chaotic dynamics*. Singapore: World Scientific.

Mira, C., Gardini, L., Barugola, A., & Cathala, J. C. (1996). *Chaotic Dynamics in two-dimensional noninvertible maps*. Singapore: World Scientific.

Mosekilde, E., Zhusubaliyev, Z. T. (2003). *Bifurcations and chaos in piecewise-smooth dynamical systems*. World Scientific.

Nusse, H. E., & Yorke, J. A. (1992). Border-collision bifurcations including period two to period three for piecewise smooth systems. *Physica D, 57*, 39–57.

Nusse, H. E., & Yorke, J. A. (1995). Border-collision bifurcation for piecewise smooth one-dimensional maps. *International Journal of Bifurcation Chaos, 5*, 189–207.

Okuguchi, K. (1964). The stability of the Cournot oligopoly solution: A further generalization. 287. *Journal of Economic Studies, 31*, 143–146.

Okuguchi, K., & Szidarovszky, F. (1999). *The theory of oligopoly with multi-product firms* (2nd ed.). Berlin: Springer.

Puu, T. (1991). Chaos in duopoly pricing. *Chaos, Solitons & Fractals, 1*(6), 573–581.

Puu, T., & Norin, A. (2003). Cournot duopoly when the competitors operate under capacity constraints. *Chaos, Solitons & Fractals, 18*, 577–592.

Rand, D. (1978). Exotic phenomena in games and duopoly models. *Journal of Mathematical Economics, 5*, 173–184.

Sushko, I., Avrutin, V., & Gardini, L. (2015). Bifurcation structure in the skew tent map and its application as a border collision normal form. *Journal of Difference Equations and Applications*. doi:10.1080/10236198.2015.1113273.

Sushko, I., Gardini, L., & Avrutin, V. (2016). Nonsmooth One-dimensional maps: Some basic concepts and definitions. *Journal of Difference Equations and Applications*, 1–56. doi:10.1080/10236198.2016.1248426.

Sushko, I., Gardini, L., & Matsuyama, K. (2014). Superstable credit cycles and u-sequence. *Chaos, Solitons & Fractals, 59*, 13–27.

Szidarovszky, F., & Okuguchi, K. (1997). On the existence and uniqueness of pure Nash equilibrium in rent-seeking games. *Games and Economic Behavior, 18*, 135–140.

Szidarovszky, F. (1999). Adaptive expectations in discrete dynamic oligopolies with production adjustment costs. *Pure Mathmatics and Application, 10*(2), 133–139.

Teocharis, R. D. (1960). On the stability of the Cournot solution on the oligopoly problem. *The Review of Economic Studies, 27*, 133–134.

Tramontana, F., & Gardini, L. (2011). Border collision bifurcations in discontinuous one-dimensional linear-hyperbolic maps. *Communications in Nonlinear Science and Numerical Simulation, 16*, 1414–1423.

Tramontana, F., Gardini, L., & Puu, T. (2011). Mathematical properties of a discontinuous Cournot-Stackelberg model. *Chaos, Solitons & Fractals, 44*, 58–70.

Van Huyck, J., Cook, J., & Battalio, R. (1984). Selection dynamics, asymptotic stability, and adaptive behavior. *Journal of Political Economy, 102*, 975–1005.

Van Witteloostuijn, A., & Van Lier, A. (1990). Chaotic patterns in Cournot competition. *Metroeconomica, 2*, 161–185.

Zhusubaliyev, Z. T., Mosekilde, E., Maity, S., Mohanan, S., & Banerjee, S. (2006). Border collision route to quasiperiodicity: Numerical investigation and experimental confirmation. *Chaos, 16*, 1–11.

Zhusubaliyev, Z. T., Soukhoterin, E., & Mosekilde, E. (2007). Quasiperiodicity and torus breakdown in a power electronic dc/dc converter. *Mathematics and Computers in Simulation, 73*, 364–377.

An Adaptive Learning Model for Competing Firms in an Industry

Haiyan Qiao

Abstract In an industry of competing firms the market price function is not completely known, however based on repeated price information the firms are able to continuously adjust their beliefs. Under simplifying conditions, it is shown that these beliefs converge to the true price function as time goes to infinity, that is, successful learning can be achieved. The same result holds if there is continuously distributed delay in the price information, if the weighting function is exponential, however in the case of fixed delays stability may be lost if the delay is sufficiently large.

1 Introduction

Learning models play an important role in knowledge engineering. In most cases the model is dynamic, when the values of the unknown quantities are updated repeatedly leading to special dynamic systems. If the time scales are discrete, then difference equations describe the dynamic development in the learning process. If the time scales are continuous, then ordinary differential equations model the process. A necessary condition of successful learning is the existence of a steady state which has to be equal to the true values of the quantities being the subjects of learning. As time progresses, the updated values also should converge to the steady state meaning the asymptotic stability of the learning dynamism.

In this paper a special learning process is analyzed in which competing firms learn about the price function in an industry. In the economic literature, this economic situation is called oligopoly, and it is assumed that the firms sell their products in a homogeneous market, so their competition is through the price function which is a decreasing function of the total industry output. A comprehensive summary of the early developments in oligopoly is given in Okuguchi (1976) and their multi-product generalizations are discussed in Okuguchi and Szidarovszky (1999). The most recent results focusing on nonlinear models are reported in Bischi et al. (2010). The liter-

H. Qiao (✉)
School of Computer Science and Engineering, California State University
San Bernardino, San Bernardino, CA, USA
e-mail: hqiao@csusb.edu

A. Matsumoto (ed.), *Optimization and Dynamics with Their Applications*,
DOI 10.1007/978-981-10-4214-0_6

ature of learning in games is also very rich. For example, Fudenberg and Levine (1998) introduces the main concepts and methods. The asymptotical properties of discrete and continuous dynamic systems are discussed in many textbooks, the reader can find the most important facts for example, in Szidarovszky and Bahill (1998). If time delay is introduced into dynamic models due to information lag, then the dynamic behavior of the processes becomes more complicated. If fixed delay is considered, then difference-differential equations are introduced (Bellman and Cooke 1956). In the cases of continuously distributed delays the processes are described by Volterra-type integro-differential equations (Cushing 1977).

This paper is organized as follows. First a special adaptive learning process is introduced, and then its asymptotical behavior is examined. Fixed delays are then introduced into the model and their effects on stability are investigated. This section is followed by the discussion on continuously distributed delays. The last section draws conclusions and future research directions.

2 The Learning Model

Consider an industry with N firms producing the same product and selling it to a homogeneous market. If x_k denotes the production (output) level of firm k, then the total production level of the industry is $s = \sum_{k=1}^{N} x_k$. Assume that the cost function of firm k is $C_k(x_k) = c_k x_k$, the linearity of which is assumed for mathematical convenience. The unit price function is decreasing in s, so we assume that it is $p(s) = B - As$ with both A and B being positive. The maximum price is B and the marginal price is $-A$. The profit of firm k is the difference of its revenue and cost:

$$\varphi_k = x_k(B - Ax_k - As_k - c_k), \tag{1}$$

where $s_k = \sum_{l \neq k} x_l$ is the output of the rest of the industry. In this way, an N-person non-cooperative game is defined, where the firms are the players, x_k is the strategy and φ_k is the payoff of firm k. It is also assumed that the firms know the technologies of the competitors and the marginal price is a common knowledge. Therefore the values of parameters A, c_k $(k = 1, 2, \ldots, N)$ are known by all firms, however they are uncertain in the value of the maximum price B, which is the subject of the learning process. Let $B_k(t)$ denote firm k's current estimate of the maximum price. Then firm k believes that the payoff of any firm l is given as

$$\varphi_l^{(k)} = x_l(B_k(t) - Ax_l - As_l - c_l), \tag{2}$$

so the believed best response of firm l is obtained from the first order condition,

$$B_k(t) - Ax_l - A(s - x_l) - c_l - Ax_l = 0$$

implying that

$$x_l = \frac{B_k(t) - As - c_l}{A}. \tag{3}$$

The firm then can assess its belief of the industry output by adding Eq. (3) for all values of l,

$$s^{(k)} = \frac{1}{A}\left(NB_k(t) - NAs^{(k)} - \sum_{l=1}^{N} c_l\right)$$

from which

$$s^{(k)} = \frac{NB_k(t) - \sum_{l=1}^{N} c_l}{(N+1)A}, \tag{4}$$

and therefore firm k believes that the equilibrium price will be

$$p^{(k)} = B_k(t) - As^{(k)} = \frac{B_k(t) + \sum_{l=1}^{N} c_l}{N+1}. \tag{5}$$

However every firm thinks in the same way independently of the others, so in reality the industry output is as follows:

$$s = \sum_{k=1}^{N} x_k = \frac{1}{(N+1)A}\left(\sum_{l=1}^{N} B_l(t) - \sum_{l=1}^{N} c_l\right), \tag{6}$$

since each firm k produces its equilibrium output level

$$x_k = \frac{B_k(t) - As^{(k)} - c_k}{A} = \frac{B_k(t) + \sum_{l=1}^{N} c_l - (N+1)c_k}{(N+1)A}.$$

Consequently the actual market price becomes

$$p = B - As = B - \frac{1}{N+1}\left(\sum_{l=1}^{N} B_l(t) - \sum_{l=1}^{N} c_l\right). \tag{7}$$

Based on the discrepancy between the believed and actual prices, firm k adjusts its belief as

$$\dot{B}_k(t) = K_k(p - p^{(k)}),\ (K_k > 0) \tag{8}$$

since if the actual price is higher than the believed price, then the firm wants to increase its believed price by increasing the value of $B_k(t)$. If the actual price is the lower, then the firm wants to decrease its belief of the price function. If the two prices are equal, then the firm does not want to make changes in its belief. Based on Eqs. (5), (7) and (8) we obtain the following system of ordinary difference equations:

$$\dot{B}_k(t) = \frac{K_k}{N+1}\left((N+1)B - \sum_{l=1}^{N} B_l(t) - B_k(t)\right). \tag{9}$$

Clearly, the only steady state is $B_k = B$ for all k. The asymptotical stability of the system is examined by investigating the location of the eigenvalues of the coefficient matrix

$$\frac{1}{N+1}\begin{pmatrix} -2K_1 & -K_1 & \cdots & -K_1 \\ -K_2 & -2K_2 & \cdots & -K_2 \\ \vdots & \vdots & & \vdots \\ -K_N & -K_N & \cdots & -2K_N \end{pmatrix} = \boldsymbol{D} - \boldsymbol{a}\,\boldsymbol{1}^T, \tag{10}$$

where with the notation $\overline{K}_k = K_k/(N+1)$, $\boldsymbol{D} = diag(-\overline{K}_1, -\overline{K}_2, \ldots, -\overline{K}_N)$, $\boldsymbol{a} = (\overline{K}_1, \overline{K}_2, \ldots, \overline{K}_N)^T$ and $\boldsymbol{1}^T = (1, 1, \ldots, 1)$.

The characteristic polynomial of this matrix can be written as

$$\begin{aligned} &det(\boldsymbol{D} - \boldsymbol{a}\,\boldsymbol{1}^T - \lambda \boldsymbol{I}) = det(\boldsymbol{D} - \lambda \boldsymbol{I}) det(\boldsymbol{I} - (\boldsymbol{D} - \lambda \boldsymbol{I})^{-1} \boldsymbol{a}\,\boldsymbol{1}^T) \\ &= \prod_{k=1}^{N} \left(-\overline{K}_k - \lambda\right) \left[1 - \sum_{k=1}^{N} \frac{\overline{K}_k}{-\overline{K}_k - \lambda}\right], \end{aligned} \tag{11}$$

where we used the simple fact that if $\boldsymbol{a}$ and $\boldsymbol{b}$ are N-element real column vectors and $\boldsymbol{I}$ is the $N \times N$ identity matrix, then

$$det(\boldsymbol{I} + \boldsymbol{a}\,\boldsymbol{b}^T) = 1 + \boldsymbol{b}^T \boldsymbol{a}.$$

For a simple proof the reader is referred to Bischi et al. (2010, Appendix E). The eigenvalues are $\lambda = -\overline{K}_k$ and the roots of equation

$$\sum_{k=1}^{N} \frac{\overline{K}_k}{\overline{K}_k + \lambda} = -1. \tag{12}$$

In order to show asymptotic stability it is sufficient to prove that the roots of Eq. (12) are real and negative. If $g(t)$ denotes the left hand side, then $\lim_{\lambda \to \pm\infty} g(\lambda) = 0$, $\lim_{\lambda \to -\overline{K}_k + 0} g(\lambda) = \infty$ and $\lim_{\lambda \to -\overline{K}_k - 0} g(\lambda) = -\infty$, furthermore

$$g'(\lambda) = \sum_{k=1}^{N} \frac{-\overline{K}_k}{\left(\overline{K}_k + \lambda\right)^2} < 0.$$

That is, $g(\lambda)$ is strictly decreasing and its poles are the $-\overline{K}_k$ values. If we assume that they are different and $\overline{K}_1 > \overline{K}_2 > \cdots > \overline{K}_N$, then there is a root before $-\overline{K}_1$ and one root inside each interval $(-\overline{K}_k, -\overline{K}_{k+1})$, so we found N real negative roots. Since Eq. (12) is equivalent with an Nth-degree polynomial equation, there are no additional roots. Hence system (9) is globally asympototically stable implying that as $t \to \infty$, all estimates $B_k(t)$ converge to the true value B. A similar argument can be used if some of the K_k values are equal. That is, the firms are successful in learning.

3 Learning with Fixed Delays

Assume that at time period t the firms cannot obtain simultaneous prince information from the market, since the actual information has a fixed delay τ. Then the received price is

$$B - \frac{1}{N+1}\left(\sum_{l=1}^{N} B_l(t-\tau) - \sum_{l=1}^{N} c_l\right),$$

so the dynamic system (9) becomes a system of difference-differential equations

$$\dot{B}_k(t) = \overline{K}_k\left((N+1)B - \sum_{l=1}^{N} B_l(t-\tau) - B_k(t)\right) \quad (k = 1, 2, \ldots, N). \tag{13}$$

For mathematical simplicity assume that the firms have identical speed of adjustments, $K_k \equiv \overline{K}$, and their initial estimates of the maximum price are also identical. Then system (13) reduces to a single-dimensional equation

$$\dot{B}(t) = \overline{K}\left((N+1)B - NB(t-\tau) - B(t)\right). \tag{14}$$

The characteristic equation is obtained by substituting the exponential solution $B(t) = e^{\lambda t}u$ into the homogeneous equation,

$$\lambda e^{\lambda t}u = \overline{K}\left(-Ne^{\lambda(t-\tau)}u - e^{\lambda t}u\right)$$

which can be rewritten as an exponential-polynomial equation

$$\lambda + \overline{K} + \overline{K}Ne^{-\lambda\tau} = 0. \tag{15}$$

At $\tau = 0$, the only eigenvalue is negative, so the system is asymptotically stable. At any stability switch $\lambda = iv$, $v > 0$, so

$$iv + \overline{K} + \overline{K}N\left(\cos(v\tau) - i\sin(v\tau)\right) = 0$$

implying that

$$1 + N\cos(v\tau) = 0 \tag{16}$$

and

$$v - \overline{K}N\sin(v\tau) = 0. \tag{17}$$

Since

$$\cos(v\tau) = -\frac{1}{N}$$

and $\sin(v\tau)$ is positive,

$$v\tau = \arccos\left(-\frac{1}{N}\right) + 2n\pi \qquad (n = 0, 1, 2, \ldots) \tag{18}$$

Furthermore we have

$$1 = \sin^2(v\tau) + \cos^2(v\tau) = \frac{1}{N^2} + \frac{v^2}{\overline{K}^2 N^2}$$

implying that

$$v = \overline{K}\sqrt{N^2 - 1}, \tag{19}$$

so stability switches might occur at delays

$$\tau = \frac{1}{\overline{K}\sqrt{N^2 - 1}}\left(\arccos\left(-\frac{1}{N}\right) + 2n\pi\right). \tag{20}$$

In order to check if stability switches actually occur or not, we select τ as the bifurcation parameter and assume $\lambda = \lambda(\tau)$. Implicitly differentiating the characteristic Eq. (15) with respect to τ we have

$$\dot{\lambda} + \overline{K}Ne^{-\lambda\tau}\left(-\dot{\lambda}\tau - \lambda\right) = 0$$

implying that

$$\dot{\lambda} = \frac{\lambda\overline{K}Ne^{-\lambda\tau}}{1 - \tau\overline{K}Ne^{-\lambda\tau}} = \frac{-\lambda(\lambda + \overline{K})}{1 + (\lambda + \overline{K})\tau}.$$

If $\lambda = iv$, then

$$\begin{aligned} Re\dot{\lambda} &= Re\frac{v^2 - iv\overline{K}}{1 + \overline{K}\tau + iv\tau} \\ &= \frac{v^2(1 + \overline{K}\tau) - v\overline{K}(v\tau)}{(1 + \overline{K}\tau)^2 + v^2\tau^2} > 0 \end{aligned}$$

implying that the real part of an eigenvalue becomes positive, so stability is lost. It is lost first at

$$\tau^* = \frac{1}{\overline{K}\sqrt{N^2-1}} \arccos\left(-\frac{1}{N}\right), \tag{21}$$

and stability cannot be regained later. At this critical value Hopf bifurcation occurs giving the possiblity of the birth of limit cycles. In summary, the learning process converges to true knowledge if $\tau < \tau^*$, at $\tau = \tau^*$ it shows cyclic behavior and at $\tau > \tau^*$ there is no convergence to the true value, so learning is not possible.

4 Learning with Continuously Distributed Delays

If the delay is uncertain, then it is assumed to be a random variable. If the largest probability is assigned to the most current data and the probability decreases afterwards, then an exponential density function describes the probabilistic nature of the situation. In this case Eq. (14) is modified as

$$\dot{B}(t) = \overline{K}\left((N+1)B - N\int_0^t w(t-s)B(s)ds - B(t)\right) \tag{22}$$

where

$$w(t-s) = \frac{1}{T}e^{-\frac{t-s}{T}} \ (t > s)$$

is the weighting function. Looking again for the solution in the exponential form $B(t) = e^{\lambda t}u$ and substituting it into the homogeneous version of Eq. (22) we have

$$\lambda e^{\lambda t} = -\overline{K}N \int_0^t \frac{1}{T}e^{-\frac{t-s}{T}} e^{\lambda s} ds - \overline{K}e^{\lambda t}. \tag{23}$$

Introduce the new integration variable $z = t - s$. The integral becomes

$$\int_0^t \frac{1}{T}e^{-\frac{z}{T}+\lambda(t-z)}dz,$$

and then simplify Eq. (23) by $e^{\lambda t}$ to get

$$\lambda + \overline{K}N \int_0^t \frac{1}{T}e^{-z(\lambda+\frac{1}{T})}dz + \overline{K} = 0.$$

Introduce again a new integration variable $v = z(\lambda + 1/T)$, then we have

$$\lambda + \overline{K}N \int_0^{t(\lambda+\frac{1}{T})} \frac{1}{T}e^{-v}\frac{T}{1+\lambda T}dv + \overline{K} = 0$$

and finally as $t \to \infty$, the characteristic equation of (22) is obtained:

$$\lambda + \overline{K} + \overline{K}N\frac{1}{1+\lambda T} = 0.$$

Here we assumed that $\lambda > -\frac{1}{T}$, since negative eigenvalues cannot destroy stability. Notice that this is a quadratic equation,

$$\lambda^2 T + \lambda(1 + T\overline{K}) + \overline{K}(1 + N) = 0.$$

Since all coefficients are positive, both eigenvalues have negative real parts, and therefore the system is asymptotically stable, so successful learning is possible.

5 Conclusions

An adaptive learning process was investigated, when firms adjust their beliefs on the maximum price based on the discrepancies between the believed and actual market prices. For the sake of simplicity linear price and cost functions were assumed. The nonlinear case can be treated by linearization in a similar manner, however in the case of nonlinear systems only local stability can be guaranteed.

Without information delay the dynamic model is globally asymptotically stable which guarantees that the firms can successfully learn about the maximum price. In the case of continuously distributed delays with exponential weighting function the asymptotical stability of the system is preserved, however the presence of fixed delay might destroy stability, if the delay is sufficiently large.

Further study is needed to examine the cases of bellshaped weighting functions and learning the values of other parameters, perhaps learning several parameter values simultaneously.

An addtional extention would be to introduce multiple delays by assuming that the delays for different firms are different. For two and three delays Gu et al. (2005) and Gu and Naghnaeian (2011) offered the methodology.

References

Bellman, R., & Cooke, K. L. (1956). *Differential-difference equations*. New York: Academic Press.

Bischi, G. I., Chiarella, C., Kopel, M., & Szidarovszky, F. (2010). *Nonlinear oligopolies: Stability and Bifurcations*. Berlin, Heidelberg, New York: Springer.

Cushing, J. (1977). *Integro-differential equations and delay models in population dynamics*. Berlin, Heidelberg, New York: Springer.

Fudenberg, D., & Levine, D. (1998). *The theory of learning in games*. Cambridge, MA: MIT Press.

Gu, K., Niculescu, S.-I., & Chen, J. (2005). On stability crossing curves for general systems with two delays. *Journal of Mathematical Analysis and Applications*, *311*, 231–253.

Gu, K., & Naghnaeian, M. (2011). Stability crossing set for systems with three delays. *IEEE Transactions on Automatic Control*, *56*(1), 11–26.
Okuguchi, K. (1976). *Expectations and stability in oligopoly models*. Berlin, Heidelberg, New York: Springer.
Okuguchi, K., & Szidarovszky, F. (1999). *The theory of oligopoly with multi-product firms* (2nd ed.). Berlin, Heidelberg, New York: Springer.
Szidarovszky, F., & Bahill, T. (1998). *Linear systems theory* (2nd ed.). Boca Raton, FL: CRC Press.

The Coordination and Dynamic Analysis of Industrial Clusters: A Multi-agent Simulation Study

Jijun Zhao

Abstract An agent based simulation model is presented to investigate the long-term behavior of firms in an industrial district. The firms are interconnected with each other through input-output relations, product markets, labor, and innovation spillover. The prices of the products depend on the supply-demand balance of the market as well as on the innovation levels of the firms. Dynamic strategies of the firms are examined and conditions for successful industrial cluster formation are developed.

Keywords Industrial clusters · Agent based simulation · Oligopoly theory · Innovation

1 Introduction

Industrial clusters are important examples of coordinated multi-agent systems in which the industrial firms are the agents that are interconnected to each other by their inputs and outputs as well as to the markets through inverse demand functions. The high complexity and the large sizes of industrial clusters make their analytical investigation impossible. In this paper agent-based simulation is used to examine the coordination and dynamic properties of industrial clusters. The interrelation of the firms is modeled as an extended oligopoly, when in addition to the competition of the firms on the product markets we are able to consider their competition for labor as well.

Traditional literature on industrial clusters has mainly focused on their identification, driving forces and policies. These studies try to answer the fundamental questions such as how the firms can benefit from belonging to a cluster. A very important problem is the evolution of industrial clusters. Investigations of the evolution mainly focused on the life cycle, entry, exit and growth of the clusters (Maskell 2001; Swann

J. Zhao (✉)
Institute of Complexity Science, Qingdao University,
Qingdao 266071, Shandong, China
e-mail: jijunzhao@yahoo.com

A. Matsumoto (ed.), *Optimization and Dynamics with Their Applications*,
DOI 10.1007/978-981-10-4214-0_7

et al. 1998). Results are drawn mostly from case studies or from empirical studies. Most of these studies analyzed the industrial clusters only after they became successful, and not during the transformation period. In addition, case studies can lead to special results from individual clusters, which cannot be generalized.

A new strand of study has recently emerged in which the evolution dynamics of industrial clusters are analyzed by using agent-based simulation. With two versions, spatial and non-spatial, these studies focus on the formation, development and coordination of artificial industrial clusters. In our paper, we will follow this strand, with the additional question: how the decisions and the behavior of industrial firms will promote the formation of a cluster when these firms are already in a given system structure of a district.

This question is very practical. It has been already discussed in literature that the initiation and support of public policies may be successful in the formation of clusters (Bresnahan et al. 2001). It is also known that clusters could grow through some types of network structure. However, how to ensure the formation, effectiveness and growth of clusters is a crucial question. Local government might help to build a structure or to introduce policies to promote the important local industries. For example, in the developing areas of some regions in China, the government plays a crucial role to initiate the development of certain industries. However, with similar policies and perhaps with similar environments, some districts were promoted to clusters while some others were not. There must be many other factors to explain this difference in the result. In this paper and in our future works, we are interested in the firm level influencing factors: what firms should do to help remove the barriers to the cluster formation and in exploring their own opportunities.

In this study the district structure means the topologies defined by Markusen (1996), who identified several types of system structures of industrial clusters. In this paper, we assume the particular structure, which is called 'Hub and Spoke' by Markusen. It consists of several large anchor companies and several small companies. (Hence we are not going to study industrial clusters with a large number of small and medium sized firms). In reality, some emerging clusters have similar system structure like this. Take again Chinese regional industrial clusters as an example, several foreign invested global companies were attracted into the developing district, and then many relatively small suppliers and accessorial firms moved in.

As mentioned earlier, we will propose an agent-based simulation model to show how an industrial cluster could emerge in a location which already includes several firms. Agent based simulation is a flexible tool to investigate emerged behaviors of complex systems from individuals. Researchers are already using this tool to examine industrial clusters. The reputation dynamics (Giardini et al. 2008) and the growth of clusters (Zhang 2003) are good examples. In the case of most studies, the individual-level decision rules are relatively simple, and the topologies of the district are never considered. In our study, by considering the environment of the designed system structure, we will adopt the Hub and Spoke topology and express it as a two-layer network. Firms will be modeled as bounded rational agents. Each agent has its own production input factors, labor, and production. During each period of time, each agent will make decisions based on its former behavior, the other firms former behavior and its own decision rules. For the decision making process of the agents, we will integrate oligopoly theory into the agent-based model. Hence our model will be an agent-based and game theory integrated model.

In the spatial version of the agent-based models of industrial clusters, moving and relocating agents are basic elements. However we will not consider these features, since firms cannot move easily like residents. Our primary model is a non-spatial one, and the distance of locations is not our concern. This is a reasonable assumption since it is more important to decide if a given firm is in the cluster or not. When we consider firms only in a specific location, spatial distance is not an important factor.

The methodology of this paper might have further applications. A potential study area is the examination of the change of behavior and decision patterns of firms that can transform declining clusters into new ones. In addition, with the relaxation of some assumptions, we may study more general situations. This paper is only a starting point of a long-term research project.

This paper develops as follows. Section 2 presents the related literature review. In Sect. 3, we will outline the fundamentals of agent-based models and oligopoly theory. Simulation methodology and numerical results will be reported in Sect. 4. Final conclusions will be drawn in Sect. 5.

2 Related Literature Review

In this section, we will briefly review the history of the two main tools that will be used in our study: oligopoly theory and agent-based social simulation.

2.1 Oligopoly Theory

The classical oligopoly theory dates back to the pioneering work of Cournot (1838). It examines an industry in which several firms produce identical product or offer identical service to a homogeneous market. Since then a significant number of researchers focused on the different extensions and generalizations of Cournots classical model. Comprehensive summaries of the earlier works and multi-product models are given in Okuguchi (1976), Okuguchi and Szidarovszky (1999). In the early stages, oligopolies were considered as noncooperative games in which the firms are the players, their output levels are the strategies, and the profit functions are the payoffs. The existence and uniqueness of the equilibrium was first the main issue, under certain monotonicity and convexity assumptions the existence and uniqueness of the equilibrium was proved. This important result was later extended to more realistic model variants including single product models with product differentiation, multi-product oligopolies, labor-managed and rent-seeking games among others.

The main focus of the studies in oligopoly theory has later turned into dynamic extensions. Models were developed with discrete and continuous time scales and the resulting difference and differential equation systems were investigated. The main issue was the asymptotical stability of the equilibrium; conditions were derived to guarantee that the output trajectories converge to the equilibrium in the long run. Most models were linear, where local and global stability are equivalent and very little attention was given to nonlinear dynamics until the late 80s. In developing dynamic models there are usually two alternative ways. In the case of best response dynamics it is assumed that each firm adjusts its output into the direction toward its best response. This approach requires the knowledge of the best response functions of the firms, which needs the solution of usually nonlinear optimization problems based on global information on the payoff functions. In the case of gradient adjustments it is assumed that the firms adjust their outputs in proportion to their marginal profits. This idea has a lot of sense, since in the case of positive (negative) gradient value the firms interest is to increase (decrease) its output level. This concept requires only local information about the payoff functions, so it is much more realistic than the use of best response dynamics. A comprehensive summary of the recent developments in this area can be found in Bischi et al. (2009).

Most studies in oligopoly theory considered only the market as a link between the firms; the unit price was always a function of the total output level of the industry due to the demand-supply balance. However in realistic economies the firms are linked together in much more complicated ways. First, they use common supply of energy, raw material, labor, capital etc., and therefore they also compete on this secondary market in addition to the market of their products. This idea was elaborated in the studies of oligopsonies (Szidarovszky and Okuguchi 2001). In multiproduct oligopolies on the other hand the firms might buy and use the products of other firms, so a network of firms develops. Network oligopolies were introduced and some results were reported in Szidarovszky (1997).

It has been also demonstrated that partial or complete cooperation of the firms in oligopolies will benefit the firms similarly to the well-known prisoners dilemma game (Chiarella and Szidarovszky 2005). Even by any increase in the cooperation level of the firms their benefit also increases.

In most models analytic results could be derived under only very special conditions, which are not the case in realistic economies. Instead of investigating very limited cases theoretically, it is much more important and practical to use computer simulation under realistic conditions and examine the evolution of more advanced production systems such as the industrial clusters.

2.2 The Agent-Based Industrial Cluster Model

In agent-based models, individuals are modeled as heterogeneous agents. Agents have goals and decision rules, and they interact with each other and with the environment. Agent-based model is a bottom up modeling method; it studies a system as an interaction evolving system. It can explicitly explain the decision process of the micro individuals, and the macro emergence from the individuals' interaction.

Agent-based models have been widely used in the analysis of complex economic and social systems (Tesfatsion and Judd 2006). Some initial attempts use agent-based simulation to study some special aspects of industrial clusters (Giardini et al. 2008; Zhang 2003; Albino et al. 2003, 2006a, b; Brenner 2001; Dawid and Wersching 2006; Fioretti 2005).

Fioretti (2005) explained what agent-based models are, the advantages of using agent-based model to study industrial clusters, and introduced some possible simulation tools. Fioretti also reviewed some connectionist models of industrial clusters that are related to agent-based modeling.

Brenner (2001) studied the spatial dynamics of entry, exit and growth of firms. Functions for productivity of firms, innovations, exit and entry of firms, public opinions etc. are modeled and then parameters' impact are analyzed by computer simulations.

Zhang considered a 100×100-lattice environment, on the lattice, agents are born and could choose whether to start a firm or not (Zhang 2003). Production functions and profit functions are adopted for firms. The emergence of a firm in a landscape could inspire its neighbors to choose to start firms; hence industrial clusters might emerge. Computer simulation was adopted to analyze dynamics of market price, firm size distribution, location of clusters, etc.

Giardini et al. (2008) modeled social evaluations as social links, and examined the effects of the reputation of the firms and the quality of the products in a cluster. Their simulation results show that higher reputation of the suppliers and information sharing will result in higher profit for the producers.

Albino et al. (2003) proposed a model to study the multiple forms of the cooperative and competitive relationships among agents and to prove the benefits of the selected type of interaction. In their model, firms and coordination mechanisms are agents; computer simulations were used to evaluate the benefit of cooperation. In the simulations, 3 buyers and 3 sellers were simulated and simple interaction rules were adopted. Albino et al. (2006a, b) introduced the concept of complex adaptive systems into agent-based model and the study mainly focused on innovation dynamics. Their simulation elements and efforts were very similar to their previous works.

3 The Agent-Based and Oligopoly Integrated Model

3.1 The Structure of the System

An industry of a region usually consists of several types of firms. In our model, we consider the situation in which there are several large firms and many smaller suppliers. The large firms produce final products that are sold directly to the market; their products could be substitutes or not. Small firms produce materials, parts, components that large firms buy and build in their final products; their products could be also substitutes or not. Therefore there is a complicated input-output relation between the large and small firms. For example, household appliances are manufactured in a certain location. Relatively large firms produce one or more of the following products: refrigerator, washing machine, television and air conditioner; and smaller firms provide resources to these larger firms.

Firms' interactions are in the form of networks. We establish the inter-firm network as a 2-layer network: one layer of all producers and one layer of all suppliers. The connections between firms in the producer layer and firms in the supplier layer are defined by input-output relations. Firms in the same layer compete for the resources and prices. Firms in the producer layer also compete with each other for new knowledge: the R and D investment of any firm spills over to others who can also benefit from the innovation. We assume that formal systematic R and D is performed only in large firms; this is based on the study of Santarelli and Sterlacchini (1990). All firms also compete in the secondary market. In the secondary market we consider only labor pool. The interactions of the firms in the system can be described therefore as the interaction among producers, the interaction among suppliers, the interaction between producers and suppliers (through supplies), and the interaction through the secondary market (the labor).

For the sake of simplicity, we assume that if a producer needs more supplies than the suppliers can produce, then it will buy them from outside the system with the same price; and when a supplier produces more than the producers need, it will sell the surplus outside the system for the same price. These assumptions will be relaxed in our next study.

3.2 Agents, Interactions and Environment

Individual firms in the system are modeled as agents. There are two types of agents: 'suppliers' who produce and offer their products to producers, 'producers' who produce final products to an open market. There are m supplier agents and n producer agents in the system. Producer agents have innovation ability, with relatively high technical advances, and they are linked together through the open market, the secondary market and by innovation spillovers. In this first model, only the size growth of the existing firms will be considered, the entry of new firms for the growth of the cluster will be studied in our future research. Hence the number of suppliers, m, and the number of producers, n, are considered fixed in the simulation model.

Agents have states and decisions. For any supplier i and any producer j, the main state variables are listed in Table 1, and notations related to the innovation of producer agents are given in Table 2. All variables of these tables vary with time according to state updating rules that will be introduced in the next subsection. In this simple model, we consider only the firms' productions and their innovation investments as factors influencing the formation of cluster. The decision variable of a supplier agent is its production level. The decision variables of a producer agent are its purchases from the suppliers, its output and innovation investment. The 4 types of interactions among agents and with environment are as follows: (1) interaction between suppliers and producers: supply demand balance; (2) interaction among suppliers: competition

Table 1 State variables of suppliers and producers

	Supplier i	Producer j
Number of firms	m	n
Productivity	s_i	z_j
Product price	p_i^s	p_j^p
Labor usage	L_i^s	L_i^p
Profit	φ_i^s	φ_j^p

Table 2 Notations related to the innovation of producers

Innovation development step	I_j
Total cumulative innovation level	$\widetilde{I}_j$
Impact of innovation level on sale price	$F(\widetilde{I}_j)$
Cost function of innovation	$D_j(I_j)$

without product interaction; (3) interaction among producers: competition, information transmission, possible relation in products; (4) interaction among all agents: competition for labor among all firms.

The environment supplies energy, raw material and labor, and it has its rule to change the labor price. The final products of the large firms are sold to the consumers in the environment. From the environment, all agent gathers information: the outputs of their competitors, market prices, spillover of innovation from its cooperators, and the price of the labor pool. Depending on the information from the environment and the agents own state, each agent will make its decision. The details of the decision rules are discussed in the next subsection.

3.3 Agents State Updating Rules Based on Oligopoly Theory

The agents decisions are based on their decision rules.

Let x_{ij} be the amount of the product that producer j purchased from supplier i, then the total physical product of a producer is represented by a production function which is assumed to be linear

$$z_j = \sum_{i=1}^{m} a_{ij}x_{ij} + a_{0j}, \tag{1}$$

where $a_{ij} \geq 0$, $a_{0j} \geq 0$. The marginal productivity of x_{ij} is denoted by a_{ij}. If $a_{ij} > 1$, then an increase in x_{ij} will result in a more than proportionate increase in the output of producer j; for $a_{ij} < 1$, the proportionate increase in the output of producer j is less than that of input x_{ij}; for $a_{ij} = 1$, the proportionate increases are equal.

The price of any supply is a decreasing function of the supplier's own output and the outputs of all other suppliers:

$$p_i^s(s_1, \ldots, s_m) = A_i - B_i s_i - \sum_{l \neq i} b_{il} s_l, \tag{2}$$

where $A_i > 0$, $1 \geq B_i > 0$ and $1 > b_{il} \geq 0$. Larger value of b_{il} represents higher level of similarity between the supplies, or higher level of competition among them. Similarly, the prices of the final products are also linear. It is also assumed that the final products are substitutes:

$$p_j^p(z_1, \ldots, z_n) = \overline{A}_j - \overline{B}_j z_j - \sum_{l \neq j} \overline{b}_{jl} z_l, \tag{3}$$

where $\overline{A}_j > 0$, $1 \geq \overline{B}_j > 0$ and $1 > \overline{b}_{jl} \geq 0$.

The revenue of a supplier is the product of its output and supply's price $s_i p_i^s$. For the revenue of a producer, we also have to consider the innovation effect. The innovation development and spillover of producer j are modeled as

$$\widetilde{I}_j(t+1) = \widetilde{I}_j(t) + I_j + \sum_{l \neq j} k_{jl} I_l, \tag{4}$$

that is, each producer invests in innovation development by increasing its technology level by a step I_j and can utilize the knowledge spillover from other producers. The spillover $k_{jl} I_l$ from agent l is proportional to agent l's innovation investment, where $1 > k_{jl} \geq 0$. The price of any final product is affected by the technology level dependent factor

$$F_j(\widetilde{I}_j) = 1 + (F_j^{max} - 1)(1 - e^{-\omega_j \widetilde{I}_j}). \tag{5}$$

In this function form we model the fact that with higher technological level, better and more expensive final products are produced. If $\widetilde{I}_j = 0$, then this factor equals 1, then it increases in $\widetilde{I}_j$ and converges to a maximum value F_j^{max} as $\widetilde{I}_j$ tends to infinity. The graph of function $F_j(\widetilde{I}_j)$ is shown in Fig. 1. With this innovation dependent factor, the revenue of producer j is given as $z_j p_j^p(z_1, \ldots, z_n) F_j(\widetilde{I}_j)$.

We assume that larger production level requires more labor, so the labor usage of supplier i is

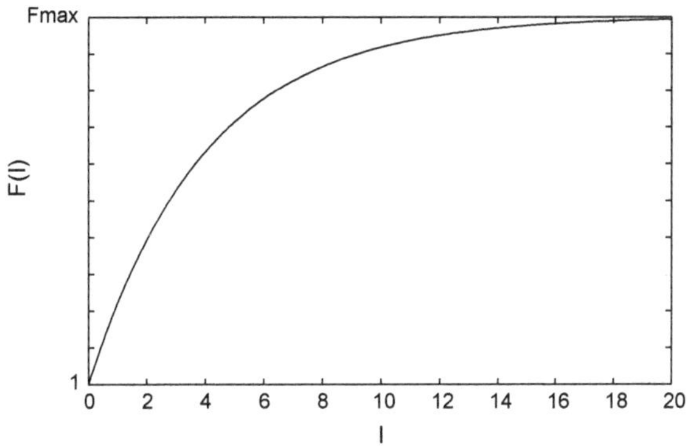

Fig. 1 The graph of function of $F_j(\widetilde{I}_j)$

$$L_i^s(s_i) = \gamma_i + \delta_i s_i. \tag{6}$$

The need of labor of producer j depends on its production and technical levels:

$$L_j^p(z_j, \widetilde{I}_j) = (\overline{\gamma}_j + \overline{\delta}_j z_j) e^{-\overline{\overline{\omega}}_j \widetilde{I}_j}, \tag{7}$$

that is, innovation decreases the labor need of the producers.

The price function of labor in the whole cluster is denoted by p^L, which depends on the total demand of labor. The price of labor is a linear function of the total labor usage:

$$p^L = c - d\left(\sum_i L_i^s + \sum_j L_j^p\right), \tag{8}$$

where $c > 0$ and $d > 0$. In this decreasing function form we model the fact that higher labor force usage decreases the ratio of skilled workers, so the average wage decreases.

The profit of a supplier is modeled as the difference of its revenue and labor cost:

$$\varphi_i^s = s_i p_i^s(s_1, \ldots, s_m) - L_i^s(s_i) p^L\left(\sum_{i=1}^m L_i^s(s_i) + \sum_{j=1}^n L_j^p(z_j, \widetilde{I}_j)\right), \tag{9}$$

For simplicity, we set all other costs to zero. The profit function of the producers is the following:

$$\varphi_j^p = z_j p_j^p(z_1, \dots, z_n) F_j(\widetilde{I}_j) - L_j^p(z_j, \widetilde{I}_j) p^L(\sum_{i=1}^m L_i^s(s_i)$$
$$+ \sum_{j=1}^n L_j^p(z_j, \widetilde{I}_j)) - \sum_{i=1}^m x_{ij} p_i^s(s_1, \dots, s_m) - D_j(I_j), \tag{10}$$

where the innovation development cost is also assumed to be linear:

$$D_j(I_j) = u_j + v_j I_j. \tag{11}$$

In this model the basic decision variables of the suppliers are their output levels s_i, those of the producers are the x_{ij} flows from the suppliers to the firms and the innovation investment I_j. We assume in our model that extra supplies can be sold outside the cluster for the same price, and in the case of supply shortages they can be purchased from sources outside the cluster.

3.4 The Decision Rules of Productions and the Innovation Step

In dynamic oligopoly models, there are two alternative ways to study the evolution of the system. In the case of best response dynamics it is assumed that each firm adjusts its output into the direction toward its best response. In the case of gradient adjustments it is assumed that the firms adjust their outputs in proportion to their marginal profits, which requires only local information about the payoff functions. So it is much more realistic than the use of best response dynamics.

In our earlier papers (Szidarovszky and Zhao 2009; Zhao and Szidarovszky 2008), we assumed gradient adjustment with constant speed of adjustment as updating rules. For producers, they adjust their inputs as

$$x_{ij}(t+1) = x_{ij}(t) + \frac{\varphi_j^p(x_{ij}(t) + \Delta^x) - \varphi_j^p(x_{ij}(t))}{\Delta^x} \varepsilon^x \tag{12}$$

and then the output of producer j at time period $t + 1$ becomes $z_j(t+1) = \sum_{i=1}^m a_{ij} x_{ij}(t+1) + a_{0j}$.

The output of the suppliers is updated according to

$$s_i(t+1) = s_i(t) + \frac{\varphi_i^s(s_i(t) + \Delta^s) - \varphi_i^s(s_i(t))}{\Delta^s} \varepsilon^s. \tag{13}$$

In our earlier models (Szidarovszky and Zhao 2009; Zhao and Szidarovszky 2008), we selected $\Delta^x = 10$, $\Delta^s = 10$, $\varepsilon^x = 1$ and $\varepsilon^s = 0.1$. In this paper, we will also investigate the effects of these parameters on the behaviors of the agents and compare this linear decision updating rules to a special nonlinear rule, which is introduced next:

$$x_{ij}(t+1) = x_{ij}(t) + K_j^p \cdot \frac{2}{\pi} \arctan(\frac{\varphi_j^p(x_{ij}(t) + \Delta^x) - \varphi_j^p(x_{ij}(t)}{\Delta^x}) \tag{14}$$

$$s_i(t+1) = x_i(t) + K_i^s \cdot \frac{2}{\pi} \arctan(\frac{\varphi_i^s(s_i(t) + \Delta^s) - \varphi_i^s(s_i(t))}{\Delta^s}), \tag{15}$$

where $K_j^p = r^p x_{ij}(t)$, $r^p < 1$, $K_i^s = r^s s_i(t)$ and $r^s < 1$.

In the case of large marginal profits the adjustment schemes (12) and (13) might lead to large fluctuations of the output levels of the firms, which make the system unstable. By introducing the inverse tangent function into the adjustment rules we make all output changes bounded, so large fluctuations become impossible.

In our former papers (Szidarovszky and Zhao 2009; Zhao and Szidarovszky 2008), we assumed a constant step in innovation increase $I_j(t) = 0.001$. In this paper however, we will study the effect of innovation step on the behavior of the firms, so we selected a similar updating rule of innovation:

$$I_j(t+1) = K_j^I \cdot \frac{2}{\pi} \arctan(\frac{\varphi_j^p(I_j(t) + \Delta^I) - \varphi_j^p(I_j(t))}{\Delta^I}) \tag{16}$$

with $K_j^I = r^I \cdot \widetilde{I}_j(t)$, $r^I < 1$.

4 The Simulation Process

4.1 *Parameters of the Model*

We have a total population of 25 agents, including 20 suppliers and 5 producer ($m = 20$ and $n = 5$). For any supplier i, we have the maximum price of $A_i = 300$ and marginal price $B_i = 1$ in Eq. (2). To represent the relatively low level of interaction between the suppliers in their prices, b_{il} is selected as 0.1 for $l \neq i$. That is, the suppliers specialize in different supplies, so their prices do not interfere with each other much. The parameters of the labor function (6) of the suppliers are chosen as $\gamma_i = 10$ and $\delta_i = 0.4$. For any producer j, the parameters of its production function are chosen as $a_{0j} = 20$, $a_{ij} = 0.1$ in Eq. (1). We also assumed much higher prices for final products than those of the supplies, so we select the common maximum price of the producers as $\overline{A}_j = 1300$ and similarly to the situation of the supplier agents, we have $\overline{B}_j = 1$ and $\overline{b}_{jl} = 0.1$. As the knowledge spillover is concerned, we consider 10% of the innovation as spillover, hence $k_{jl} = 0.1$ for $l \neq j$. Besides, $F_j^{max} = 2$ and $\omega_j = 0.1$ for the innovation dependent factor of Eq. (5). The parameters of the innovation development cost in Eq. (11) are selected as $u_j = 50$ and $v_j = 0.1$. The sizes of the producers are assumed to be larger than those of the suppliers, hence $\overline{\gamma}_j = 50$, $\overline{\delta}_j = 0.3$ and $\overline{\overline{\omega}}_j = 0.05$. For the labor market, we have the maximum labor price $c = 300$ and d is selected as 0.2 in Eq. (8).

In this paper, we will analyze only the effect of the decision rules. At the beginning of the simulation process, the initial values $x_{ij}(0)$ were generated randomly by using uniform distribution from the interval [0, 20]. The initial value of s_i is $s_i(0) = \sum_j x_{ij}(0)$; the corresponding values of z_j are calculated according to Eq. (1). The same set of the initial $x_{ij}(0)$ values was used in the same simulation group for comparison purposes. The initial value of technology level of all producers was chosen as 1.

There might be situations when prices, labors might become negative in the process, therefore these variables will be bounded from below. It is reasonable to assume that final products are sold for higher prices than supplies. The prices of final products are bounded from below by 5, those of the suppliers are bounded from below by 0. Usually, government has minimum wage policy; hence, for the whole system the price of labor is bounded by 10 from below.

4.2 *Simulation Results*

4.2.1 The Effect of Parameters of Gradient Adjustment

First we fixed the values of Δ^x and Δ^s as 10, changed ε^s from 0.1 to 1.1 with the step size of 0.2, and with each value of ε^s, ε^x varies from 0.1 to 2 with varying step sizes depending on the pattern changes in the behavior of the agents.

1. $\epsilon^s = 0.1$

When ε^x varies gradually from 0.1 to 1.79, the patterns of the behavior of the agents remain the same: fast increase or decrease at the beginning (this can be interpreted as the primitive formation of clusters) then the patterns converge or increase (decrease) slowly (Fig. 2). If we consider a time cross section with increasing value of ε^x, the output, labor usage and profit of both the suppliers and the producers increase and the average price of all firms and the labor price decrease. In the long run, the output and profit of the producers always increase, but when ε^x is small (< 1 in the figure) the labor usage is actually decreasing slowly in time. In this situation, the producers will not increase their firm sizes. As $\varepsilon^x < 1$, the profits of the suppliers also decrease slowly and when t is large enough, the profit might drop down to a negative value. Hence the supplier firms might shut down or sell their products outside the

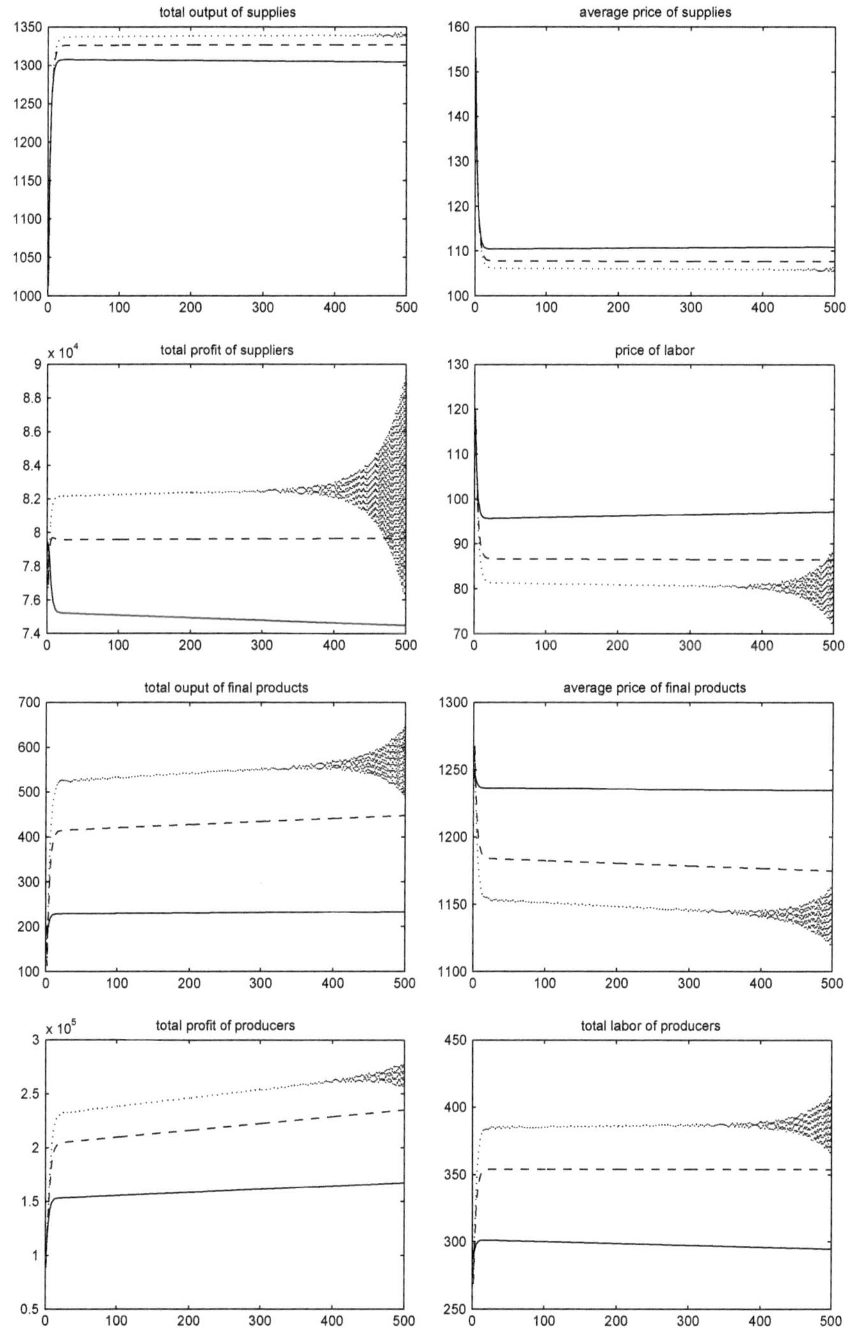

Fig. 2 Firms' behaviors when $\varepsilon^s = 0.1$, $\varepsilon^x = 0.1$ (*solid line*), 1 (*dashed line*), and 1.8 (*dotted line*)

cluster. In such situations, the cluster will never survive. Until ε^x is increased above 1, the labor of the producers and the profit of the suppliers keep a steady value after an increasing period. In this situation, it is hard to say that the exiting firms will expand.

From $\varepsilon^x = 1.795$, an oscillating behavior can be observed between two states (Fig. 2). That is, a two-period cycle emerges. The amplitudes of the oscillations become larger with time until they become stable. The oscillation starts after a linear pattern and its amplitude increases. Since we will have later other kinds of oscillation patterns, in order to distinguish between them, we call the oscillations just described as tail oscillations. Larger value of ε^x makes the oscillations start earlier in time and when ε^x becomes large enough, oscillations start almost at the beginning of the time scale, and the cycles will have more points. Hence for a stable system, when suppliers update their outputs as slowly as $\varepsilon^s = 0.1$, the producers should not choose large value of $\varepsilon^x (\varepsilon^x \geq 1.795)$. For a stable cluster that could stay, the range of ε^x should be $1 \leq \varepsilon^x < 1.795$.

2. $\varepsilon^s = 0.3$

Unlike higher values of ε^x which produce tail oscillations, higher values of ε^s induce behavior oscillation from the beginning of the time scale (we call this type oscillation as head oscillation) (Fig. 3). For $\varepsilon^x = 0.1$, there is a small oscillation at the beginning of time but the trajectories converge later. When ε^x increases, the amplitude and the length of the oscillating period become larger. If ε^x is larger than 1.4, then the trajectories do not converge anymore, and the shape of the time series looks like a dog bone as the result of the combination of the head oscillation and the tail oscillation. When $\varepsilon^s = 0.3$ and $\varepsilon^x < 1$, the long term behavior patterns are the same as those with $\varepsilon^s = 0.1$ and $\varepsilon^x < 1.795$.

Another impact of the higher value $\varepsilon^s = 0.3$ is that the tail oscillation patterns, which were induced by increasing values of ε^x, appear earlier than in the case of $\varepsilon^x = 0.1$. The two types of oscillations (head and tail) are combined again to the dog bone shape when ε^x is slightly larger than 1.4, and if ε^x becomes even larger, then the behavior oscillates between two stable states, forming a two-period cycle.

For $\varepsilon^s = 0.3$, smaller value of ε^x should be used to avoid the large amplitude oscillations. However, like in the case of $\varepsilon^s = 0.1$, with small value of ε^x, the small decreasing labor usage and decreasing profits imply that the cluster will not survive. Hence, for a surviving stable cluster, the range of ε^x should be $1 \leq \varepsilon^x \leq 1.3$.

3. $\varepsilon^s = 0.5$

The situation of $\varepsilon^s = 0.5$ is very similar to the case of $\varepsilon^s = 0.3$, however with larger amplitude of oscillation. The possible range of ε^x for a stable system is very narrow.

4. $\varepsilon^s \geq 0.7$

When $\varepsilon^s = 0.7$, the behaviors of both the suppliers and the producers oscillate irregularly, profits might drop down to negative values (Fig. 4). When ε^x increases, the producers production levels and profits also increase. When ε^x is increased to 1, the behaviors of the two types of agents converge, however the corresponding profits of suppliers become negative.

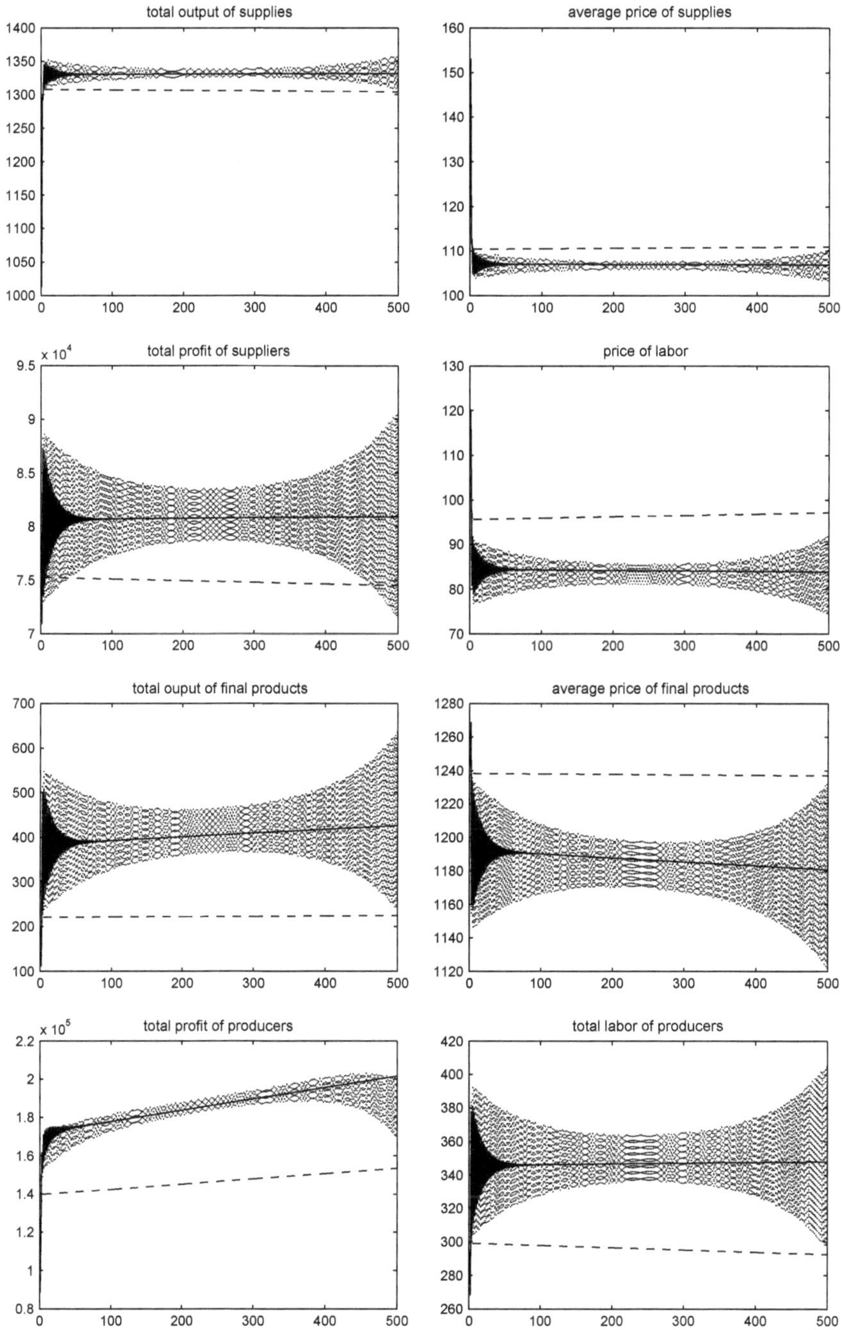

Fig. 3 Firms' behaviors when $\varepsilon^s = 0.3$, $\varepsilon^x = 0.1$ (*dashed line*), 1.3 (*solid line*), 1.4 (*gray dotted line*)

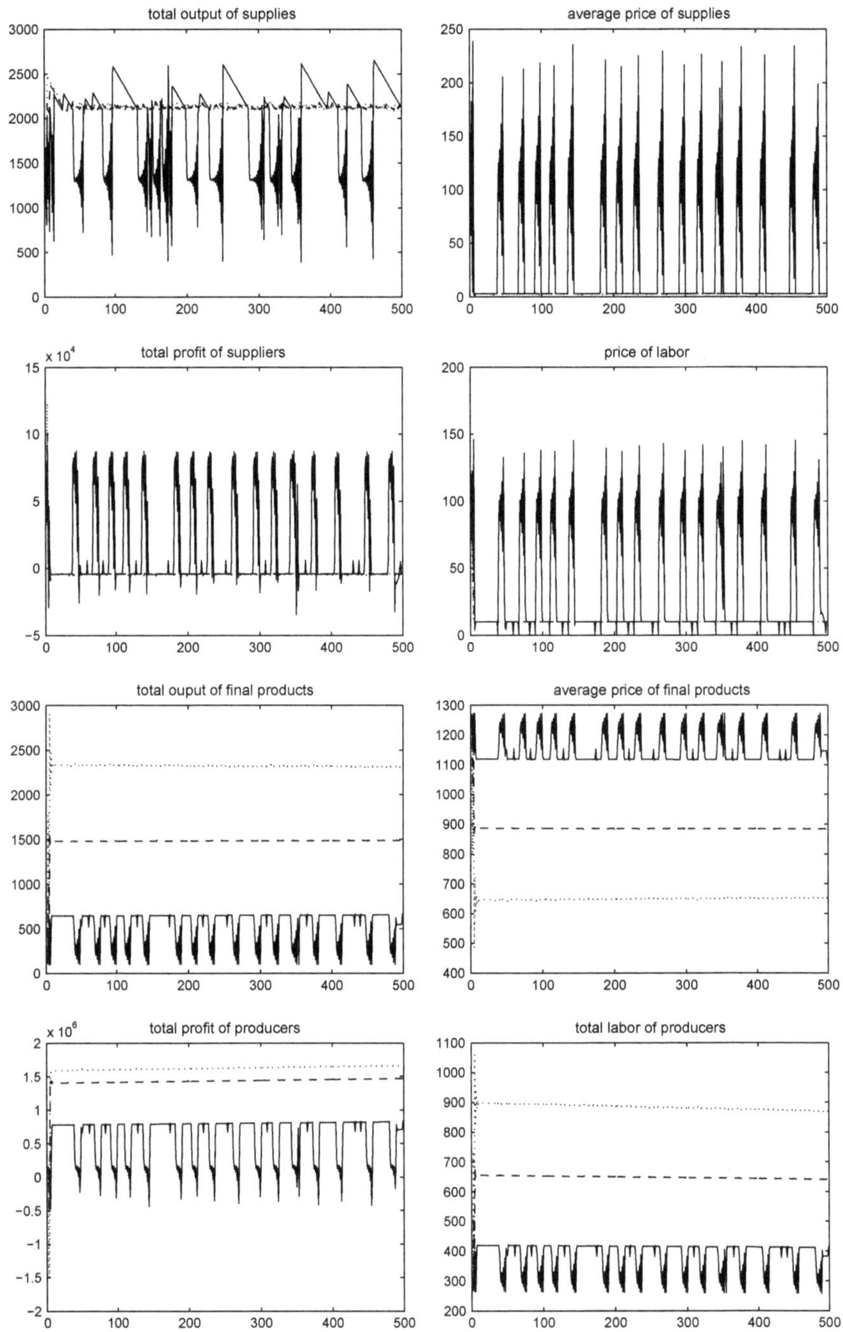

Fig. 4 Firm's behaviors when $\varepsilon^s = 0.7$, $\varepsilon^x = 0.3$ (*solid*), 1 (*dashed*), 2 (*dotted*)

When $\varepsilon^s = 0.9$, the patterns are similar to the previous case with the difference that when ε^x increased to 0.5, the behaviors of both types of agents converge, however the profits of the suppliers become negative.

When $\varepsilon^s = 1.1$, there are oscillations and sparks in the behaviors, they are never stable regardless of the value of ε^x (Fig. 5).

Overall, changes in the values of ε^x and ε^s have significant influence on the behaviors of the suppliers and the producers. The combination of the different values of ε^s and ε^x will generate many different patterns. Larger value of ε^s and larger value of ε^x induce unstable systems.

It is interesting to analyze the reason why oscillation is observed. When ε^s is very small, as 0.1, any increase of ε^x in a certain range will benefit the suppliers and the producers in the short-term, all make more profit even with decreased average prices. When ε^x is increased, then the behavior oscillates between two values. The amplitude of the oscillation increases in time through many iterations and then becomes stable. Our time scale is $0 \leq t \leq 500$. When $\varepsilon^x = 1.795$, oscillations emerge at the end of the time scale, and when $\varepsilon^x = 1.85$, oscillations emerge before time period 200. The reason is the following. When ε^x is increased, the outputs of the producers should also increase; this brings more labor to the cluster and decreases the labor price. This benefits the suppliers, increases their profits. Since the outputs and profits of the producers increase in time, until a certain time period, more and more outputs are produced, and when it accumulates to a certain value (when ε^x is large enough), the producers will over adjust their outputs, their profits decrease, then they adjust to the opposite direction. This drives the oscillations; the oscillation amplitude increases gradually until stable cycles occur.

If the updating step is too large, it will generate unstable behaviors, however if it is too small, then the firms development is also too slow. To form a stable cluster and also to keep the cluster for a longer time period, the values of ε^x and ε^s should be selected properly.

We also repeated the simulations for $\Delta^x = 1$. The pattern changes in the behavior of the firms were similar to those observed for $\Delta^x = 10$, only the critical values for pattern changes were slightly different.

4.2.2 Simulation Results with New Updating Rules

The combination of $\varepsilon^s = 0.1$ and $\varepsilon^x = 1$ is chosen as a benchmark for the comparison of the two different updating rules, the old rule (12)–(13), and the new rule (14)–(15). Figure 6 shows the results of the four different updating strategies: both types of agents use old updating rules and the innovation step is constant (dashed line); both types of agents use new updating rules and the innovation step is constant (dotted line); both types of agents use old updating rules but innovation increase uses new rule (dash-dot line); both types of agents use new updating rules and innovation increase uses new rule (solid line). In the new updating rules, $\Delta^x = 1$, $r^p = r^s = r^l = 0.1$. Even though the selected value of Δ^x is large, say 10, the behaviors of the firms become much smoother than before, only small oscillations within a small range can

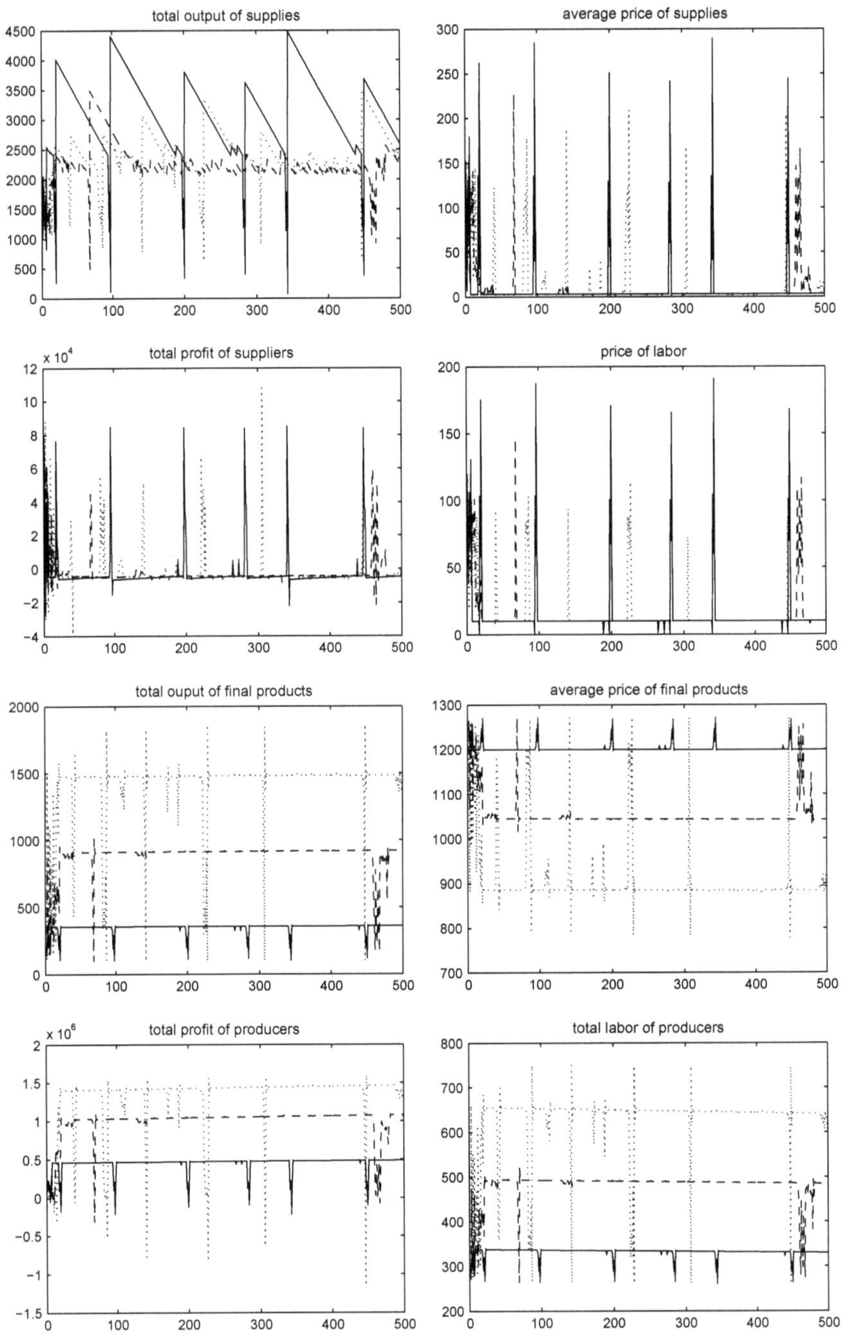

Fig. 5 Firms' behaviors when $\varepsilon^s = 1.1$, $\varepsilon^x = 0.1$ (*dashed line*), 0.5 (*solid line*), 1(*gray dotted line*)

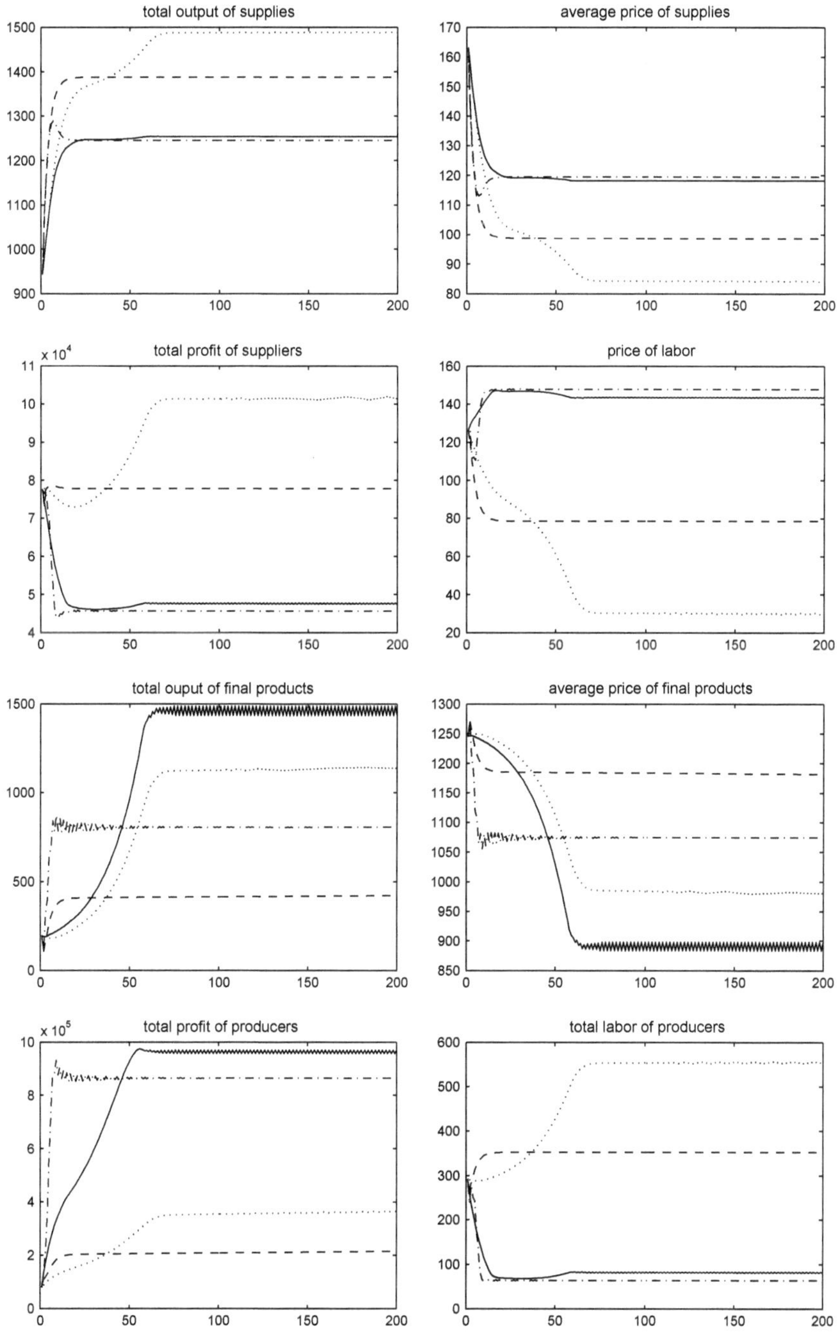

Fig. 6 Comparison of patterns of aggregated behavior by using different updating rules. *Dashed lines* old update rule; *dotted lines* x and z updated using new rule; *dash-dot lines* only I was updated using new rule; *solid lines* all use new rule

be observed. It is surprising to see that the new updating rule of the innovation step will harm the suppliers no matter the agents adopt new output strategy or not (since the suppliers' profits are decreased and the average prices are increased). Hence a stable innovation step is a relatively good choice. From the simulation results we have the main conclusion that to bring existing firms into a stable cluster, both types of agents should adopt new output strategies.

5 Conclusion

This paper presents an integrated model of agent-based simulation and network oligopoly to study the evolution of a group of local firms for the possibility of forming a long lasting industrial cluster. Agent-based simulation model is used to study the effect of firms' decisions on the formation of the cluster, and network oligopoly theory is used to model the decisions and interaction rules of the agents. We studied the very simple situation when the firms only concern is their marginal profits and their decisions are their productivity and innovation investments. Firms interact through the product market and the secondary market of labor. The structure of the system is similar to Markusen's 'Hub and Spoke' type of cluster. From the simulation results we demonstrated that under some production decision rules the group could have the potential to involve into a surviving cluster. This paper offers a starting point to study the cluster formation from exiting firms. More complicated situations will be considered in our future research. We considered only fixed network with existing firms. For investigating the growth of the cluster, innovations disperse and relationship establishment, dynamic spatial networks have to be used. This task will be the topic of our future work.

References

Albino, V., Carbonara, N., & Giannoccaro, I. (2003). Coordination mechanisms based on cooperation and competition within industrial districts: An agent-based computational approach. *Journal of Artificial Societies and Social Simulation, 6*(4). http://jasss.soc.surrey.ac.uk/6/4/3.html.

Albino, V., Carbonara, N., & Giannoccaro, I. (2006a). The competitive advantage of geographical clusters as complex adaptive systems: an exploratory study based on case studies and network analysis. In *International Conference on Complex Systems*. Boston, MA, USA, June 25–30.

Albino, V., Carbonara, N., & Giannoccaro, I. (2006b). Innovation in industrial districts: An agent-based simulation model. *International Journal of Production Economics, 104*(1), 30–45.

Bischi, G. I., Chiarella, C., Kopel, M., & Szidarovszky, F. (2009). *Nonlinear oligopolies: Stability and bifurcations*. Berlin/New York: Springer.

Brenner, T. (2001). Simulating the evolution of localised industrial clusters an identification of the basic mechanisms. *Journal of Artificial Societies and Social Simulation, 4*(3). http://www.soc.surrey.ac.uk/JASSS/4/3/4.html.

Bresnahan, T., Gambardella, A., & Saxenian, A. (2001). Old economy inputs for new economy outcomes: Cluster formation in the new silicon valleys. *Industrial and Corporate Change*, *10*(4), 835–860.
Chiarella, C., & Szidarovszky, F. (2005). The complex asymptotic behavior of dynamic oligopolies with partially cooperating firms. *Pure Mathematics and Applications*, *16*(4), 365–375.
Cournot, A. (1838). Recherches sur les principes mathmatiques de la thorie de richesses, hachett, paris (English translation, 1960. Researches into the Mathematical Principles of the Theory of Wealth, Kelley, New York)
Dawid, H., & Wersching, K. (2006). On technological specialization in industrial clusters: an agent-based analysis. In J. P. Rennard (Eds.), *Handbook of Research on Nature Inspired Computing for Economic and Management* (pp. 367–378). Idea Group Publishers.
Fioretti, G. (2005). Agent-based models of industrial clusters and districts. In F. Columbus (Eds.), *Contemporary Issues in Urban and Regional Economics*. Nova Science Publishers.
Giardini, F., Tosto, G. D., & Conte, R. (2008). A model for simulating reputation dynamics in industrial districts. *Simulation Modelling Practice and Theory*, *16*, 231–241.
Markusen, A. (1996). Sticky places in slippery space: A typology of industrial districts. *Economic Geography, 72*, 293–313.
Maskell, P. (2001). Towards a knowledge-based theory of the geographical cluster. *Industrial and Corporate Change*, *10*(4), 921–943.
Okuguchi, K. (1976). *Expectations and stability in oligopoly models*. Berlin/New York: Springer.
Okuguchi, K., & Szidarovszky, F. (1999). *The theory of oligopoly with multi-product firms*. Berlin/New York: Springer.
Santarelli, F., & Sterlacchini, A. (1990). Innovation, formal vs. informal R and D and Firm size: Some evidence from italian manufacturing firms. *Small Business Economics*, *2*(3), 223–228.
Swann, G. M. P., Prevezer, M., & Stout, D. (1998). *The dynamics of industrial clustering*. New York: Oxford University Press.
Szidarovszky, F. (1997). Network oligopolies. *Pure Mathematics and Applications*, *8*(1), 117–123.
Szidarovszky, F., & Okuguchi, K. (2001). Dynamic analysis of oligopsony under adaptive expectations. *Southwest Journal of Pure and Applied Mathematics*, *2*, 53–60.
Szidarovszky, F., & Zhao, J. (2009). The dynamic evolution of industrial clusters. *Cubo A Mathematical Journal*, *11*(2), 37–54.
Tesfatsion, L., & Judd, K. L. (2006). *Handbook of computational economics: Agent-based computational economics* (Vol. 2). Oxford, UK: Elsevier.
Zhang, J. (2003). Growing silicon valley on a landscape: An agent-based approach to high-tech industrial clusters. *Journal of Evolutionary Economics*, *13*, 529–548.
Zhao, J., & Szidarovszky, F. (2008). A dynamic model and simulation of industrial clusters. In *32nd Annual IEEE International Computer Software and Applications* (pp. 890–895).

Approximation of LPV-Systems with Constant-Parametric Switching Systems

Sandor Molnar and Mark Molnar

Abstract A common problem in systems and control theory is to provide an approximation to non-linear systems. We provide a novel approach as a general solution to this problem originally conceived by Gamkrelidze. We consider and solve a general approximation problem which provides the fundamentals for various switching-type systems thus encompassing a wide range of systems theory problems.

1 Introduction

Vertical integration is one of the key elements of modern industrial production (Molnár and Szigeti 1994; Molnár 1989). We witness the widespread penetration of industrial sensors and monitoring devices embedded in the production chain resulting in a vast amount of digital data on plant level activity. This, together with the control and system theory provides a unique chance to implement a decentralised automated production system resulting in improved efficiency and higher output as production optimisation requires optimal systems. In the following we provide an important element in optimal system control applicable to a wide range of state-of-the-art systems.

Optimising linear control systems shows similarities with linear programming as it was noticed by Pontryagin and others. Starting with the qualitative analysis of optimal control, Gamkrelidze (1978) considered the

S. Molnar (✉)
Faculty of Mechanical Engineering, Department of Mathematics and Informatics, Szent Istvan University, Gödöllő, Hungary
e-mail: molnar.sandor@gek.szie.hu

M. Molnar
Faculty of Economics and Business Management, Department of Macroeconomics, Szent Istvan University, Gödöllő, Hungary
e-mail: molnar.mark@gtk.szie.hu

A. Matsumoto (ed.), *Optimization and Dynamics with Their Applications*,
DOI 10.1007/978-981-10-4214-0_8

$$\dot{x}(t) = A(t)x(t) + B(t)u(t), x(0) = \xi,$$

linear, time varying (LTV) system for controls over the $u(t) \in \times_1^k [0, 1]$ cube. He proved that if u: $[0, T] \rightarrow \times_1^k [0, 1] = U$ is piecewise continuous, then for every $\varepsilon > 0$ there exists a v: $[0, T] \rightarrow U$ piecewise continuous control, which directs the system to the vertices of the cube and the solution

$$\dot{y}(t) = A(t)y(t) + B(t)v(t), \quad y(0) = \xi$$

satisfies the condition

$$\|x(t) - y(t)\| < \varepsilon.$$

Thus, even though the *u(t)* and *v(t)* controls might show significant pointwise differences, the respective trajectories will still remain uniformly close.

We will use a simple application in the following sections to demonstrate the novel approach presented in the approximation theorems. This application is a simple Buck-Boost converter and is widely applied (Sira-Ramírez 2015; Sira-Ramírez and Agrawal 2004).

2 Applications

We will demonstrate the proposed approximation problem through a well-known engineering application. Let's consider the so called Buck-Boost converter circuit:

The function *v(t)* describes the state of the switch and can take discrete values $\{0, 1\}$, as it is visible on Fig. 1. The ideal behaviour can be characterised by a piecewise continuous function u: $[0, T] \rightarrow [0, 1]$ which is described by the following system of differential equations

$$\begin{aligned} L\dot{x}_1 &= (1 - u)x_2 + uE \\ C\dot{x}_2 &= -(1 - u)x_1 - \tfrac{x_2}{R} \end{aligned} \quad (1)$$

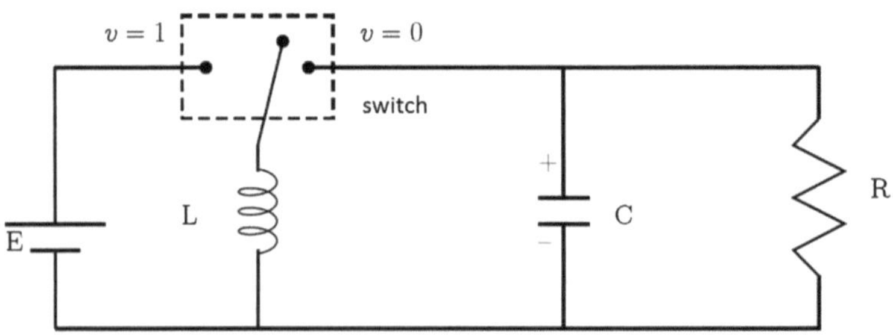

Fig. 1 A Buck-Boost converter circuit

where x_1 denotes current flowing through the coil and x_2 the voltage drop on the capacitor.

Function u which describes the ideal behaviour cannot be generated by turning the switch on and off, only a good approximation can be achieved. For example, for a given precision $\varepsilon > 0$ a function $v(t)$ can be assigned describing a set of switchings,

$$\begin{aligned} L\dot{y}_1 &= (1-v)y_2 + vE \\ C\dot{y}_2 &= -1(1-v)y_1 - \frac{y_2}{R} \end{aligned} \tag{2}$$

where the resulting y_1 and y_2 fulfill the conditions

$$|x_1(t) - y_1(t)| < \varepsilon, \; |x_2(t) - y_2(t)| < \varepsilon.$$

It is visible that our simple example is nonlinear, does not have any connection with the optimal control, and is a typical problem in systems theory. Mathematically, however it is of a very similar nature as the approximation problem considered by Gamkrelidze.

3 Problem Formulation

Herewith we consider and solve a general approximation problem which provides the fundamentals for further switching-type systems encompassing a wide range of systems theory problems. The Buck-Boost converter is considered as the paradigm of switching systems.

3.1 Introducing LPV Systems from a Buck-Boost Converter

The system is non-linear, thus it can not be fitted in the framework of the Gamkrelidze approximation theorem. Instead of linearising we introduce a parameter to replace the $\begin{pmatrix} x_1 \\ x_2 \end{pmatrix}$ state making the system formally linear. After dividing by the physical constant constants L and C, (1) can be rewritten as

$$\begin{aligned} \dot{x}_1 &= \frac{1}{L}x_2 + \left(\frac{E}{L} - p_2\frac{1}{L}\right)u, \\ \dot{x}_2 &= -\frac{1}{C}x_1 - \frac{1}{RC}x_2 + \frac{1}{C}p_1 u, \end{aligned} \tag{3}$$

or in matrix form,

$$\begin{pmatrix} \dot{x}_1 \\ \dot{x}_2 \end{pmatrix} = \begin{pmatrix} 0 & \frac{1}{L} \\ -\frac{1}{C} & \frac{1}{RC} \end{pmatrix} \begin{pmatrix} x_1 \\ x_2 \end{pmatrix} + \begin{pmatrix} \frac{1}{L} p_2 \\ \frac{1}{C} p_1 \end{pmatrix} u. \tag{4}$$

Considering the $\begin{pmatrix} p_1 \\ p_2 \end{pmatrix}$ parameter vector as the $\begin{pmatrix} x_1 \\ x_2 \end{pmatrix}$ state variable, we arrive to the non-linear system in (2). We can immediately raise the problem of adopting a Gamkrelidze-type approximation theorem to LPV or LPTV systems, while in the same time giving a generalisation.

It is well visible that the outlined problem is purely of system theory and not of optimisation theory. We will call the approximating systems which will be piece-wise continuous constant parametric linear systems, similarly to the Buck-Boost switches switching systems.

Let $U \subset R^{k_1}$ convex polyhedron and $P \subset R^{k_1}$ a compact set. We assume that

$$A: P \times [0, T] \to \mathbb{R}^{n \times n}$$
$$B: P \times [0, T] \to \mathbb{R}^{n \times k_1}$$

are satisfying the uniform Lipschitz-condition in t with L_1 and L_2 parameters, respectively.

We substitute the p parameter of the

$$\dot{x} = A(p, t)x + B(p, t)u \tag{5}$$

LPTV system with a state-time-dependent variable. For this consider an open set $D \subset \mathbb{R}^n \times [0, T]$, and the $p: D \to P$ parameter-function. We assume that function p is uniformly Lipschitz-continuous with an L_3 Lipschitz-constant. To have an easy picture of the structure of such a set D, consider the following Fig. 2.

The basis and top shape of the cylinder-like object belongs to the D-domain, but the constituents of the cylinder do not, due to the openness in $\mathbb{R}^2 \times [0, T]$

Fig. 2 Structure of set D

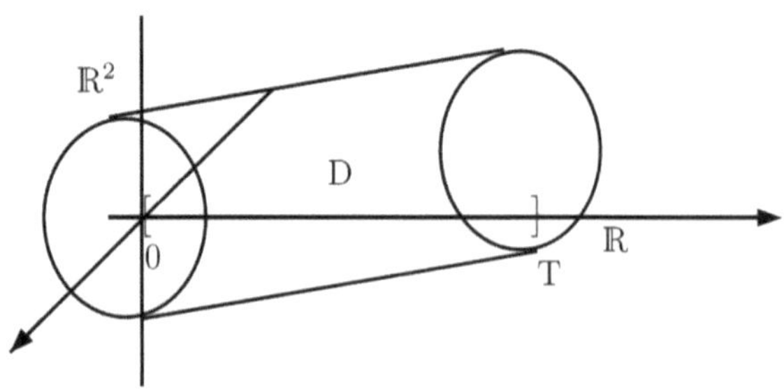

4 Approximation Theorems

In the followings we outline two approximation theorems. In the case of the first theorem the approximation of control u can contain large discrete-point errors, since the approximating v control considers the vertices of the U convex polyhedron, but the respective trajectories are uniformly close. The second theorem states significantly more. Not just that the above mentioned statement holds, but also parameter function q (which is used for approximating parameter function p) considers values of the vertices of the P convex polyhedron and despite this, the approximation of the trajectories is uniform.

Approximation Theorem 1 *Assume that for a piecewise continuous* $u\colon [0,T] \to U$ *control, and for a* $p\colon D \to P$ *uniformly global Lipschitz-continuous state-time-dependent parameter function in t, and for a* $\xi \in \mathbb{R}^n$ *initial condition there is a solution for the*

$$\begin{aligned} \dot{x}(t) &= A(p(x(t),t),t)x(t) + B(p(x(t),t),t)u(t), \\ x(0) &= \xi \end{aligned} \tag{6}$$

initial value problem. Then, there is a $\varepsilon_0 > 0$ *for which for every* $0 < \varepsilon < \varepsilon_0$ *the following exist:*

(1) $\delta > 0$,
(2) *a piecewise constant* $v\colon [0,T] \to U$ *which takes the values of the vertices of the U convex polyhedron,*
(3) $q\colon D \to P$ *piecewise constant state-time-dependent parameter function, that for all* $\eta \in \mathbb{R}^n$ *initial conditions satisfying* $\|\xi - \eta\| < \delta$ *the solution of the initial value problem of*

$$\begin{aligned} \dot{y}(t) &= A(q(y(t),t),t)y(t) + B(q(y(t),t),t)v(t), \\ y(0) &= \eta \end{aligned} \tag{7}$$

has the whole [0, T] interval as a domain, and there $\|x(t) - y(t)\| < \varepsilon$ *holds.*

For the next interpolation theorem let's assume that $P \subset \mathbb{R}^{k_2}$ is also a convex polyhedron. We assume that functions $A\colon P \to \mathbb{R}^{n\times n}$ és a $B\colon P \to \mathbb{R}^{n\times k_1}$ are linear in p, that is, functions $p \to A(p,t)$ and $p \to B(p,t)$ are linear for all t. From this, for a fixed t the Lipschitz-condition holds, thus we need only assume continuity and uniformity in t. With these additional assumptions we can improve our previous theorem.

Approximation Theorem 2 *Assume that for a piecewise continuous control* $u\colon [0,T] \to U$ *and for a time-state-dependent function* $p\colon D \to P$ *satisfying the uniform global Lipschitz-condition in t, and for an initial state* $\xi \in \mathbb{R}^n$ *there is a* *solution* x: $[0, T] \rightarrow \mathbb{R}^n$ *for the initial value problem on the whole* $[0, T]$ *domain. Then there exists* $\varepsilon_0 > 0$ *so that for all* $0 < \varepsilon < \varepsilon_0$ *the followings exist:*

(1) $\delta > 0$,
(2) *a piecewise constant* v: $[0, T] \rightarrow U$, *which takes its values from the vertices of the* U *convex polyhedron,*
(3) *a* $q: D \rightarrow P$ *piecewise constant state-time parameter function which takes its values from the* P *convex polyhedron, so that for all* $\eta \in \mathbb{R}^n$, *for which* $\|\xi - \eta\| < \delta$, *if the initial value condition* $y(0) = \eta$ *is satisfied the solution of the initial value problem is available on the whole domain* $[0, T]$, *and there*

$$\|x(t) - y(t)\| < \varepsilon$$

holds.

While the Gamkrelidze-type of optimisation cannot be applied to the Buck-Boost switch, our approximation theorems are applicable. As we discuss a system which is linear in parameters we can endeavour to apply the second theorem. For this we have to ensure that at least on the finite [0, T] interval we can keep the $p = \begin{pmatrix} P_1 \\ P_2 \end{pmatrix}$ parameters in a convex polyhedron if $p = \mathbf{x}$. Let's consider Eq. (1) and integrate on the interval $[0, t) \subset [0, T]$. Consider the following system of equations

$$\begin{aligned} x_1(t) &= \xi_1 + \frac{E}{L}\int_0^t u(\tau)d\tau + \frac{1}{L}\int_0^t (1 - u(\tau))\, x_2(\tau)d\tau \\ x_2(t) &= \xi_2 + \frac{1}{C}\int_0^t (u(\tau) - 1)\, x_1(\tau)d\tau - \frac{1}{RC}\int_0^t x_2(\tau)d\tau \end{aligned} \tag{8}$$

and substitute $x_1(t)$ from the first equation into the second and from this equation substitute $x_2(t)$ to get:

$$\begin{aligned} x_2(t) = {} & \xi_2 + \frac{1}{C}\int_0^t (u(\tau_1) - 1)\left[\xi_1 + \frac{E}{L}\int_0^{\tau_1} u(\tau_2)d\tau_2 + \frac{1}{L}\int_0^{\tau_1}(1 - u(\tau_2))x_2(\tau_2)d\tau_2\right]d\tau_1 - \\ & - \frac{1}{RC}\int_0^t x_2(\tau)d\tau = \frac{1}{C}\int_0^t (u(\tau_1) - 1)d\tau_1\xi_1 + \xi_2 + \frac{E}{LC}\int_0^t (u_1(\tau_1) - 1)\int_0^{\tau_1} u(\tau_2)d\tau_2 d\tau_1 + \\ & + \frac{1}{LC}\int_0^t\left((u(\tau_1) - 1)\int_0^{\tau_1}(1 - u(\tau_2))x_2(\tau_2)d\tau_2\right)d\tau_1 - \frac{1}{RC}\int_0^t x_2(\tau)d\tau = \\ = {} & \frac{1}{C}\int_0^t (u(\tau_1) - 1)d\tau_1\xi_1 + \xi_2 + \frac{E}{LC}\int_0^t (u_1(\tau_1) - 1)\int_0^{\tau_1} u(\tau_2)d\tau_2 d\tau_1 + \\ & + \int_0^t\left[\frac{1}{LC}(1 - u(\tau_1))\int_{\tau_1}^t (u(\tau_2) - 1)\, d\tau_2 - \frac{1}{RC}\right]x_2(\tau_1)d\tau_1. \end{aligned}$$

Let's introduce the following notations

$$\phi_1(t) = \frac{1}{C}\int_0^t (u(\tau_1) - 1)d\tau_1,$$

$$\phi_2(t) = \frac{E}{LC}\int_0^t (u_1\tau_{-1})\int_0^{\tau_2} u(\tau_2)d\tau_2 d\tau_1, \quad \text{és}$$

$$\phi_3(\tau_1) = \frac{1}{LC}(1 - u(\tau_1))\int_{\tau_1}^t (u(\tau_2) - 1)d\tau_2 - \frac{1}{RC}.$$

Then

$$x_2(t) = \phi(t)\xi_1 + \xi_2 + \phi_2(t) + \int_0^t \phi_3(\tau)x_2(\tau)d\tau.$$

From this follows

$$x_2(t) = \varphi_1(t)\xi_1 + \xi_2 + \varphi_2(t) + \int_0^t \varphi_3(\tau_1)\left[\varphi_1(\tau_1)\xi_1 + \xi_2 + \varphi_2(\tau_1) + \int_0^{\tau_1} \varphi_3(\tau_2)x_2(\tau_2)d\tau_2\right]d\tau_1 =$$

$$= \left(\varphi_1(t) + \int_0^t \varphi_3(\tau)\varphi_1(\tau)d\tau\right)\xi_1 + \left(1 + \int_0^t \varphi_3(\tau_1)d\tau_1\right)\xi_2 + \varphi_2(t) + \int_0^t \varphi_3(\tau)\varphi_2(\tau)d\tau +$$

$$+ \int_0^t \varphi_3(\tau_1)\int_0^{\tau_1} \varphi_3(\tau_2)x_2(\tau_2)d\tau_2 d\tau_1 = \left(\varphi_1(t) + \int_0^t \varphi_3(\tau)\varphi_1(\tau)d\tau\right)\xi_1 +$$

$$+ \left(1 + \int_0^t \varphi_3(\tau)d\tau\right)\xi_2 + \left(\varphi_2(t) + \int_0^t \varphi_3(t)\varphi_2(\tau)d\tau + \right) +$$

$$+ \int_0^t \varphi_3(\tau_1)\int_0^{\tau_1} \varphi_3(\tau_2)\left[\varphi_1(\tau_2)\xi_1 + \xi_2 + \varphi_2(\tau_2) + \int_0^{\tau_2} \varphi_3(\tau_3)x_2(\tau_3)d\tau_3\right]d\tau_2 d\tau_1 =$$

$$= \left[\varphi_1(t) + \int_0^t \varphi_3(\tau_1)\varphi_1(\tau_1)d\tau_1 + \int_0^t \varphi_3(\tau_1)\int_0^{\tau_1} \varphi_3(\tau_2)\varphi_1(\tau_2)d\tau_2 d\tau_1\right]\xi_1 +$$

$$+ \left[1 + \int_0^t \varphi_3(\tau_1)d\tau_1 + \int_0^t \varphi_3(\tau_1)\int_0^{\tau_1} \varphi_3(\tau_2)d\tau_2 d\tau_1\right]\xi_2 +$$

$$+ \left[\varphi_2(t) + \int_0^t \varphi_3(\tau_1)\varphi_2(\tau_1)d\tau_1 + \int_0^t \varphi_3(\tau_1)\int_0^{\tau_1} \varphi_3(\tau_2)\varphi_2(\tau_2)d\tau_2 d\tau_1\right] +$$

$$+ \int_0^t \varphi_3(\tau_1)\int_0^{\tau_1} \varphi_3(\tau_2)\int_0^{\tau_2} \varphi_3(\tau_3)x_2(\tau_3)d\tau_3 d\tau_2 d\tau_1.$$

Continuing with this procedure we get

$$x_2(t)=\left[\varphi_1(t)+\sum_{i=1}^{k}\int_0^t \varphi_3(\tau_1)\int_0^{\tau_1}\varphi_3(\tau_2)\ldots\int_0^{\tau_{k-2}}\varphi_3(\tau_{k-1})\int_0^{\tau_{k-1}}\varphi_3(\tau_k)\varphi_1(\tau_k)d\tau_k\ldots d\tau_1\right]\xi_1+$$
$$+\left[1+\int_0^t \varphi_3(\tau_1)+\ldots+\int_0^t \varphi_3(\tau_2)\ldots\int_0^{\tau_{k-2}}\varphi_3(\tau_{k-1})\int_0^{\tau_{k-1}}\varphi_3(\tau_k)\varphi_1(\tau_k)d\tau_k\ldots d\tau_1\right]\xi_2+$$
$$+\left[\varphi_2(t)+\sum_{i=1}^{k}\int_0^t \varphi_3(\tau_1)\int_0^{\tau_1}\varphi_3(\tau_2)\ldots\int_0^{\tau_{k-2}}\varphi_3(\tau_{k-1})\int_0^{\tau_{k-1}}\varphi_3(\tau_k)\varphi_2(\tau_k)d\tau_k\ldots d\tau_1\right]+$$
$$+\int_0^t \varphi_3(\tau_1)\int_0^{\tau_1}\varphi_3(\tau_2)\int_0^{\tau_2}\varphi_3(\tau_3)\ldots\int_0^{\tau_{k-1}}\varphi_3(\tau_k)\int_0^{\tau_k}\varphi_3(\tau_{k+1})x_2(\tau_{k+1})d\tau_{k+1}d\tau_k\ldots d\tau_1.$$

Repeating partial integration allows for the following transformation of integrals:

$$\int_0^t \varphi(\tau_1)\int_0^{\tau_1}\varphi(\tau_2)\ldots\int_0^{\tau_{k-2}}\varphi(\tau_{k-1})\int_0^{\tau_{k-1}}\varphi(\tau_k)\psi(\tau_k)d\tau_k d\tau_{k-1}\ldots d\tau_1=$$
$$=\frac{1}{k!}\int_0^t\left(\int_{\tau_1}^t \varphi(\tau_2)d\tau_2\right)^k \psi'(\tau_1)d\tau_1$$

and

$$\int_0^t \varphi(\tau_1)\int_0^{\tau_1}\varphi(\tau_2)\ldots\int_0^{\tau_{k-2}}\varphi(\tau_{k-1})\int_0^{\tau_k}\varphi(\tau_k)d\tau_k d\tau_{k-1}\ldots d\tau_1=$$
$$=\frac{1}{k!}\int_0^t\left(\int_{\tau_1}^t \varphi(\tau_2)d\tau_2\right)^k d\tau_1 \text{ and } \psi(t)=\int_0^t \psi'(t)d\tau.$$

This yields

$$x_2(t)=\left(\sum_{i=0}^{k}\frac{1}{i!}\int_0^t\left(\int_{\tau_1}^t \phi_3(\tau_2)d\tau_2\right)^i \phi_1'(\tau_1)d\tau_1\right)\xi_1+\left(\sum_{i=0}^{k}\frac{1}{i!}\int_0^t\left(\int_0^t \phi_3(\tau_2)d\tau_2\right)^i d\tau_1\right)\xi_2$$
$$+\left(\sum_{i=0}^{k}\frac{1}{i!}\int_{\tau_1}^t\left(\int_{\tau_1}^t \phi_3(\tau_2)d\tau_2\right)^i \phi_2'(\tau_1)d\tau_1\right)+\frac{1}{k+1}\int_0^t\left(\int_{\tau_1}^t \phi_3(\tau_2)d\tau_2\right)^{k+1} x_2'(\tau_1)d\tau_1,$$

Taking the right limit when $k \to \infty$, we get

$$x_2(t) = \int_0^t \exp\left(\int_{\tau_1}^t \phi_3(\tau_2)d\tau_2\right)\phi_1'(\tau_1)d\tau_1\xi_1 + \int_0^t \exp\left(\int_{\tau_1}^t \phi_3(\tau_2)d\tau_2\right)d\tau_1\xi_2 + \\ + \int_0^t \exp\left(\int_{\tau_1}^t \phi_3(\tau_2)d\tau_2\right)\phi_2'(\tau_1)d\tau_1.$$

From this, substituting $x_2(t)$ from (12) in the first equation of (8) yields:

$$x_1(t) = \xi_1 + \frac{E}{L}\int_0^t u(\tau)d\tau + \frac{1}{L}\int_0^t (1-u(\tau_1))\int_0^{\tau_1} \exp\left(\int_{\tau_2}^{\tau_1} \phi_3(\tau_3)d\tau_3\right)\phi_1'(\tau_2)d\tau_2d\tau_1\xi_1 + \\ + \frac{1}{L}\int_0^t (1-u(\tau_1))\int_0^{\tau_1} \exp\left(\int_{\tau_2}^{\tau_1} \phi_3(\tau_3)d\tau_3\right)d\tau_2d\tau_1\xi_2 + \\ + \frac{1}{L}\int_0^t (1-u(\tau_1))\int_0^{\tau_1} \exp\left(\int_{\tau_2}^{\tau_1} \phi_3(\tau_3)d\tau_3\right)\phi_2'(\tau_2)d\tau_2d\tau_1, \quad \text{azaz}$$

$$x_1(t) = \left[1 + \frac{1}{L}\int_0^t (1-u(\tau_1))\int_0^{\tau_1} \exp\left(\int_{\tau_2}^{\tau_1} \phi_3(\tau_3)d\tau_3\right)\phi_1'(\tau_2)d\tau_2d\tau_1\right]\xi_1 + \\ + \frac{1}{L}\int_0^t (1-u(\tau_1))\int_0^{\tau_1} \exp\left(\int_{\tau_2}^{\tau_1} \phi_3(\tau_3)d\tau_3\right)d\tau_2d\tau_1\xi_2 + \\ + \frac{E}{L}\int_0^t u(\tau)d\tau + \frac{1}{L}\int_0^t (1-u(\tau_1))\int_0^{\tau_1} \exp\left(\int_{\tau_2}^{\tau_1} \phi_3(\tau_3)d\tau_3\right)\phi_2'(\tau_2)d\tau_2d\tau_1.$$

We can estimate a $P \subset \mathbb{R}^2$ polyhedron (in our case a square $P = [-M, M] \times [-M, M]$) using the physical constants, the definitions of ϕ_1, ϕ_2, ϕ_3, and the condition $0 \le u(t) \le 1$, which will contain the $p = \mathbf{x}$ parameter function all along the interval $[0, T]$. Our second theorem guarantees an interesting but physically not implemented circuit approximation model for the Buck-Boost converter.

Instead of the model in (1) we can use models defined over discrete sections where in the bilinear element we put $\pm M$ instead of x_1 as the current in the coil, and $\pm M$ instead of x_2 for the potential in the capacitor.

$$\begin{aligned} L\dot{x}_1 &= x_2 + (E \pm M)u, \\ C\dot{x}_2 &= -x_1 - \frac{1}{R}x_2 \pm Mu. \end{aligned} \tag{9}$$

This means that the sectionwise switching of the four linear system models corresponding to the four vertices of the square will be the "linearisation" of the bilinear system, nonetheless having not too many common features with conventional linearisation.

Before providing the proofs for these two theorems, we prove an approximation lemma which will replace matrix-norm estimations during the proofs of the theorems.

Approximation Lemma *Let* $U \subset \mathbb{R}^{k_1}$ *be a convex finite polyhedron,* $u: [0, T] \to U$ *a piecewise continuous function, and* $\mathbf{B}: [0, T] \to \mathbb{R}^{n \times k_1}$ *a piecewise continuous matrix function. Then for any* $\bar{\varepsilon} > 0$ *there exists a* $v: [0, T] \to U$ *piecewise continuous approximation function, which takes values in the vertices of* U*, and*

$$\left\| \int_0^t B(\tau)(u(\tau) - v(\tau))\, d\tau \right\| < \bar{\varepsilon} \tag{10}$$

Proof Let d_U be the diameter of the U polyhedron, that is $d = \max \|u_{l_1} - u_{l_2}\|$, where u_{l_1}, u_{l_2} run along the vertices of the U polyhedron, furthermore

$$|B| = \sup_{t \in [0, T]} \|B(t)\|$$

Since u and B are piecewise continuous, thus for all $\bar{\varepsilon} > 0$ there exist such $0 = t_0 < t_1 < \ldots < t_{k-1} < t_k = T$ points in $[0, T]$ that on the $[t_{i-1}, t_i) \subset [0, T]$ intervals both u and B are continuous, and

$$\begin{aligned} &\|u(t) - u(t_{i-1})\| < \frac{\bar{\varepsilon}}{3T|B|}, \quad \text{if } t \in [t_{i-1}, t_i) \\ &\|B(t) - B(t_{i-1})\| < \frac{\bar{\varepsilon}}{3Td_U} \text{if } t \in [t_{i-1}, t_i) \\ &(t_i - t_{i-1})|B| d_u < \frac{\bar{\varepsilon}}{3T} \end{aligned}$$

Denote the vertices of the U polyhedron with $u_1, u_2, \ldots, u_L$. Since U is convex, therefore there exists at least one convex combination $u_1, u_2, \ldots, u_L$ of the vertices for which

$$u(t_{i-1}) = \sum_{l=1}^{L} \lambda_l^i u_l \quad \left(\sum_{l=1}^{L} \lambda_l^i = 1, \lambda_l^i \geq 0 \right).$$

Let's partition the $[t_{i-1}, t_i)$ interval in $\lambda_1^i, \lambda_2^i \ldots \lambda_L^i$ proportions to sub-intervals, allowing for 0 length: $t_{i1} = t_{i0} \le t_{i1} \le \cdots \le t_{iL} = t$ where

$$t_{il} = t_{i-1} + (t_i - t_{i-1}) \left(\sum_{l_1=1}^{l} \lambda_{l_1}^i \right).$$

Let's define the approximating function $v: [0, T] \to U$ on the sub-interval $t \in [t_{il-1}, t_{il}) \subset [t_{i-1}, t_i)$ with the value $v(t) = u_l$ from the vertex. Let $t \in [0, T)$ be arbitrary, then $t \in [t_{i-1}, t_i)$ for an ith sub-interval.

Then, for $t < T$,

$$\left\| \int_0^t B(\tau)(u(\tau) - v(\tau))d\tau \right\| \le \left\| \int_{t_{i-1}}^t B(\tau)(u(\tau) - v(\tau))d\tau \right\| + \sum_{j=1}^{i-1} \left\| \int_{t_{j-1}}^{t_j} B(\tau)(u(\tau) - v(\tau))d\tau \right\|.$$

We estimate seperately the first element and the sum of the right side:

$$\left\| \int_0^t B(\tau)(u(\tau) - v(\tau))d\tau \right\| \le \int_{t_{i-1}}^t \|B(\tau)\| \|(u(\tau) - v(\tau))\| d\tau \le$$
$$\le (t - t_{i-1})|B| d_U \le (t_i - t_{i-1})|B| d_U \le \frac{\bar{\varepsilon}}{3}.$$

Simple calculations yield the estimation

$$\left\| \int_0^t B(\tau)(u(\tau) - v(\tau))d\tau \right\| \le \bar{\varepsilon}$$

proving our approximation lemma.

Approximation Theorem Proof
Assume that

$$\xi \in \mathbb{R}^n \text{ and } \|\xi - \varsigma\| < \frac{\varepsilon}{6} \exp(-LT)$$

where the L constant can be defined using

$$|A| = \sup \|A(p, t)\|, \ |B| = \sup \|B(p, t)\|, \ |u| = \sup \|u(t)\|, \ |x| = \sup \|x(t)\|$$

as follows:

$$L = L_1 L_3 |x| + L_2 L_3 |u| \cdot |A|,$$

where L_1, L_2, L_3 are the Lipschitz-constans introduced to the A, B, P as global functions of t, respectively.

According to our conditions the initial value problem has a solution on the complete $[0, T]$. Since $D \subset \mathbb{R}^n \times [0, T]$ is an open set, therefore there exists a $\varepsilon_0 > 0$, for which the closed ε_0 radius neighbourhood of $x\colon [0, T] \to \mathbb{R}^n$ is a subset of D. Then an $\varepsilon > 0$ can be selected arbitrarily under the condition $\varepsilon < \varepsilon_0$. Let v be an arbitrary control for now, on the following problem:

$$\begin{aligned} &\dot{z}(t) = A(p(z(t), t), t)z(t) + B(p(z(t), t), t)v(t), \\ &z(0) = \zeta. \end{aligned} \tag{11}$$

We can compare the z solution of this problem on the semi-open domain $[0,\ T_0)$ of the interval $[0,\ T]$ with the solution of $x\colon [0,\ T] \to \mathbb{R}^n$. For this reason we integrate Eqs. (6) and (11) and estimate the deviations of the solutions,

$$\begin{aligned} \|x(t) - z(t)\| \le \|\xi - \varsigma\| &+ \left\| \int_0^t (A(p(x(\tau), \tau), \tau)x(\tau) - A(p(z(\tau), \tau), \tau)z(\tau))d\tau \right\| + \\ &+ \left\| \int_0^t (B(p(x(\tau),\ \tau),\ \tau)u(\tau) - B(p(z(\tau),\ \tau),\ \tau)v(\tau))d\tau \right\| \le \\ &\le \|\xi - \varsigma\| + \left\| \int_0^t (A(p(x(\tau),\ \tau),\ \tau)x(\tau) - A(p(z(\tau),\ \tau),\ \tau)x(\tau))d\tau \right\| + \\ &+ \left\| \int_0^t (A(p(z(\tau),\ \tau),\ \tau)(x(t) - z(\tau)))d\tau \right\| + \\ &+ \left\| \int_0^t (B(p(x(\tau),\ \tau),\ \tau)u(\tau) - B(p(z(\tau),\ \tau),\ \tau)u(\tau))d\tau \right\| + \\ &+ \left\| \int_0^t (B(p(z(\tau),\ \tau))(u(\tau) - v(\tau)))d\tau \right\|. \end{aligned}$$

Let us estimate each of the last four elements seperately.

$$\left\| \int_0^t (A(p(x(\tau),\tau),\tau) - A(p(z(\tau),\tau),\tau))x(\tau)d\tau \right\| \leq$$

$$\leq \int_0^t \|(A(p(x(\tau),\tau),\tau) - A(p(z(\tau),\tau),\tau))\| \|x(\tau)\| d\tau \leq L_1 L_3 |x| \int_0^t \|x(\tau) - z(\tau)\| d\tau.$$

$$\left\| \int_0^t (A(p(z(\tau),\tau),\tau))(x(\tau) - z(\tau))d\tau \right\| \leq |A| \int_0^t \|x(\tau) - z(\tau)\| d\tau +$$

$$\left\| \int_0^t (B(p(x(\tau),\tau),\tau) - B(p(z(\tau),\tau),\tau))u(\tau)d\tau \right\| \leq$$

$$\leq \int_0^t \|B(p(x(\tau),\tau),\tau) - B(p(z(\tau),\tau),\tau)\| \|u(\tau)\| d\tau \leq L_2 L_3 |u| \int_0^t \|x(\tau) z(\tau)\| d\tau.$$

The last element can not be easily estimated to justify our approximation theorem. Therefore, we introduce the following notation:

$$L = L_1 L_3 |x| + L_2 L_3 |u| + |A|,$$

$$\psi(t) = \left\| \int_0^t B(p(z(\tau),\tau),\tau)(u(\tau) - v(\tau))d\tau \right\|,$$

$$\varphi(t) = \|x(t) - z(t)\|.$$

Then, we get

$$\phi(t) \leq \phi(0) + L \int_0^t \phi(\tau)d\tau + \psi(t) \tag{12}$$

We substitute (12) for $\phi(\tau)$:

$$\phi(t) \leq \phi(0) + L\int_0^t \left(\phi(0) + L\int_0^t \phi(\tau_2)d\tau_2 + \psi(\tau_1) \right) d\tau_1 + \psi(t).$$

We repeat this with the resulting inequality, to get that

$$\begin{aligned} \|x(t) - z(t)\| \leq\ & \exp(LT)\|\xi - \zeta\| + \left\| \int_0^t B(p(z(t), \tau), \tau)(u(\tau) - v(\tau))d\tau \right\| + \\ & + L\int_0^t \exp(L(t - \tau_1)) \left\| \int_0^{\tau_1} B(p(z(\tau_2), \tau_2))(u(\tau_2) - v(\tau_2))d\tau_2 \right\| d\tau_1. \end{aligned} \tag{13}$$

Assuming that

$$\|\xi - \zeta\| < \frac{\delta}{2} < \frac{\varepsilon}{6}\exp(-LT), \quad \text{that is } \delta < \frac{\varepsilon}{3}\exp(-LT)$$
$$\psi(t) = \left\| \int_0^t B(p(z(\tau), \tau), \tau)(u(\tau) - v(\tau))d\tau \right\| < \frac{\varepsilon}{6},$$

and, for having the third element to be smaller than $\frac{\varepsilon}{6}$, it is necessary, that

$$\psi(t) = \left\| \int_0^t B(p(z(\tau), \tau)\tau)(u(\tau) - v(\tau))d\tau \right\| \leq \frac{\varepsilon}{6(\exp LT - 1)}$$

also holds. Applying our approximation lemma using the function

$$B(t) = B(p(z(t), t), t)$$

there exists a control v: $[0, T] \to U$ which can have values only from the vertices of the U convex polyhedron, is piecewise constant and $\psi(t) < \frac{\varepsilon}{6}$.

Thus

$$\psi(t) < \min\left\{ \frac{\varepsilon}{6(\exp LT - 1)}, \quad \frac{\varepsilon}{6} \right\},$$

from which follows, that

$$\|x(t) - z(t)\| < \frac{\varepsilon}{2}$$

From this follows from the bound-to-bound continuity of the solutions that for the piecewise constant control v: $[0, T] \to$ U taking its range in the vertices of the U convex polyhedron, the respective $z\colon [0, T] \to \mathbb{R}^n$ has the complete $[0, T]$ as its domain. Therefore we need to examine if such a v: $[0, T] \to U$ exists, as stated in the approximation theorem, for which

$$\psi(t) = \left\| \int_0^t B(p(z(\tau), \tau), \tau)\,(u(\tau) - v(\tau))\,d\tau \right\| \leq$$
$$\leq \min\left\{ \frac{\varepsilon}{6(\exp LT - 1)}, \ \frac{\varepsilon}{6} \right\}$$

holds. For this, apply the approximation lemma on the function

$$t \to B(p(z(t), t), t) = B(t)$$

and on the number

$$\bar{\varepsilon} = \min\left\{ \frac{\varepsilon}{6(\exp LT - 1)}, \ \frac{\varepsilon}{6} \right\}$$

which asserts that with the constructed v control the solutions x and z of the initial value problems (6) and (11) satisfy $\|x(t) - z(t)\| < \frac{\varepsilon}{2}$, thus z also has its domain on the whole $[0, T]$.

Let us compare now the solutions y and z of initial value problems (7) and (11). We would like to choose the piecewise constant parameter-function $q\colon D \to P$ in Eq. (7) that $\|x(t) - z(t)\| < \frac{\varepsilon}{2}$ would hold assuming $\|\eta - \xi\| < \bar{\bar{\delta}}$ for some $\bar{\bar{\delta}} > 0$ which would be smaller than $\bar{\delta}$. Then, $\delta > 0$ in the first approximation theorem could be chosen as $\delta = \min\{\bar{\delta}, \bar{\bar{\delta}}\}$. Stating the estimation the usual way and assuming that $\|\zeta - \eta\| < \bar{\bar{\delta}}$, and $\|p(x,t) - q(x,t)\| < \bar{\varepsilon}$, $\quad \forall (x,t) \in D$ we find that for the function $\phi(t) = \|z(t) - y(t)\|$ and the constants

$$L = L_1 L_3 \left(|x| + \frac{\varepsilon}{2} \right) + L_2 L_3 |v| + |A|,$$
$$M = \left(L_1 (|x| + \frac{\varepsilon}{2}) + L_2 |v| \right) T$$

the inequality

$$\phi(t) \le \phi(0) + L \int_0^t \phi(\tau) d\tau + M$$

holds. The iterative process applied previously multiple times yields

$$\phi(t) \le (\phi(0) + M\xi) \exp(-LT).$$

This is almost the original form of the Gronwall-Bellman-lemma. If we want this deviation to be smaller than $\frac{\varepsilon}{2}$, then we have to choose the $\bar{\delta} < \frac{\varepsilon}{4} \exp(-LT)$, and

$$\bar{\varepsilon} < \frac{\varepsilon}{4M} \exp(-LT)$$

constraints. This, together with the inequality

$$\|x(t) - y(t)\| \le \|x(t) - z(t)\| + \|z(t) - y(t)\| < \frac{\varepsilon}{2} + \frac{\varepsilon}{2} = \varepsilon$$

proves our first approximation theorem.

Proof Approximation Theorem 2
We structure this similarly to Theorem 1. From the proof of Theorem 1 it can be easily seen that in estimating the deviation $\|x(t) - z(t)\|$ the function $q: D \to P$ does not play any role, therefore it is sufficient to refer to this part of the proof of Theorem 1.

For the estimation of $\|z(t) - y(t)\|$ in Theorem 2 we need to construct $q: D \to P$ taking into account our additional conditions. Thus $P \subset \mathbb{R}^{k_2}$ is a convex polyhedron, and $p \to A(p,t)$ and $p \to B(p,t)$ are linear functions for all fixed t. The Lipschitz-condition has to hold uniformly in t in this case, too. We will also use the notations introduced in Theorem 1.

We start with a geometrical construction. The basic idea behind this is that we will create a "toothed" domain inside D with the union of smaller cubicles which also has its closure also D.

We defined the desired piecewise constant q state-time-variable function which has its values in the P convex polyhedron. Now let FrD denote the boundary of set D. Let $FrD_{(0,T)} = FrD \cap (\mathbb{R}^n \times (0,T))$. For an arbitrary $\delta_0 > 0$ let $N_{\delta_0}(FrD_{(0,T)}) \subset \mathbb{R}^n \times [0,T]$ denote the δ_0 radius neighbourhood of the boundary in the $\mathbb{R}^n \times [0,T]$ band. Let now be $\delta > 0$ a value satisfying only one condition, $(n+1)\delta < \delta_0$. Consider a $\mathbf{t} = (t_0, t_1, \ldots, t_{l-1}, t_l)$ partition of the $[0, T]$ interval, $0 = t_0 < t_1 < \cdots t_{l-1} < t_l = T$, which satisfies the condition $t_i - t_{i-1} < \delta$ for all i.

Consider the dot-grid $\mathbb{Z}^n = \{\mathbf{m}: \mathbf{m} = (m_1, m_2, \ldots, m_n),\ m_j \in \mathbb{Z}\}$. Let $Q_{\delta\mathbf{m}}$ denote the above-open cube with edge length 2δ, and centre $\delta\mathbf{m} \in \mathbb{R}^n$:

$$Q_{\delta\mathbf{m}} = \{\mathbf{x} \in \mathbb{R}^n, \mathbf{x} \in [(m_j - 1)\delta, (m_j + 1)\delta], \quad j = 1, 2, ..n\}.$$

The definition of the forementioned $D_{\delta_0,\delta,\mathbf{t}} \subset D$ "toothed" set is as follows:

$$D_{\delta_0,\delta,\mathbf{t}} = \mathbb{R}^n \times [0, T] \cap \left(\cup_{Q_{\delta\mathbf{m}} \times [t_{i-1}, t_i] \cap N_{\delta_0}(FrD_{(0,T)}) \neq \varnothing} Q_{\delta\mathbf{m}} \times [t_{i-1}, t_i] \right)$$

Consider the state-time variable p parameter function's values in the $(\delta\mathbf{m},\ t_{i-1})$ locations: $p(\delta\mathbf{m}, t_{i-1}) \in \mathbf{P}$. Since $\mathbf{P}$ is a convex polyhedron with $P_1, P_2, \ldots, P_M$ vertices, therefore there is at least one convex combination, for which

$$p(\delta\mathbf{m}, t_{i-1}) = \sum_{m=1}^{M} \lambda_m^{\mathbf{m},i} P_m.$$

Now let's consider the $\lambda_m^{\mathbf{m},i}$ proportioned partitioning of the t_{i-1}, t_i interval:

$$\begin{aligned} &t_{i-1} = t_{i0} \le t_{i1} \le t_{i2} \le \cdots . \le t_{i(M-1)} \le t_{iM} = t_i; \\ &t_{i1} - t_{i0} : t_{i2} - t_{i1} : \cdots : t_{iM} - t_{iM-1} = \lambda_1^{\mathbf{m},i} : \lambda_2^{\mathbf{m},i} : \cdots : \lambda_M^{\mathbf{m},i} \end{aligned}$$

or

$$t_{im} = t_{i-1} + (t_i - t_{i-1}) \left(\sum_{\eta=1}^{m} \lambda_\eta^{\mathbf{m},i} \right).$$

Now we can define the $q: D_{\delta_0,\delta,\mathbf{t}} \to P$ function as follows:

If $(x, t) \in D_{\delta_0,\delta,\mathbf{t}}$, then there is $Q_{\delta\mathbf{m}} \times [t_{i-1}, t_i)$ for which $(x, t) \in Q_{\delta\mathbf{m}} \times [t_{i-1}, t_i)$ for all $t < T$, and for (x, T) exists such $Q_{\delta\mathbf{m}}$, which satisfies $(x, T) \in Q_{\delta\mathbf{m}} \times [t_{I-1}, t_I]$. Taking $p(\delta\mathbf{m}, t_{i-1}) \in P$ and assigning a convex combination

$$p(\delta_{\mathbf{m}}, t_{i-1}) = \sum_{m=1}^{M} \lambda_m^{\mathbf{m},i} P_m$$

for all $t \in [t_{im-1}, t_{im})$ intervals where $t_{im} - t_{im-1} > 0$, and $t_{im} \neq T$ so that

$$q(x, t) = P_m, \quad \text{if} \quad (x, t) \in Q_{\delta\mathbf{m}} \times [t_{im-1}, t_{im}),$$

and

$$q(x, T) = P_m, \quad \text{if} \quad (x, T) \in Q\delta_m$$

hold.

Our conditions assure that the initial value problem has a solution on the whole $[0, T]$ interval. We put another constraining condition on the solution x, connecting the $\delta_0, \delta, \mathbf{t}$ in the construction of $q\colon D_{\delta_0}, \delta, \mathbf{t} \to P$ and the ε_0 -radius neighbourhood of x in the "toothed" domain of $D_{\delta_0, \delta, \mathbf{t}}$. This is a common condition for the x solution and the $\delta_0, \delta, \mathbf{t}$, ε_0. So we prove the first part of Theorem 2 under this condition. For all $0 < \varepsilon < \varepsilon_0$ there is a $\overline{\delta} > 0$, which, for all $\xi \in \mathbb{R}^n \|\xi - \varsigma\| < \overline{\delta}$ and for any v: $[0, T] \to$ U piecewise constant control taking its range on the vertices of the U satisfies that the deviation of the solutions of (6) and (11) on the complete $[0, T]$ is $\|x(t) - z(t)\| < \frac{\varepsilon}{2}$.

Let us compare now the solutions of the initial value problems in (7) and (11) perhaps putting more stringent constraints on the choice of $\mathbf{t}$, δ_0, δ than in the first proof. We start the comparison the usual way:

$$\|z(t) - y(t)\| \leq \|\xi - \eta\| + \left\| \int_0^t A(p(z(\tau), \tau), \tau) z(\tau) - A(q(y(\tau), \tau), \tau) y(\tau) d\tau \right\| +$$

$$+ \left\| \int_0^t B(p(z(\tau), \tau), \tau) - B(q(y(\tau), \tau), \tau) v(\tau) d\tau \right\| \leq \|\xi - \eta\| +$$

$$+ \left\| \int_0^{t_{i-1}} A(p(z(\tau), \tau), \tau) z(\tau) - A(q(y(\tau), \tau), \tau) y(\tau) d\tau \right\| +$$

$$+ \left\| \int_{t_{i-1}}^{t_i} A(p(z(\tau), \tau), \tau) z(\tau) - A(q(y(\tau), \tau), \tau) y(\tau) d\tau \right\|$$

$$+ \left\| \int_0^{t_{i-1}} B(p(z(\tau), \tau), \tau) - B(q(y(\tau), \tau), \tau) v(\tau) d\tau \right\| +$$

$$+ \left\| \int_{t_{i-1}}^{t_i} B(p(z(\tau), \tau), \tau) - B(q(y(\tau), \tau), \tau) v(\tau) d\tau \right\|$$

for all $t \in [t_{i-1}, t_i)$. For the remaining $[t_{i-1}, t)$ interval both elements are estimated separately. Elements containing A and B are estimated individually, breaking up the $[0, t_{i-1})$ interval to the union of the $[t_{j-1}, t_j), j = 1, \ldots, i-1$ intervals:

$$[0, t_{i-1}) = \cup_{j=1}^{i-1} [t_{j-1}, t_j).$$

$$\left\| \int_{t_{j-1}}^{t_j} (A(p(z(\tau),\tau),\tau)z(\tau) - A(q(y(\tau),\tau),\tau)y(\tau))d\tau \right\| \leq$$

$$\leq \left\| \int_{t_{j-1}}^{t_j} (A(p(z(\tau),\tau),\tau)z(\tau) - A(p(y(\tau),\tau),\tau)z(\tau))d\tau \right\| +$$

$$+ \left\| \int_{t_{j-1}}^{t_j} (A(p(y(\tau),\tau),\tau)z(\tau) - A(p(y(\tau),\tau),\tau)z(t_{j-1}))d\tau \right\| +$$

$$+ \left\| \int_{t_{j-1}}^{t_j} (A(p(y(\tau),\tau),\tau)z(t_{j-1}) - A\big(p(y(\tau),\tau),t_{j-1}\big)z(t_{j-1}))d\tau \right\| +$$

$$+ \left\| \int_{t_{j-1}}^{t_j} (A(p(y(\tau),\tau),t_{j-1})z(t_{j-1}) - A\big(q(y(\tau),\tau),t_{j-1}\big))z(t_{j-1})d\tau \right\| +$$

$$+ \left\| \int_{t_{j-1}}^{t_j} (A(q(y(\tau),\tau),t_{j-1})z(t_{j-1}) - A(q(y(\tau),\tau),\tau))z(t_{j-1})d\tau \right\| +$$

$$+ \left\| \int_{t_{j-1}}^{t_j} (A(q(y(\tau),\, \tau),\, \tau)z(t_{j-1}) - A(q(y(\tau),\, \tau),\, \tau)\, z(\tau))d\tau \right\| +$$

$$+ \left\| \int_{t_{j-1}}^{t_j} (A(q(y(\tau),\, \tau),\, \tau)z(\tau) - A(q(y(\tau),\, \tau),\, \tau)\, y(\tau))d\tau \right\|.$$

We break up elements containing B similarly:

$$\left\| \int_{t_{j-1}}^{t_j} (B(p(z(\tau),\tau),\tau) - B(q(y(\tau),\tau),\tau))v(\tau)d\tau \right\| \leq$$

$$\leq \left\| \int_{t_{j-1}}^{t_j} (B(p(z(\tau),\tau),\tau) - B(p(y(\tau),\tau),\tau))v(\tau)d\tau \right\| +$$

$$+ \left\| \int_{t_{j-1}}^{t_j} \big(B(p(y(\tau),\tau),\tau) - B\big(p(y(\tau),\tau),t_{j-1}\big)\big)v(\tau)d\tau \right\| +$$

$$+ \left\| \int_{t_{j-1}}^{t_j} \big(B(p(y(\tau),\tau),t_{j-1}) - B\big(q(y(\tau),\tau),t_{j-1}\big)\big)v(\tau)d\tau \right\| +$$

$$+ \left\| \int_{t_{j-1}}^{t_j} \big(B(q(y(\tau),\tau),t_{j-1}) - B(q(y(\tau),\tau),\tau)\big)v(\tau)d\tau \right\|.$$

The estimation of the remaining integrals yields (after a few simple steps):

$$\left\| \int_{t_{i-1}}^{t_i} A(p(z(\tau),\tau),\tau)z(\tau) - A(q(y(\tau),\tau),\tau)y(\tau)d\tau \right\| \leq$$

$$\leq L_1 L_3 (t_i - t_{i-1}) \left(|x| + \frac{\varepsilon}{2} \right) \int_{t_{j-1}}^{t_j} \|z(\tau) - y(\tau)\| d\tau + |A| \int_{t_{i-1}}^{t_i} \|z(\tau) - y(\tau)\| d\tau =$$

$$= \left(L_1 L_3 (t_i - t_{i-1}) \left(|x| + \frac{\varepsilon}{2} \right) + |A| \right) \int_{t_{i-1}}^{t_i} \|z(\tau) - y(\tau)\| d\tau.$$

On the other hand

$$\left\| \int_{t_{j-1}}^{t_j} (B(p(z(\tau),\tau),\tau)v(t) - B(q(y(\tau),\tau),\tau))v(\tau)d\tau \right\| \leq$$

$$\leq \int_{t_{j-1}}^{t_j} \|(B(p(z(\tau),\tau),\tau)v(t) - B(q(y(\tau),\tau),\tau))\| \|v(\tau)\| d\tau \leq$$

$$\leq L_2 L_3 |v| \int_{t_{i-1}}^{t_i} \|z(\tau) - y(\tau)\| d\tau.$$

The estimation for $\|z(\tau) - z(t_{j-1})\|$ can be simply derived from the integro-differential equation:

$$\|z(\tau) - z(t_{j-1})\| \leq \int_{t_{j-1}}^{t_j} \|A(p(z(\tau),\tau),\tau)\| \|z(\tau)\| + \|B(p(z(\tau),\tau),\tau)\| \|v(\tau)\| d\tau \leq$$

$$\leq \left(|A| + \left(x + \frac{\varepsilon}{2} \right) + |B||v| \right) (t_j - t_{j-1})$$

Since A and B are uniformly continuous therefore for all $\bar{\bar{\varepsilon}} > 0$ exists a $\bar{\bar{\delta}} > 0$, for which

$$\|A(p,t) - A(p,t_{j-1})\| < \bar{\bar{\varepsilon}}, \quad \text{if} \quad t - t_{j-1} < \bar{\bar{\delta}}$$
$$\|B(p,t) - B(p,t_{j-1})\| < \bar{\bar{\varepsilon}}, \quad \text{if} \quad t - t_{j-1} < \bar{\bar{\delta}}$$

When estimating elements containing A and B we can use simple norm estimation with the exceptions of the third and fourth elements, respectively. This yields the following inequalities (omitting some simplifications):

(a)

$$\left\| \int_{t_{j-1}}^{t_j} (A(p(z(\tau),\tau),\tau)z(\tau) - A(p(y(\tau),\tau),\tau))z(\tau)d\tau \right\| \le$$
$$\le \int_{t_{j-1}}^{t_j} \|(A(p(z(\tau),\tau),\tau) - A(p(y(\tau),\tau),\tau))\| \|z(\tau)\| d\tau \le$$
$$\le L_1 L_3 \left(|x| + \frac{\varepsilon}{2}\right) \int_{t_{j-1}}^{t_j} \|z(\tau) - y(\tau)\| d\tau.$$

(b)

$$\left\| \int_{t_{j-1}}^{t_j} A(p(y(\tau),\tau),\tau)z(\tau) - A(p(y(\tau),\tau),\tau)z(t_{j-1})d\tau \right\| \le$$
$$\le |A| \left(|A| \left(|x| + \frac{\varepsilon}{2} \right) + |B||v| \right) (t_j - t_{j-1}) \bar{\delta}.$$

(c)

$$\left\| \int_{t_{j-1}}^{t_j} A(p(y(\tau),\tau),\tau) - A(p(y(\tau),\tau),t_{j-1})z(t_{j-1})d\tau \right\| \le$$
$$\le \left(|x| + \frac{\varepsilon}{2}\right)(t_j - t_{j-1})\bar{\bar{\varepsilon}} \quad \text{if} \quad t_j - t_{j-1} < \bar{\bar{\delta}}.$$

(d)

$$\left\| \int_{t_{j-1}}^{t_j} A(q(y(\tau),\tau),t_{j-1})z(t_{j-1}) - A(q(y(\tau),\tau),t_{j-1})z(\tau)d\tau \right\| \le$$
$$\le \left(|x| + \frac{\varepsilon}{2}\right) \int_{t_{j-1}}^{t_j} \|A(q(y(\tau),\tau),t_{j-1}) - A(q(y(\tau),\tau),\tau)\| d\tau \le$$
$$\le \left(|x| + \frac{\varepsilon}{2}\right)(t_j - t_{j-1})\bar{\bar{\varepsilon}} \quad \text{if} \quad t_j - t_{j-1} < \bar{\bar{\delta}}.$$

(e)

$$\left\| \int_{t_{j-1}}^{t_j} A(q(y(\tau),\tau),\tau)z(t_{j-1}) - A(q(y(\tau),\tau),\tau)z(\tau)d\tau \right\| \le$$
$$\le |A| \left(|A| \left(|x| + \frac{\varepsilon}{2} \right) + |B||v| \right) (t_j - t_{j-1}) \bar{\delta}.$$

(f)

$$\left\| \int_{t_{j-1}}^{t_j} A(q(y(\tau),\tau),\tau)z(\tau) - A(q(y(\tau),\tau),\tau)y(\tau)d\tau \right\| \leq$$

$$\leq \int_{t_{j-1}}^{t_j} \|A(q(y(\tau),\tau),\tau)\| \|z(\tau) - y(\tau)\| d\tau \leq$$

$$\leq |A| \int_{t_{j-1}}^{t_j} \|z(\tau) - y(\tau)\| d\tau.$$

(g)

$$\left\| \int_{t_{j-1}}^{t_j} (B(p(z(\tau),\tau),\tau) - B(p(y(\tau),\tau),\tau))v(\tau)d\tau \right\| \leq$$

$$\leq |v| \int_{t_{j-1}}^{t_j} \|B(p(z(\tau),\tau),\tau) - B(p(y(\tau),\tau),\tau)\| d\tau \leq$$

$$\leq |v| L_2 L_3 \int_{t_{j-1}}^{t_j} \|z(\tau) - y(\tau)\| d\tau.$$

(h)

$$\left\| \int_{t_{j-1}}^{t_j} \left(B(p(y(\tau),\tau),\tau) - B\left(p(y(\tau),\tau),t_{j-1}\right)\right)v(\tau)d\tau \right\| \leq$$

$$\leq |v| \int_{t_{j-1}}^{t_j} \|B(p(z(\tau),\tau),\tau) - B\left(p(y(\tau),\tau),t_{j-1}\right)\| d\tau \leq$$

$$\leq |v|(t_j - t_{j-1})\bar{\bar{\varepsilon}}, \quad \text{ha} \quad t_j - t_{j-1} < \bar{\bar{\delta}}.$$

Similarly,

(i)

$$\left\| \int_{t_{j-1}}^{t_j} \left(B(q(y(\tau),\tau),t_{j-1}) - B(q(y(\tau),\tau),\tau)\right)v(\tau)d\tau \right\| \leq$$

$$\leq |v|(t_j - t_{j-1})\bar{\bar{\varepsilon}}, \quad \text{ha} \quad t_j - t_{j-1} < \bar{\bar{\delta}}.$$

Estimation of the fourth member of A elements and the third member of B elements is undertaken similarly, using the definitions of v and q further partitioning the $[t_{j-1}, t_j)$ interval. We assume that in defining the two piecewise constant functions we use the same $\mathbf{t} = (t_0, t_1, \ldots, t_{I-1}, t_I)$ partitioning. Omitting the usual calculations we get

(j)
$$\begin{aligned}&\left\|\int_{t_{j-1}}^{t_j} A(p(y(\tau),\tau),t_{j-1})z(t_{j-1}) - A(q(y(\tau),\tau),\tau)z(t_{j-1})d\tau\right\| \leq \\ &\leq \int_{t_{j_{m-1}}}^{t_{jm}} \left\|A(p(y(\tau),\tau),t_{j-1})z(t_{j-1}) - A(p(y(t_{j-1}),t_{j-1}),t_{j-1})d\tau\right\| \left\|z(t_{j-1})\right\| \\ &\leq \left(|x| + \frac{\varepsilon}{2}\right)(t_j - t_{j-1})\bar{\bar{\varepsilon}}, \quad \text{if} \quad t_j - t_{j-1} < \bar{\delta}.\end{aligned}$$

due to the uniform continuity of A.

We proceed in the same manner in estimating the third element containing B. But in the approximation of this element we use two piecewise constant functions; notably

$$v\colon [0, T] \to U \text{ and}$$
$$q\colon D_{\delta_0,\,\delta,\,\mathbf{t}} \to P$$

are the approximating functions.

Let the initial partitioning be $\mathbf{t} = (t_0, t_1, \ldots, t_I)$, which we can choose to any given $\varepsilon > 0$ due to the uniform continuity of A, B and the piecewise uniform continuity of u. Let us define the range of the piecewise constant v function with the values taken in the vertices of the U convex polyhedron. For this reason we further partition the $[t_{i-1}, t_i)$ sub-interval according to the

$$u(t_{i-1}) = \sum_{l=1}^{L} \lambda_l u_l, \qquad \sum_{l=1}^{L} \lambda_l = 1, \quad \lambda_l \geq 0$$

convex combination with

$$t_{il} = t_{i-1} + (t_i - t_{i-1})\left(\sum_{k=1}^{l} \lambda_k\right)$$

division points. Then

$$v(t) = u_l, \quad \text{if} \quad t \in [t_{il-1}, t_{il}), \quad \text{and} \quad v(T) = u_2.$$

Now let's further divide the $[t_{il-1}, t_{il})$ interval.
If $(x, t) \in Q_{\delta\mathbf{n}} \times [t_{il-1}, t_{il})$, then according to the convex combination

$$p(\delta_{mn, t_{i-1}}) = \sum_{m=1}^{M} \mu_m^{\delta\mathbf{n}} P_m, \ \sum_{m=1}^{M} \mu_m^{\delta\mathbf{n}} = 1, \quad \mu_m^{\delta\mathbf{n}} \geq 0$$

this division points will be

$$t_{ilm} = t_{il-1} + (t_{il} - t_{il-1}) \left(\sum_{k=1}^{m-1} \mu_k^{\delta\mathbf{n}} \right)$$

used for $[t_{il-1}, t_{il})$. According to this partition

$$q(x, t) = P_m, \quad \text{if} \quad (x, t) \in Q_{\delta\mathbf{n}} \times [t_{i-1}, t_i) \quad \text{and} \quad t \in [t_{ilm-1}, t_{ilm}),$$
$$q(x, T) = P_m, \quad \text{if} \quad x \in Q_{\delta\mathbf{n}}.$$

Following this, a few simplifications yield.

$$\begin{aligned} & \left\| \int_{t_{i-1}}^{t_i} (B(p(y(\tau), \tau), t_{i-1}) - B(q(y(\tau), \tau), t_{i-1}))v(\tau)d\tau \right\| \\ \text{(k)} \qquad & \leq \int_{t_{i-1}}^{t_i} \|(B(p(y(\tau), \tau), t_{i-1}) - B(p(y(t_{i-1}), t_{i-1}), t_{i-1}))d\tau\| \|v(\tau)\| \\ & \leq L_2 |v| \int_{t_{i-1}}^{t_i} \|p(y(\tau), \tau) - p(y(t_{i-1}), t_{i-1})\| d\tau \leq L_2 |v| \bar{\bar{\varepsilon}} (t_i - t_{i-1}), \end{aligned}$$

where for $\bar{\bar{\varepsilon}} > 0$ there is $\bar{\bar{\delta}} > 0$, so that due to the uniform continuity of p and y when $t_i - t_{i-1} < \bar{\bar{\delta}}$ holds

$$p(y(\tau), \tau) - p(y(t_{i-1}), t_{i-1}) < \bar{\bar{\varepsilon}}, \quad \text{if} \quad \tau \in [t_{i-1}, t_i).$$

We can conclude the above estimations using the alphabetical notations as follows.

$$
\begin{aligned}
&\|z(t)-y(t)\| \le \|\varsigma-\eta\| + \\
&\sum_{j=1}^{i-1}\left\|\int_{t_{j-1}}^{t_j} A(p(z(\tau),\tau),\tau)z(\tau)-A(q(y(\tau),\tau),\tau)y(\tau)d\tau\right\| \\
&+\sum_{j=1}^{i-1}\left\|\int_{t_{j-1}}^{t_j} B(p(z(\tau),\tau),\tau)-B(q(y(\tau),\tau),\tau)v(\tau)d\tau\right\| \\
&+\left\|\int_{t_{i-1}}^{t_j} A(p(z(\tau),\tau),\tau)z(\tau)-A(q(y(\tau),\tau),\tau)y(\tau)d\tau\right\| \\
&+\left\|\int_{t_{i-1}}^{t_j} B(p(z(\tau),\tau),\tau)-B(q(y(\tau),\tau),\tau)v(\tau)d\tau\right\| \\
&\le \overbrace{L_1L_3\left(|x|+\frac{\varepsilon}{2}\right)\sum_{j=1}^{i-1}\int_{t_{j-1}}^{t_j}}^{a} \|z(\tau)-y(\tau)\|d\tau + \\
&+\overbrace{|A|\left(|A|\left(|x|+\frac{\varepsilon}{2}\right)+|B||v|\right)\bar{\delta}\sum_{j=1}^{i-1}\left(t_j-t_{j-1}\right)}^{b} + \overbrace{\left(|x|+\frac{\varepsilon}{2}\right)\bar{\bar{\varepsilon}}\sum_{j=1}^{i-1}\left(t_j-t_{j-1}\right)}^{c} \\
&+\overbrace{\left(|x|+\frac{\varepsilon}{2}\right)\bar{\bar{\varepsilon}}\sum_{j=1}^{i-1}\left(t_j-t_{j-1}\right)}^{d} + \overbrace{|A|\left(|A|\left(|x|+\frac{\varepsilon}{2}\right)+|B||v|\right)\bar{\delta}\sum_{j=1}^{i-1}\left(t_j-t_{j-1}\right)}^{e} \\
&+\overbrace{|A|\sum_{i=1}^{j-1}\int_{t_{j-1}}^{t_j}\|z(\tau)-y(\tau)\|d\tau}^{f}+\overbrace{L_2L_3|v|\sum_{i=1}^{j-1}\int_{t_{j-1}}^{t_j}\|z(\tau)-y(\tau)\|d\tau}^{g} \\
&+\overbrace{|v|\bar{\bar{\varepsilon}}\sum_{j=1}^{i-1}\left(t_j-t_{j-1}\right)}^{h} + \overbrace{|v|\bar{\bar{\varepsilon}}\sum_{j=1}^{i-1}\left(t_j-t_{j-1}\right)}^{i} + \overbrace{\left(|x|+\frac{\varepsilon}{2}\right)\bar{\bar{\varepsilon}}\sum_{j=1}^{i-1}\left(t_j-t_{j-1}\right)}^{j} \\
&+L_2|v|\bar{\bar{\varepsilon}}\sum_{j=1}^{i-1}\left(t_j-t_{j-1}\right)=\|\xi-\eta\| \\
&+\left(L_1L_3\left(|x|+\tfrac{\varepsilon}{2}\right)+L_2L_3|v|+|A|\right)\int_{t_{i-1}}^{t}\|z(\tau)-y(\tau)\|d\tau + \\
&+2|A|\left(|A|\left(|x|+\tfrac{\varepsilon}{2}\right)+|B||v|t_{i-1}\right)\bar{\delta} \\
&+\left(3\left(|x|+\tfrac{\varepsilon}{2}\right)+2|v|+L_2|v|\right)t_{i-1}\bar{\bar{\varepsilon}}+L_2|v|t_{i-1}\bar{\bar{\varepsilon}} \\
&=\|\varsigma-\eta\|+\left(L_1L_3\left(|x|+\tfrac{\varepsilon}{2}\right)+L_2L_3|v|+|A|\right)\int_{0}^{t}\|z(\tau)-y(\tau)\| \\
&+2|A|\left(|A|\left(|x|+\tfrac{\varepsilon}{2}\right)+|B||v|t_{i-1}\right)\bar{\delta}+L_2|v|t_{i-1}\bar{\bar{\varepsilon}}+\left(3\left(|x|+\tfrac{\varepsilon}{2}\right)+2|v|+L_2|v|\right)t_{i-1}\bar{\bar{\varepsilon}}.
\end{aligned}
$$

Let $|U|$ denote the maximum of the norms of the vertices of the U convex polyhedron. Let's introduce the notations

$$L = L_1 L_3 \left(|x| + \tfrac{\varepsilon}{2}\right) + L_2 L_3 |U| + |A|,$$
$$K_1 = 2|A| \left(A| \left(|x| + \tfrac{\varepsilon}{2}\right) + |B|\ |U| T\right)$$
$$K_2 \left(3\left(|x| + \tfrac{\varepsilon}{2}\right) + 2|U| + L_2 |U|\right) T$$
$$K_3 = L_2 |U| T,$$

and

$$\phi(t) = \|z(t) - y(t)\|.$$

Then the inequalities can be rewritten to the following inequality:

$$\phi(t) \le \phi(0) + L \int_0^t \phi(\tau) d\tau + K_1 \bar{\bar{\delta}} + K_2 \bar{\bar{\varepsilon}} + K_3 \bar{\bar{\varepsilon}}$$

Proceeding in a similar manner as before this yields a Gronwall-Bellman-type inequality:

$$\|z(t) - y(t)\| \le \|\varsigma - \eta\| \exp LT + K_1 \bar{\bar{\delta}} \exp LT + K_2 \bar{\bar{\varepsilon}} \exp LT + K_3 \bar{\bar{\varepsilon}} \exp LT$$

We would like all members of the right side to be smaller than $\frac{\varepsilon}{8}$, so that the sum would be smaller than $\frac{\varepsilon}{2}$.

1. Due to the uniform continuity of A, p and y the function $t \to A(p(y(t,t)), t_{i-1})$ is also uniformly continuous, therefore δ has to be small enough, for the following condition to hold.
   If $t_i - t_{i-1} < \delta$, then

   $$\|A(p(y(t), t), t_{i-1}) - A(p(y(t_{i-1}), t_{i-1}), t_{i-1})\| < \frac{\varepsilon}{8K_2} \exp(-2T)$$

   for all $t \in [t_{i-1}, t_i)$.
2. Since p and y are uniformly continuous therefore $t \mapsto p(y(t), t)$ is also uniformly continuous, so if $t_i - t_{i-1} < \delta$ then

   $$\|p(y(t), t) - p(y(t_{i-1}), t_{i-1})\| < \frac{\varepsilon}{8K_3} \exp(-LT),$$

   has to hold for all $t \in [t_{i-1}, t_i)$.

Further elements will be smaller than $\frac{\varepsilon}{8}$, if for δ

$$\delta < \frac{\varepsilon}{8}\exp(-LT) \text{ and } \delta < \frac{\varepsilon}{8K_1}\exp(-LT).$$

will hold. If all these hold then

$$\|z(t) - y(t)\| < \frac{\varepsilon}{2}$$

so

$$\|x(t) - y(t)\| \leq \|x(t) - z(t)\| + \|z(t) - y(t)\| < \frac{\varepsilon}{2} + \frac{\varepsilon}{2} = \varepsilon.$$

This concludes the proof of the second approximation theorem. □

5 Conclusions

Let us now notice that although we required that the A and B matrix functions to be linearly dependent from the parameter, we can, for all $(x, t) \to A(x(t), t)$, $(x, t) \to B(x(t), t)$, satisfying the smoothness condition transform the

$$\dot{x} = A(x, t)x + B(x, t)n$$

system to an LPV system. For this consider the following

$$A(x, t) = \left(a_{ij}(x, t)\right)_{i,j=1}^{n}, \quad B(x, t) = \left(b_{ij}(x, t)\right)_{i=1,j=1}^{n,k_1}$$

matrices. Replace $a_{ij}(x,t)$ with the p_{ij} parameters, and $b_{ij}(x,t)$ with the q_{ij} parameters. Thus, we defined the $A(p)$ and $B(q)$ parameter-variable matrices, which are obviously linear in p, q parameters, which is nothing else but the linearity of matrix addition. With this we get the

$$\dot{x} = A(p)x + B(q)u$$

LPV system. Using the substitutions $p_{ij} = a_{ij}(x,t)$ and $q_{ij} = b_{ij}(x,t)$ trivially the original system is recovered. Obviously, not too much is gained with this LPV-ification, under such general circumstances the system will not show any interesting qualities.

Our example with the converter well illustrates the range of applicability of our approximation theorems specific to real life applications (Hermes and Lasalle 1969; Berkovitz 1974).

In systems theory analysis these theorems are also quite promising, e.g. in stability, controllability and observability (Pontryagin et al. 1962; Warga 1972), especially Theorem 2, where we can approximate arbitrary system qualities by switching constant parametric linear systems assigned to the vertices of the P convex polyhedron.

References

Berkovitz, L. D. (1974). Optimal control theory (Applied Mathematical Sciences, Vol. 12). Springer, New York.

Gamkrelidze, R V. (1978). *Principles of optimal control theory.* (Mathematical concepts and methods in science and engineering) Translation of Osnovy optimal'nogo upravleniia. Based on lectures presented in 1974 at Tbilisi State University. (New York: Plenum Press).

Hermes, H., & Lasalle, J. P. (1969). *Functional analysis and time-optimal control.* New York: Academic Press.

Molnár, S. (1989). A special decomposition of linear systems. *Belgian Journal of Operations Research, Statistics and Computer Science, 29*(4), 1–19.

Molnár, S., & Szigeti, F. (1994). On "verticum"-type linear systems with time-dependent linkage. *Applied Mathematics and Computation, 60,* 89–102.

Pontryagin, L. S., Boltyanskii, V. G., Gamkrelidze, R. V., & Mishchenko, E. F. (1962). *The mathematical theory of optimal processes.* New York: Wiley.

Sira-Ramírez, H. (2015). *Sliding mode control* (p. 258). Birkhauser Verlag Gmbh.

Sira-Ramírez, H., Agrawal, S.K. (2004). Differentially flat systems (automation and control engineering) (p. 450). New York, Basel: Marcel Dekker Inc.

Warga, J. (1972). *Optimal control of differential and functional equations.* New York: Academic Press.

Love Affairs Dynamics with One Delay in Losing Memory or Gaining Affection

Akio Matsumoto

Abstract A dynamic model of a love affair between two people is examined under different conditions. First the two-dimensional model is analyzed without time delays in the interaction of the lovers. Conditions are derived for the existence of a unique as well as for multiple steady states. The nonzero steady states are always stable and the stability of the zero steady state depends on model parameters. Then a delay is assumed in the mutual-reaction process called the Gaining-affection process. Similarly to the no-delay case, the nonzero steady states are always stable. The zero steady state is either always stable or always unstable or it is stable for small delays and at a certain threshold stability is lost in which case the steady state bifurcates to a limit cycle. When delay is introduced to the self-reaction process called the Losing-memory process, then the asymptotic behavior of the steady state becomes more complex. The stability of the nonzero steady state is lost at a certain value of the delay and bifurcates to a limit cycle, while the stability of the zero steady state depends on model parameters and there is the possibility of multiple stability switches with stability losses and regains. All stability conditions and stability switches are derived analytically, which are also verified and illustrated by using computer simulation.

1 Introduction

The dynamics of love affairs has been modeled in various ways since Strogatz (1988) has proposed a 2D system of linear differential equations to describe the time evolution of a love affair between two individuals called Romeo and Juliet. Strogatz's

---

The author highly appreciate the financial supports from the MEXT-Supported Programe for the Strategic Research Foundation at Private Universities 2013-2017, the Japan Society for the Promotion of Science (Grant-in-Aid for Scientific Research (C), 24530202, 25380238, 26380316) and Chuo University (Joinet Research Grant). The usual disclaimer apply.

A. Matsumoto (✉)
Department of Economics, Senior Researcher, International Center for further Development of Dynamic Economic Research, Chuo University, 742-1, Higashi-Nakano, Hachioji, Tokyo 192-0393, Japan
e-mail: akiom@tamacc.chuo-u.ac.jp

A. Matsumoto (ed.), *Optimization and Dynamics with Their Applications*,
DOI 10.1007/978-981-10-4214-0_9

purpose was to teach harmonic oscillations by applying a topic that is already on the minds of many college students: the time evolution of a love affair between two people. The study on the love affair dynamics after Strogatz aims to explain dynamic processes of love stories in our life in a formal theoretical framework. On the one hand, real-life observations tell us that love-stories frequently develop very regularly and stay at a plateau of love affair for a long time. Reconstructing the Strogatz model with linear or nonlinear behavioral functions and secure individuals, Rinaldi and Gragnani (1998) and Rinaldi (1998a) shows that one of the model's properties concerning the dynamics of the love affair is a smoothly increasing feeling tending toward a positive stationary point. On the other hand, another real-life observations indicate that love stories often arrive at a fluctuating regime including chaotic motions. Rinaldi (1998b) models the dynamics of the real love affair between Petrarch, a poet of the 14th century, and Laura, a beautiful married lady, with three differential equations and shows the appearance of cyclical pattern ranging from ecstasy to despair. Sprott (2004) applies a 4D system of nonlinear differential equations involving Romeo, Juliet and Romeo's mistress, Guinevere and derive chaotic love regime. Introducing information delays into the Strogatz model, Liao and Ran (2007) find that the stable steady state is destabilized for a delay larger than a threshold value and then bifurcates to a limit cycle via a Hopf bifurcation when Romeo is secure and Juliet is non-secure. Son and Park (2011) investigate the effect of delay on the love dynamics and confirm a cascade of period-doubling bifurcations to chaos analytically as well as numerically. Usually a delay is believed to possess a destabilizing effect in a sense that a longer delay destabilizes a system which is otherwise stable. Bielczyk et al. (2012) reveal the stabilizing effect of the delay by showing that a unstable steady state without time delay can gain stability for certain range of delays.

In this study we follow the Liao-Ran version of the Romeo-Juliet model to investigate how the delay and nonlinearities affect love dynamics. One important issue that Liao and Ran (2007) do not examine is to investigate time evolution in the case of multiple steady state. As is seen shortly, nonlinear behavioral functions can be a source of multiple steady state. However only the unique steady state case has been considered. Our first goal is to investigate dynamics in the multiple case. The second issue we take up concerns the romantic style of Rome and Juliet. There are four specifications of the romantic style for each individual, "eager beaver", "narcissistic nerd", "cautious (or secure) lover" and "hermit".[1] The majority of the population is represented by a cautious or secure lover who loves to be loved (alternatively, hates to be

[1]See Strogatz (1994) for more precise specification.

hated) and gradually loses the emotion to the partner when the partner leaves or dies. In spite of this, most studies confine attention to the case where Romeo and Juliet are heterogeneous, one is secure and the other is non-secure. Furthermore, it is demonstrated that the Romeo-Juliet model without delays does not exhibit cyclic dynamics when both are secure lovers. Our second goal is to investigate how the delay affects love dynamics between secure Romeo and Juliet. We have one more goal. The existing studies mainly focus on the delay that exists in love stimuli sent between Romeo and Juliet. We give a detailed analysis when there is a delay in Romeo's reaction to his own emotional state, referring to the basic study conducted by Bielczyk et al. (2013).

This paper is organized as follows. Section 2 presents the basic love dynamic model that has no delays. Section 3 introduces one delay as in the Liao-Ran model and studies the dynamics of multiple steady states. Section 4 considers the case where Romeo loses the feeling for Juliet with a delay and Juliet without any delay. Section 5 concludes the paper.

2 Basic Model

Strogatz (1988) proposes a linear model of love affairs dynamics and Rinaldi (1998a) extends it to a more general model in which three aspects of love dynamics, *oblivion*, *return* and *instinct*, are taken into account. If $x(t)$ denotes Romeo's emotions for Juliet at time t while $y(t)$ denotes Juliet's feeling to Romeo at time t, then the rates of change of Romeo's love and Juliet's love are assumed to be composed of three terms,

$$\dot{x}(t) = O_x(x(t)) + R_x(y(t)) + I_x$$

$$\dot{y}(t) = O_y(y(t)) + R_y(x(t)) + I_y$$

where O_z, R_z and I_z for $z = x, y$ are specified as follows. First, O_z gives rise to a loss of interest in the partner and describes the losing-memory process that characterizes decay of love at disappearance of the partner. Second, R_z is a source of interest and describes the reaction of individual z to the partner's love in the gaining-affection process. Lastly, I_z is also a source of interest and describes the reaction of individual z to the partner's appeal reflecting physical, financial, educational, intellectual properties. We adopt the following forms of these reaction functions:

Assumption 1 $O_x(x) = -\alpha_x x,\ \alpha_x > 0$ *and* $O_y(y) = -\alpha_y y,\ \alpha_y > 0.$

Assumption 2 $R_x(y) = \beta_x \tanh(y)$ *and* $R_y(x) = \beta_y \tanh(x).$

Assumption 3 $I_x = \gamma_x A_y,\ A_y > 0$ *and* $I_y = \gamma_y A_x,\ A_x > 0.$

Assumption 1 confines attention to the case where the memory vanishes exponentially. In Assumption 2, the hyperbolic function is positive, increasing, concave

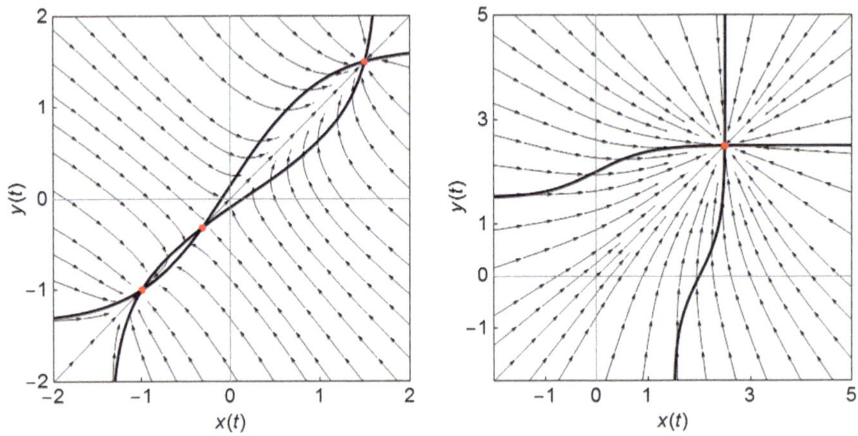

(A) Multiple equilibria (B) Unique equilibrium

Fig. 1 Orbits of system (1)

and bounded from above for positive values and is negative, increasing, convex and bounded from below for negative values. If $\beta_z > 0$. Then the feeling of individual z is encouraged by the partner and such an individual is called *secure*. On the other hand, if $\beta_z < 0$, it is discouraged and the individual is thought to be *non-secure*. Assumption 3 implies that individuals have time-invariant positive appeal. α_z is called the forgetting parameter while β_z and γ_z are the reaction coefficients of the love and appeal.

Under these assumptions, our basic model is

$$\begin{aligned} \dot{x}(t) &= -\alpha_x x(t) + \beta_x \tanh[y(t)] + \gamma_x A_y, \\ \dot{y}(t) &= \beta_y \tanh[x(t)] - \alpha_y y(t) + \gamma_y A_x. \end{aligned} \tag{1}$$

Two numerical examples are given and the directions of the trajectories are indicated by arrows. In Fig. 1a with $\alpha_x = \alpha_y = 1$, $\beta_x = \beta_y = 3/2$, $\gamma_x = \gamma_y = 1$ and $A_x = A_y = 1/7$, the isoclines, $\dot{x}(t) = 0$ and $\dot{y}(t) = 0$, intersect at three points denoted by red dots. The middle one is unstable (a saddle) while the one with positive coordinates and the other with negative coordinates are stable nodes. In Fig. 1b with $\alpha_x = \alpha_y = 1$, $\beta_x = \beta_y = 1/2$, $\gamma_x = \gamma_y = 1$ and $A_x = A_y = 2$, the steady state is unique and stable. As will be seen below, stability of system (1) is rather robust.

Assumption 3 affects the location of a steady state but does not affect dynamic properties. Since we confine our attention to dynamics of the state variables in this study, we, only for a sake of analytical simplicity, replace Assumption 3 with the following:

Assumptiom 3': $A_x = A_y = 0$.
The steady state of (1) satisfies $\dot{x}(t) = 0$ and $\dot{y}(t) = 0$. Solving $\dot{x}(t) = 0$ and $\dot{y}(t) = 0$ for y yields two functions,

$$y = \tanh^{-1}\left(\frac{\alpha_x}{\beta_x}x\right) \text{ and } y = \frac{\beta_y}{\alpha_y}\tanh(x). \tag{2}$$

Let us denote the right hand side of two equations as $u(x)$ and $v(x)$, respectively. The steady state value of x, denoted as x^*, solves

$$u(x) = v(x) \tag{3}$$

and the steady state value of y, denoted as y^*, is determined as

$$y^* = u(x^*) \text{ or } y^* = v(x^*). \tag{4}$$

We then have the following result where the proofs of this and further results are given in the Appendix:

Theorem 1 *A zero solution* (x_0^*, y_0^*) *of system (1) is a unique steady state if* $\alpha_x\alpha_y \geq \beta_x\beta_y$ *and there are three steady states* (x_i^*, y_i^*) *for* $i = 0, 1, 2$ *if* $\beta_x\beta_y > \alpha_x\alpha_y$.

Our next problem is to find out whether a solution of system (1) converges to the steady state or not. First the linearized version of system (1) is obtained by differentiating it in the neighborhood of the steady state,

$$\dot{x}(t) = -\alpha_x x(t) + \beta_x d_y^k y(t),$$

$$\dot{y}(t) = \beta_y d_x^k x(t) - \alpha_y y(t)$$

where

$$d_x^k = \left.\frac{d\tanh(x)}{dx}\right|_{x=x_k^*} \text{ and } d_y^k = \left.\frac{d\tanh(y)}{dy}\right|_{y=y_k^*}.$$

Notice that $d_x^0 = d_y^0 = 1$ at the zero steady state (x_0^*, y_0^*) and $d_x^k = d_y^k < 1$ at the nonzero steady states (x_k^*, y_k^*) for $k = 1, 2$.[2] The steady state is locally asymptotically

---

[2]By definition,

$$\tanh(x) = \frac{e^x - e^{-x}}{e^x + e^{-x}}$$

and its derivative is

$$\frac{d}{dx}\tanh(x) = \left(\frac{2}{e^x + e^{-x}}\right)^2 \leq 1.$$

It is clear that equality holds if $x = 0$. If $e^x = a$ for $x \neq 0$, then

$$e^x + e^{-x} = a + \frac{1}{a} > 2$$

implying

$$\frac{2}{e^x + e^{-x}} < 1$$

Hence the strict inequality holds if $x \neq 0$.

stable if the roots of the characteristic equation

$$\det \begin{pmatrix} \lambda + \alpha_x & -\beta_x d_y^k \\ -\beta_y d_x^k & \lambda + \alpha_y \end{pmatrix} = 0$$

or

$$\lambda^2 + (\alpha_x + \alpha_y)\lambda + (\alpha_x \alpha_y - \beta_x \beta_y d_x^k d_y^k) = 0$$

have negative real parts. It is now well-known, as a special case of the Routh-Hurwitz stability criterion, that the roots have negative real parts if the following inequality conditions hold,

$$\alpha_x + \alpha_y > 0 \text{ and } \alpha_x \alpha_y - \beta_x \beta_y d_x^k d_y^k > 0. \quad (5)$$

The first inequality always holds by assumption. Thus for the stability of the steady state, we need to check only the second inequality. The local stability results are summarized as follows:

Theorem 2 *The zero steady state* (x_0^*, y_0^*) *is*

(1) *a saddle point if* $\beta_x \beta_y > \alpha_x \alpha_y$,

(2) *a stable node if* $\alpha_x \alpha_y > \beta_x \beta_y > 0$

and in the case of $\alpha_x \alpha_y > 0 > \beta_x \beta_y$, *it is*

(3) *a stable node if* $(\alpha_x - \alpha_y)^2 + 4\beta_x \beta_y \geq 0$,

(4) *a stable focus if* $(\alpha_x - \alpha_y)^2 + 4\beta_x \beta_y < 0$

whereas the non-zero steady state (x_k^*, y_k^*) *for* $k = 1, 2$, *is always a stable node.*

3 Delay in the Gaining-Affection Process

Son and Park (2011) rise an important question on how an individual know the partner's romantic feeling. Observing a real situation in which the romantic interaction is communicated through various ways such as a talk, a phone call, an email, a letter and a rumor that "she loves you", they find that time is required for the romantic feelings of an individual to transfer to his/her partner. One delay $\tau_x > 0$ is introduced into the gaining-affection process of Juliet in system (1),[3]

[3]Liao and Ran (2007) further assume that Romeo also reacts to the delayed Juliet feeling $y(t - \tau_y)$ with $\tau_x \neq \tau_y$. Son and Park (2011) consider the special case where both individuals have the same

$$\dot{x}(t) = -\alpha_x x(t) + \beta_x \tanh[y(t)],$$
$$\dot{y}(t) = \beta_y \tanh[x(t - \tau_x)] - \alpha_y y(t). \tag{6}$$

Notice that the steady states (x_k^*, y_k^*) for $k = 0, 1, 2$ of the non-delay model are also the steady states of the delay model. The characteristic equation is obtained from the linearized version of system (6)

$$\lambda^2 + \left(\alpha_x + \alpha_y\right)\lambda + \alpha_x\alpha_y - \beta_x\beta_y d_x^k d_y^k e^{-\lambda\tau_x} = 0. \tag{7}$$

First the following result is shown:

Theorem 3 *All pure complex eigenvalues of Eq. (7) are simple.*

Suppose $\lambda = i\omega,\ \omega > 0$ is a root of (7) for some τ_x. Substituting it separates the characteristic equation into the real and imaginary parts,

$$-\omega^2 + \alpha_x\alpha_y - \beta_x\beta_y d_x^k d_y^k \cos\omega\tau_x = 0 \tag{8}$$

and

$$\left(\alpha_x + \alpha_y\right)\omega + \beta_x\beta_y d_x^k d_y^k \sin\omega\tau_x = 0. \tag{9}$$

Moving the constant terms to the right hand side and then adding the squares of the resultant equations yield a quartic equation

$$\omega^4 + \left(\alpha_x^2 + \alpha_y^2\right)\omega^2 + (\alpha_x\alpha_y)^2 - (\beta_x\beta_y d_x^k d_y^k)^2 = 0. \tag{10}$$

We first consider the stability of the nonzero steady states at which β_x and β_y have identical sign. In the proof of Theorem 1, it is shown that the second inequality condition in (5) holds. Thus all coefficients of Eq. (10) are positive, so there is no positive solution for ω^2. Therefore there is no stability switch and since they are stable at $\tau_x = 0$, they remain stable for all $\tau_x > 0$. We summarize the result:

Theorem 4 *The nonzero steady states of system (6) are stable for any $\tau_x \geq 0$.*

In Fig. 2, we illustrate the basin of attraction of the nonzero steady states, (x_1^*, y_1^*) and (x_2^*, y_2^*), taking $\alpha_x = \alpha_y = 1,\ \beta_x = 3/2, \beta_y = 3/2$ and $\tau_x = 2$. Any trajectory starting at an initial point in the light red region coverages to the positive steady state (x_1^*, y_1^*) denoted by the yellow dot and the one starting in the light blue region converges to the negative steady state denoted by the red dot. The downward-slopoing dotted line is the boundary between the two basins when there is no delay, $\tau_x = 0$.

(Footnote 3 continued)
delay $\tau_x = \tau_y$ in the gaining-affection processes. The dynamic results obtained in those studies are essentially the same as the one to be obtained in the following one delay model.

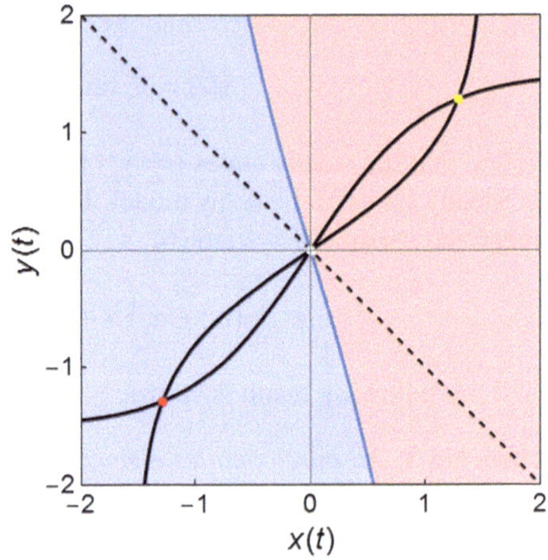

Fig. 2 Basin of attraction for system (6)

Increasing the value of the delay clockwisely rotates the boundary line. Thus the stability region of (x_1^*, y_1^*) in the fourth quadrant is enlarged and the one in the second quadrant is contracted and the same changes, but in opposite direction, occur for the stability region of the steady state (x_2^*, y_2^*). Even if the delay exists in the gaining-affection process, any trajectory converges to the positive equilibrium as far as an initial point is in the first quadrant.

Consider next the stability of the zero steady state. Solving (10) for ω^2 gives two solutions

$$(\omega_{\pm})^2 = \frac{-\left(\alpha_x^2 + \alpha_y^2\right) \pm \sqrt{D}}{2}$$

with

$$D = \left(\alpha_x^2 - \alpha_y^2\right)^2 + 4(\beta_x\beta_y)^2 > 0.$$

It is clear that $(\omega_-)^2 < 0$ and that $\left(\omega_+\right)^2$ is positive if $D > \left(\alpha_x^2 + \alpha_y^2\right)^2$ or

$$(\alpha_x\alpha_y)^2 < (\beta_x\beta_y)^2. \tag{11}$$

If there is no nonzero steady state with $\alpha_x\alpha_y > \beta_x\beta_y > 0$ or $0 > \beta_x\beta_y > -\alpha_x\alpha_y$, then inequality (11) is violated, so there is no positive solution for ω^2, and there is no stability switch in the case of $\left|\alpha_x\alpha_y\right| > \left|\beta_x\beta_y\right|$. Notice therefore that Eq. (11) might hold if, in addition to zero steady state, there are nonzero steady states or $0 > -\alpha_x\alpha_y > \beta_x\beta_y$. Substituting ω_+ into Eqs. (8) and (9) and then looking for τ_x that satisfies both equation, we have from (8)

$$\tau_x^m = \frac{1}{\omega_+}\left[\cos^{-1}\left(\frac{\alpha_x\alpha_y - \omega_+^2}{\beta_x\beta_y}\right) + 2m\pi\right] \text{ for } m = 0, 1, 2, ... \quad (12)$$

and from and (9),

$$\tau_x^n = \frac{1}{\omega_+}\left[\sin^{-1}\left(-\frac{(\alpha_x + \alpha_y)\omega_+}{\beta_x\beta_y}\right) + 2n\pi\right] \text{ for } n = 0, 1, 2, ... \quad (13)$$

Needless to say, these two solutions are different expressions for the same value when $m = n$.

To confirm direction of stability switch, we let $\lambda = \lambda(\tau_x)$ and then determine the sign of the derivative of Re $\left[\lambda(\tau_x)\right]$ at the point where $\lambda(\tau_x)$ is purely imaginary. Simple calculation shows that

$$sign\left[\text{Re}\left(\left.\frac{d\lambda(\tau_x)}{d\tau_x}\right|_{\lambda=i\omega}\right)\right] = sign\left[\omega_+^2\left(2\omega_+^2 + \alpha_x^2 + \alpha_y^2\right)\right].$$

The sign of the right hand side is apparently positive, which implies that crossing of the imaginary axis is from left to right as τ_x increases. Thus, at smallest stability switch (i.e., τ_x^m with $m = 0$), stability is lost and cannot be regained later if steady state is stable without delay. If it is unstable without delay, then it remains unstable for all $\tau_x > 0$. Concerning the stability of the zero steady state, we summarize the following results:

Theorem 5 (1) *If* $\left|\alpha_x\alpha_y\right| \geq \left|\beta_x\beta_y\right|$, *then the zero steady state is stable regardless of the values of the delay;* (2) *If* $\left|\alpha_x\alpha_y\right| < \left|\beta_x\beta_y\right|$ *and it is unstable for* $\tau_x = 0$, *then the zero steady state is unstable for any* $\tau_x > 0$; (3) *If* $\left|\alpha_x\alpha_y\right| < \left|\beta_x\beta_y\right|$ *and it is stable for* $\tau_x = 0$, *then the zero steady state is stable for* $\tau_x < \tau_x^0$, *loses stability for* $\tau_x = \tau_x^0$ *and bifurcates to a limit cycle for* $\tau_x > \tau_x^0$ *where the threshold value* τ_x^0 *is obtained from (12) with* $m = 0$.

In Fig. 3, parameter values are specified as $\alpha_x = \alpha_y = 1$, $\beta_x = 3/2$ and $\beta_y = -3/2$. Result (3) of Theorem 5 is numerically confirmed in Fig. 3a in which the bifurcation diagram with respect to τ_x is illustrated. Bifurcation parameter τ_x increases from $1/2$ to 3 with an increment $1/200$. Against each value of τ_x, the local maximum and local maximum values of $y(t)$ for $t \in [750, 800]$ are plotted. The red line starting at $y_0^* = 0$ bifurcates to two branches at $\tau_x = \tau_x^0 (\simeq 1.305)$. If the bifurcation diagram has only one point against the value of τ_x, then the system is stable and converges to the steady state. If it has two points, then one maximum and one minimum of a trajectory is plotted, that is, a limit cycle emerges. The shape of the diagram indicates that the limit cycle become larger as τ_x increases. In Fig. 3a the dotted vertical line at $\tau_x = 2.5$ intersects the diagram twice. In Fig. 3b a trajectory starting in the neighborhood of the steady state is oscillatory and converges to a limit cycle that has the maximum and minimum points corresponding to the crossing points in Fig. 3a.

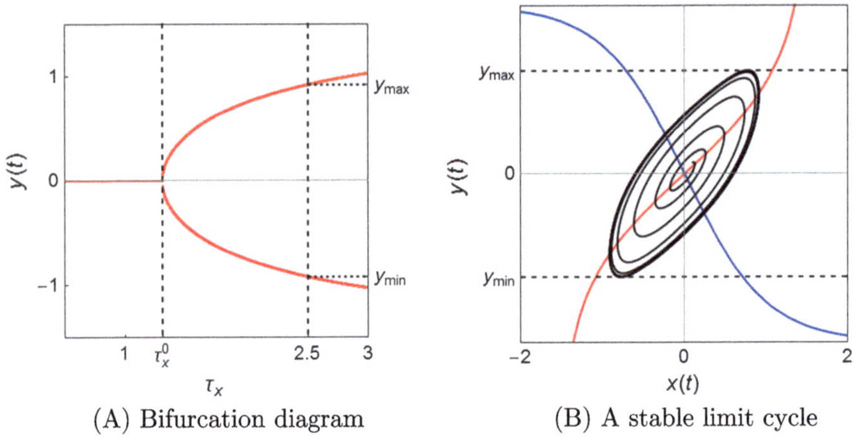

Fig. 3 Stability switch and the birth of limit cycles

4 Delay in the Losing-Memory Process

There are millions of people who can't stop loving their partners and live their life in dream of yesterday since they have been left alone. The love motions of those people may be described by a simple one delay differential equation,

$$\dot{x}(t) = -\alpha_x x(t - \tau_x), \ \alpha_x > 0. \tag{14}$$

Taking an exponential solution $x(t) = e^{\lambda t}$ and substituting it into the above equation yield a characteristic equation

$$\lambda = -\alpha_x e^{-\lambda\tau}.$$

Substituting an pure imaginary solution $\lambda = i\omega$ and then separating the resultant equation into the real and imaginary parts, we have

$$\alpha_x \cos\omega\tau = 0 \text{ and } \sin\omega\tau = \frac{\omega}{\alpha_x}.$$

Solving these equations simultaneously determines the threshold value of the delay as

$$\tau_x^0 = \frac{\pi}{2\alpha_x}.$$

If Eq. (14) is thought to be a linear approximation of the nonlinear equation preventing the possibilities of unbounded passion

$$\dot{x}(t) = -\alpha_x \tanh[x(t - \tau_x)] + A_x$$

where a positive appeal (i.e., $A_x > 0$) leads to a positive steady state. Then the steady state is stable for $\tau < \tau_x^0$ and bifurcates to a cyclic orbit for $\tau > \tau_x^0$. The memory does not vanish but keeps to oscillate around the steady state that approximates those happy hours. In this section we consider love dynamics of a Romeo who can live in memory and a Juliet who responds instantaneously. We replace Assumption 1 with the following Assumption 1',

Assumptiom 1': $O_x(x(t-\tau_x)) = -\alpha_x x(t-\tau_x)$, $\alpha_x > 0$ and $O_y(y(t)) = -\alpha_y y(t)$, $\alpha_y > 0$.

Dynamic system (6) is transformed to the following system with one delay in the losing-memory process,

$$\dot{x}(t) = -\alpha_x x(t-\tau_x) + \beta_x \tanh[y(t)],$$
$$\dot{y}(t) = \beta_y \tanh[x(t)] - \alpha_y y(t). \tag{15}$$

The characteristic equation is obtained from the linearized version of system (15)

$$\lambda^2 + \alpha_y \lambda - \beta_x \beta_y d_x^k d_y^k + \alpha_x(\lambda + \alpha_y)e^{-\lambda\tau_x} = 0. \tag{16}$$

Suppose again that the equation has a pure imaginary solution, $\lambda = i\omega$, $\omega > 0$. The characteristic equation can be broken down to the real and imaginary parts,

$$\alpha_x \alpha_y \cos\omega\tau + \alpha_x \omega \sin\omega\tau = \omega^2 + \beta_x \beta_y d_x^k d_y^k \tag{17}$$

and

$$-\alpha_x \alpha_y \sin\omega\tau + \alpha_x \omega \cos\omega\tau = -\alpha_y \omega. \tag{18}$$

Squaring both sides of each equation and adding them together yield a fourth-order equation with respect to ω,

$$\omega^4 + \left[(\alpha_y^2 - \alpha_x^2) + 2\beta_x\beta_y d_x^k d_y^k\right]\omega^2 + \left[(\beta_x\beta_y d_x^k d_y^k)^2 - (\alpha_x\alpha_y)^2\right] = 0.$$

Solving the equation with respect to ω^2 yields two solutions

$$(\omega_\pm)^2 = \frac{-\left[(\alpha_y^2 - \alpha_x^2) + 2\beta_x\beta_y d_x^k d_y^k\right] \pm \sqrt{D}}{2}$$

with

$$D = \left[(\alpha_y^2 - \alpha_x^2) + 2\beta_x\beta_y d_x^k d_y^k\right]^2 - 4\left[(\beta_x\beta_y d_x^k d_y^k)^2 - (\alpha_x\alpha_y)^2\right].$$

To simplify the analysis, we assume the following henceforth:

Assumption 4 $\alpha_x = \alpha_y = \alpha$

Then the solutions are simplified as

$$\omega_+^2 = \alpha^2 - \beta_x\beta_y d_x^k d_y^k \tag{19}$$

and

$$\omega_-^2 = -\left(\alpha^2 + \beta_x\beta_y d_x^k d_y^k\right). \tag{20}$$

Solving Eqs. (17) and (18) simultaneously presents two solutions,

$$\cos\omega\tau = \frac{\beta_x\beta_y d_x^k d_y^k}{\alpha^2 + \omega^2} \tag{21}$$

and

$$\sin\omega\tau = \frac{\omega\left(\omega^2 + \beta_x\beta_y d_x^k d_y^k + \alpha^2\right)}{\alpha(\alpha^2 + \omega^2)} \tag{22}$$

Before proceeding, we show the following:

Theorem 6 *If $\lambda = i\omega$ is a solution of Eq. (16), then it is simple.*

Concerning the direction of motion of the state variable $x(t)$ and $y(t)$ as τ is varied, we have the following result:

Theorem 7 *The stability of the steady state is lost and gained according to whether the following sign is positive or negative,*

$$sign\left[Re\left(\left.\frac{d\lambda(\tau_x)}{d\tau_x}\right|_{\lambda=i\omega}\right)\right] = \begin{cases} sign\left[2\alpha^2 - \beta_x\beta_y d_x^k d_y^k\right] & if\ \omega = \omega_+, \\ sign\left[\beta_x\beta_y d_x^k d_y^k\right] & if\ \omega = \omega_-. \end{cases}$$

4.1 Stability of Nonzero Steady State

At any nonzero steady state it is already shown that $\alpha^2 > \beta_x\beta_y d_x^k d_y^k$. So $\omega_+^2 > 0$ while $\omega_-^2 < 0$ since $\beta_x\beta_y > 0$. Then both $\cos\omega\tau$ and $\sin\omega\tau$ are positive so two threshold values of τ_x are obtained, one from Eq. (21)

$$\tau_x^m = \frac{1}{\omega_1}\left[\cos^{-1}\left(\frac{\beta_x\beta_y d_x^k d_y^k}{\alpha^2 + \omega_+^2}\right) + 2m\pi\right] \text{ for } m = 0, 1, 2, ...$$

and the other from (22)

$$\tau_x^n = \frac{1}{\omega_1}\left[\sin^{-1}\left(\frac{2\alpha\omega_1}{\alpha^2+\omega_+^2}\right)+2n\pi\right] \text{ for } n = 0, 1, 2, ...$$

where, as pointed out above, $\tau_x^m = \tau_x^n$ for $m = n$ since these describe the same relation between the delay and the parameters. Due to Theorem 7, we have

$$\text{Re}\left(\left.\frac{d\lambda(\tau_x)}{d\tau_x}\right|_{\lambda=i\omega_+}\right) > 0.$$

Then we have the following results concerning the stability switch on the nonzero steady state.

Theorem 8 *The nonzero steady state of system (15) is stable for* $\tau_x < \tau_x^0$, *loses stability for* $\tau_x = \tau_x^0$ *and bifurcates to a limit cycle for* $\tau_x > \tau_x^0$.

Figure 4a, b illustrate bifurcation diagrams with respect to τ_x. The only difference between these diagrams is the selection of the initial functions for system (15) while any other values of the parameters are the same. Simulations for the red curve is performed in the following way. The value of τ_x is increased from 1.5 to 1.825. For each value of τ_x, the delay dynamics system (15) with initial functions $x_0(t) = 0.1\cos(t)$ and $y_0(t) = 0.2\cos(t)$ runs for $0 \le t \le 5000$ and data obtained for $t \le 4950$ are discarded to get rid of transients. The local maximum and minimum from the remaining data of $y(t)$ are plotted against selected values of τ_x. The value of τ_x is increased with $1/500$ and then the same procedure is repeated until the value of τ_x arrives at 1.825. The blue curve has initial functions $x_0(t) = -0.1\cos(t)$ and $y_0(t) = -0.2\cos(t)$. Simulation has been done in the same way.

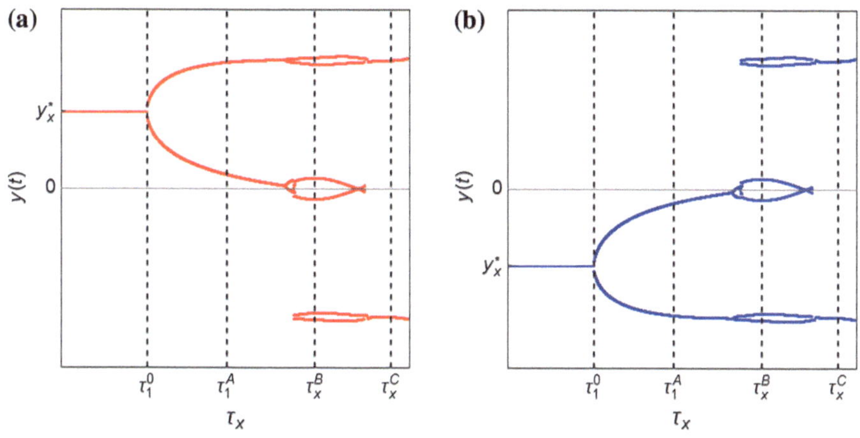

Fig. 4 Bifurcation diagrams with different initial functions

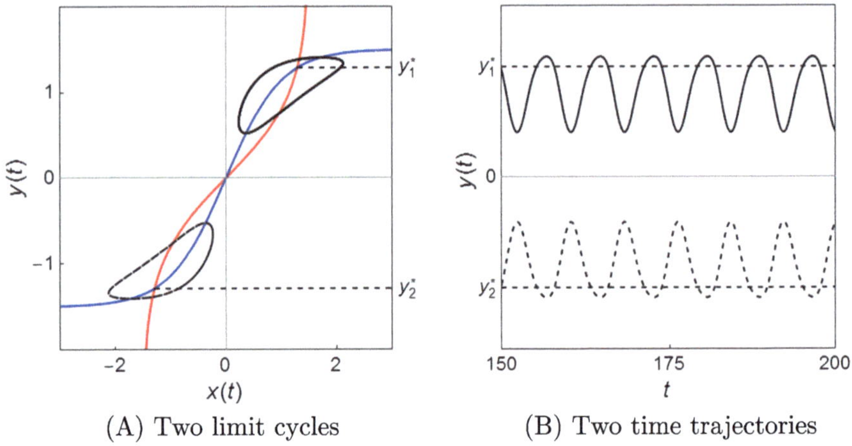

Fig. 5 Dynamics with $\tau_x = \tau_x^A$

Observing the bifurcation diagrams, we find that each diagram has four phases according to which different dynamics arises. To see what dynamics is born in each phase, we select three values of τ_x,

$$\tau_x^A = 1.68,\ \tau_x^B = 1.75 \text{ and } \tau_x^C = 1.81.$$

and then perform simulations to find dynamics in the (x, y) plane and in the $(t, y(t))$ plane. In the first phase where $\tau_x < \tau_x^0$ ($\simeq 1.617$), any trajectory converges to either $y_1^* > 0$ or $y_2^* < 0$ depending on the selection of the initial functions as each steady state is asymptotically stable. In the second phase where the diagrams have two brances and the vertical dotted line at $\tau_x = \tau_x^A$ intersects the blue diagram and the red diagram twice each. The steady state is destabilized as $\tau_x^A > \tau_x^0$. A trajectory starting in the neighborhood of the positive steady state converges to a small limit cycle surrounding the steady state. The same holds for a trajectory starting in the neighborhood of the negative steady state. The simulation results are plotted in Fig. 5a, b.

In the third phase, the diagram has six branches and the vertical dotted line at $\tau_x = \tau_x^B$ intersects the diagram six times. This implies two issues. One is that the two independent cycles are connected to form a large one cycle. Two cycles are included in the big one and each cycle has two extreme values leading to six extreme values. The other is that any trajectory converges to the same cyclic attractor regardless of the selection of the initial functions. Figure 6b indicates that a trajectory makes two small ups and downs around the positive steady state and moves down in the neighbourhood of the negative steady state within a large cycle. The real curve and dotted curve in Fig. 6b behave exactly in the same way with some phase shift.

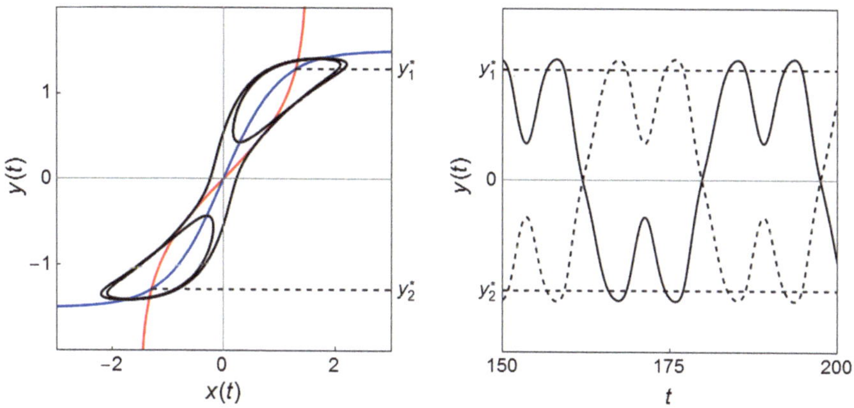

(A) Connected limit cycle (B) Time trajectories

Fig. 6 Dynamics with $\tau_x = \tau_x^B$

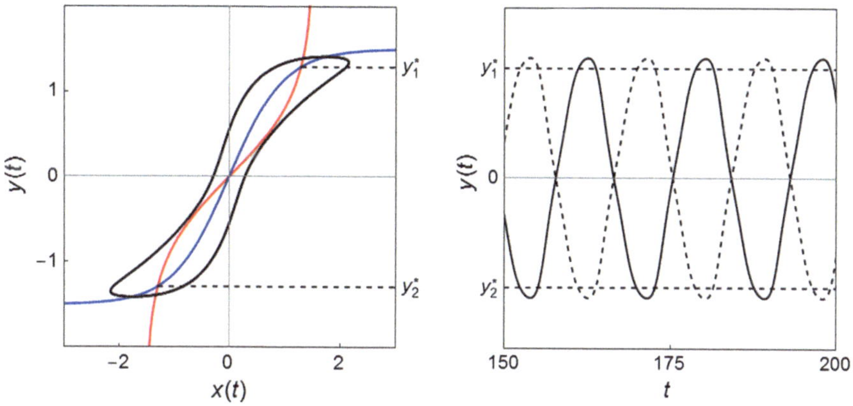

(A) Merged limit cycle (B) Time trajectories

Fig. 7 Dynamics with $\tau_x = \tau_x^C$

In the fourth phase, the diagram has two branches and thus the number of intersection of the dotted vertical line at $\tau_x = \tau_x^C$ with the bifurcation diagram decreases to two. As seen in Fig. 7a, the two small cycles are completely merged with the big cycle having one maximum and one minimum. The big limit cycle surrounds the two nonzero steady states, y_1^* and y_2^*.

4.2 *Stability of Zero Steady State*

To examine the stability switch of the zero steady state, we consider the three cases depending on the relative magnitude between $\alpha^2 = \alpha_x \alpha_y$ and $\beta_x \beta_y$.

(I) $\beta_x \beta_y \geq \alpha^2$

Under this inequality condition, Eqs. (19) and (20) indicate $\omega_+^2 \leq 0$ and $\omega_-^2 < 0$. The characteristic equation does not have a solution such as $\lambda = i\omega,\ \omega > 0$ and thus the real parts of the eigenvalues do not change their signs if τ_x increases. Hence no stability switch occurs and the stability of the zero steady state is the same as without delay. Due to (1) of Theorem 2, the zero steady state is unstable (i.e., a saddle point) for $\tau_x = 0$, it remains unstable for any $\tau_x > 0$.

(II) $\alpha^2 > \beta_x \beta_y > -\alpha^2$

Due to (3) and (4) of Theorem 2, the zero steady state is stable for $\tau_x = 0$. Equations (19) and (20) with the inequality conditions lead to $\omega_+^2 > 0$ and $\omega_-^2 < 0$, meaning that $\lambda = i\omega_+, \omega_+ > 0$ can be a solution of the characteristic equation under Assumption 4. Due to Theorem 7, we have

$$\mathrm{Re}\left(\left. \frac{d\lambda(\tau_x)}{d\tau_x} \right|_{\lambda = i\omega_+} \right) > 0$$

This implies that the solution crosses the imaginary axis from left to right as τ_x increases. We now determine the threshold value of τ_x at which the real parts of the solutions change their signs. Returning to two equations in (21), we check that the right hand side of both equations are positive. There is a unique $\omega_+ \tau_x,\ 0 < \omega_+ \tau_x < \pi/2$ for which both equations hold,

$$\tau_x^m = \frac{1}{\omega_+} \left[\cos^{-1} \left(\frac{\beta_x \beta_y}{\alpha^2 + \omega_+^2} \right) + 2m\pi \right] \quad \text{for } m = 0, 1, 2, ...$$

and

$$\tau_x^n = \frac{1}{\omega_+} \left[\sin^{-1} \left(\frac{\omega_+ (\omega_+^2 + \beta_x \beta_y + \alpha^2)}{\alpha(\alpha^2 + \omega_+^2)} \right) + 2n\pi \right] \quad \text{for } n = 0, 1, 2, ...$$

It is apparent that $\tau_x^m = \tau_x^n$ for $m = n$. It can be noticed that the zero steady state is asymptotically stable for $\tau_x < \tau_x^0$ and unstable for $\tau_x > \tau_x^0$. Thus τ_x^0 is the threshold value at which the stability switch occurs.

Numerical examples are given to confirm the analytical results. In Fig. 8a $\alpha_x = \alpha_y = 1$ and $\beta_x = \beta_y = 1/2$ are assumed and both Romeo and Juliet are secure. Stability is lost at $\tau_x = \tau_x^0 \simeq 1.648$ and a limit cycle emerges for $\tau_x > \tau_x^0$. In Fig. 8b, Romeo is still secure but Juliet is non-secure as $\beta_x = 1/2$ and $\beta_y = -1/2$. Stability is lost at $\tau_x = \tau_x^0 \simeq 1.505$ and a limit cycle emerges for $\tau_x > \tau_x^0$. It is to be noticed that the romantic syle in these examples are different, however, evolution of the emotion exhibit essentially the same.

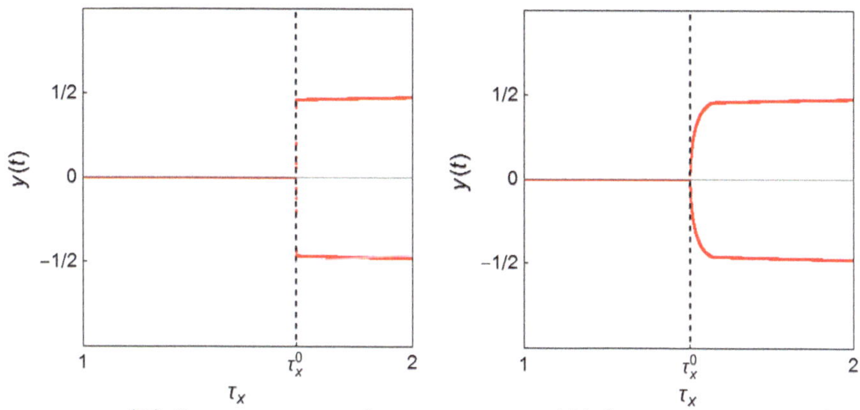

(B) Secure-secure style (A) Secure-nosecure style

Fig. 8 Bifurcation diagrams with respect to τ_x

(III) $-\beta_x\beta_y > \alpha^2 > \beta_x\beta_y$

Multiple stability switches occur in this case. Equations (19) and (20) indicate $\omega_+^2 \geq 0$ and $\omega_-^2 > 0$. It is to be noticed that (21) with (19) can be written as

$$\cos\omega_+\tau = \frac{\beta_x\beta_y}{\alpha^2 + \omega_+^2} \text{ and } \sin\omega_+\tau = \frac{2\alpha\omega_+}{\alpha^2 + \omega_+^2}$$

and (21) with (20) as

$$\cos\omega_-\tau = -1 \text{ and } \sin\omega_-\tau = 0.$$

So we have two different threshold values,

$$\tau_x^m = \frac{1}{\omega_+}\left[\cos^{-1}\left(\frac{\beta_x\beta_y}{\alpha^2 + \omega_+^2}\right) + 2m\pi\right] \text{ for } m = 0, 1, 2, ...$$

and

$$\tau_x^n = \frac{1}{\omega_-}(\pi + 2n\pi) \text{ for } n = 0, 1, 2, ...$$

Taking $\alpha_x = \alpha_y = 1$ and $\beta_y = -2$, we illustrate three τ_x^m curves for $m = 0, 1, 2$ in black and two τ_x^n curves for $n = 0, 1$ in red against values of $\beta_x \in [0, 3]$. All curves are downward-sloping and increasing the value of m (resp. n) shifts the black (resp. red) curve upward. The red curve is asymptotic to the dotted vertical line at $\beta_x = 1/2$ in Fig. 9a since ω_- goes to infinity as β_x approaches $1/2$ from above. The steady state is asymptotically stable for (β_x, τ_x) in the yellow regions and unstable otherwise. If

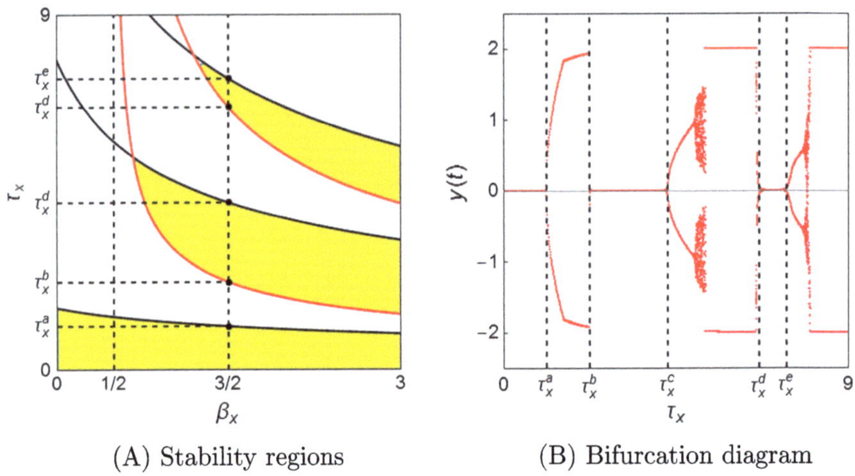

Fig. 9 Delay effect of τ_x

we fix the value of β_x at $3/2$ and increases the value of τ_x, then the dotted vertical line at $\beta_x = 3/2$ intersects the downward-sloping curves five times at

$$\tau_x^a \simeq 1.107,\ \tau_x^b \simeq 2.221,\ \tau_x^c \simeq 4.249,\ \tau_x^d \simeq 6.664 \text{ and } \tau_x^e \simeq 7.390.$$

The corresponding bifurcation diagram with respect to τ_x is illustrated in Fig. 9b. These figures illustrate multiple stability switching phenomenon from different points of view. Figure 9b indicates three Hopf bifurcation values in τ_x, $\tau_x^a < \tau_x^c < \tau_x^d$. The steady state is stable for $\tau_x = 0$ and remains stable for $\tau_x < \tau_x^a$. It loses stability at $\tau_x = \tau_x^a$ and bifurcates to a limit cycle for $\tau_x > \tau_x^a$. As the value of τ_x increases further, the steady state repeatedly passes through stability loss and gain and then eventually stays to be unstable. So as far as Fig. 9 concerns, the stability loss occurs three time and the stability gain twice for $\tau_x < 9$. Theorem 6 shows that the pure imaginary solutions are simple. Therefore at the crossing points with the stability switching curve only a pair of eigenvalues change the sign of their real part. Without delay the system is stable, all eigenvalues have negative real parts. So at the first crossing when stability is lost one pair of eigenvalues will have positive real part. If at the next crossing point stability might be regained, then the same pair of eigenvalues should change back the sign of their real part to negative, since there is no other pair with positive real parts. So all eigenvalues will have negative real parts again. In case if more than one pairs have positive real parts and the next crossing is when stability might regain, then only one pair changes back the sign of their real part to negative, the others will be still positive, so no stability regain occurs.

5 Concluding Remarks

In this paper the dynamic love affair model of Strogatz (1988) was reconsidered. First its nonlinear extension was introduced, the number of steady states was determined and the asymptotic behavior of its steady state was examined under different conditions. Conditions were derived for the existence of a unique and also for multiple steady states. First no time delay was assumed in the interaction of the lovers. In this no-delay case the nonzero steady states were always stable and conditions were derived for the stability of the zero steady state. Next a delay was assumed in the Gaining-affection process. The delay did not alter the stability of the nonzero steady states, the stability of the zero steady state was more complex. Depending on model parameter values it was either stable for all values of the delay, or always unstable, or stable for small values of the delay with stability loss at a certain threshold value of the delay. At this point the steady state bifurcated to a limit cycle. Then a delay was introduced into the Losing-memory process. The nonzero steady state was stable for small values of the delay, then stability was lost and the steady state bifurcated to a limit cycle. So this kind of delay had a destabilizing effect on the nonzero steady states. In examining the stability of the zero steady state we considered three cases depending on the relative magnitude of model parameters. In the first case the zero steady state was always unstable. In the second case stability was lost at a threshold value of the delay, and in the third case multiple stability switches could occur with repeated stability losses and regains. The stability of the steady states was analytically studied and the results were illustrated and verified by using computer simulation. In this paper we considered the cases of no or a single delay. It is a very interesting problem to see how the results of this paper change in the presence of multiple delays. This issue will be the subject of our next research project.

Appendix

Proof of Theorem 1

Proof The zero steady state, $x_0^* = 0$ and $y_0^* = 0$, is clearly a solution of (3) and (4). Thus the two isoclines intersect at least once at the origin. We investigate whether such an intersection happens only once or not. To this end, we differentiate $u(x)$ and $v(x)$,

$$u'(x) = \frac{\frac{\alpha_x}{\beta_x}}{1 - \left(\frac{\alpha_x}{\beta_x}x\right)^2}, \quad u''(x) = \frac{2x\left(\frac{\alpha_x}{\beta_x}\right)^3}{\left[1 - \left(\frac{\alpha_x}{\beta_x}x\right)^2\right]^2}$$

and

$$v'(x) = \frac{\beta_y}{\alpha_y}\left(\frac{2}{e^x + e^{-x}}\right)^2, \; v''(x) = -\frac{8\beta_y}{\alpha_y}\frac{e^x - e^{-x}}{(e^x + e^{-x})^3}.$$

Although $\alpha_x > 0$ and $\alpha_y > 0$ by assumption, the signs of β_x and β_y are undetermined. We consider three cases, depending on the signs of β_x and β_y.

(i) Assume first that β_x and β_y have different signs. Then $u'(x)$ and $v'(x)$ also have different signs, so one is strictly increasing and the other is strictly decreasing. So $x_0^* = 0$ and $y_0^* = 0$ are the only steady state if $\alpha_x\alpha_y > 0 > \beta_x\beta_y$.

(ii) Assume next that β_x and β_y are both positive. Then

$$u(0) = 0, \; u\left(\frac{\beta_x}{\alpha_x}\right) = \infty, \; u\left(-\frac{\beta_x}{\alpha_x}\right) = -\infty, \; u'(x) > 0, \; u''(x)\begin{cases} > 0 \text{ if } x > 0, \\ < 0 \text{ if } x < 0 \end{cases}$$

and

$$v(0) = 0, \; v(\infty) = \frac{\beta_y}{\alpha_y}, \; v(-\infty) = -\frac{\beta_y}{\alpha_y}, \; v'(x) > 0, \; v''(x)\begin{cases} < 0 \text{ if } x > 0, \\ > 0 \text{ if } x < 0. \end{cases}$$

Furthermore

$$u'(0) = \frac{\alpha_x}{\beta_x} \text{ and } v'(0) = \frac{\beta_y}{\alpha_y}.$$

Only zero solution is possible if $u'(0) \geq v'(0)$, that is, if

$$\frac{\alpha_x}{\beta_x} \geq \frac{\beta_y}{\alpha_y} \text{ or } \alpha_x\alpha_y \geq \beta_x\beta_y.$$

If $\alpha_x\alpha_y < \beta_x\beta_y$, then there are two nonzero solutions in addition to the zero steady state: one in the positive region $(x_1^*, y_1^*) > 0$ due to the convexity of $u(x)$ and the concavity of $v(x)$ for positive x and the other in the negative region $(x_2^*, y_2^*) < 0$ due to the concavity of $u(x)$ and the convexity of $v(x)$ for negative x.

(iii) Assume finally that $\beta_x < 0$ and $\beta_y < 0$. Equation (3) remains same if β_x and β_y are replaced by $-\beta_x$ and $-\beta_y$, so previous case may apply for existence of nonzero solutions.

■

Proof of Theorem 2

Proof We omit to prove the first four cases, (1), (2), (3) and (4). For the last case in which $\beta_x\beta_y > \alpha_x\alpha_y$, we consider two cases depending of the signs of β_x and β_y.

(i) We first assume $\beta_x > 0$ and $\beta_y > 0$. At a non-zero solution $v'(x_k^*) < u'(x_k^*)$, that is,

$$\frac{\beta_y}{\alpha_y} d_x < \frac{\frac{\alpha_x}{\beta_x}}{1 - \left(\frac{\alpha_x}{\beta_x} x\right)^2}. \tag{23}$$

Since from the first equation in (2),

$$\frac{\alpha_x}{\beta_x} x = \tanh(y),$$

the right hand side of (23) is

$$\frac{\frac{\alpha_x}{\beta_x}}{1 - \left(\frac{e^y - e^{-y}}{e^y + e^{-y}}\right)^2} = \frac{\frac{\alpha_x}{\beta_x}}{\left(\frac{2}{e^y + e^{-y}}\right)^2} = \frac{\frac{\alpha_x}{\beta_x}}{d_y}.$$

So we have

$$\frac{\beta_y}{\alpha_y} d_x < \frac{\frac{\alpha_x}{\beta_x}}{d_y} \tag{24}$$

or

$$\alpha_x\alpha_y > \beta_x\beta_y d_x d_y. \tag{25}$$

(ii) If $\beta_x < 0$ and $\beta_y < 0$, then $v'(x_k^*) > u'(x_k^*)$ for $k = 1, 2$ at any nonzero solution, so inequality (23) has opposite direction, as well as inequality (24) has opposite direction and by multiplying it by $\alpha_y\beta_x d_y < 0$, Eq. (25) remains valid. ∎

Proof of Theorem 3

Proof If any eigenvalue is multiple, then it also solves the following equation obtained by differentiating the left hand side of Eq. (7),

$$2\lambda + \left(\alpha_x + \alpha_y\right) + \beta_x\beta_y d_x^k d_y^k e^{-\lambda\tau_x}\tau_x = 0. \tag{26}$$

From Eq. (7),

$$\beta_x\beta_y d_x^k d_y^k e^{-\lambda\tau_x} = \lambda^2 + \left(\alpha_x + \alpha_y\right)\lambda + \alpha_x\alpha_y$$

that is substituted into Eq. (26),

$$2\lambda + \left(\alpha_x + \alpha_y\right) + \lambda^2 \tau_x + \left(\alpha_x + \alpha_y\right) \lambda \tau_x + \alpha_x \alpha_y \tau_x = 0$$

or

$$\lambda^2 \tau_x + \left(2 + \alpha_x \tau_x + \alpha_y \tau_x\right) \lambda + (\alpha_x + \alpha_y + \alpha_x \alpha_y \tau_x) = 0.$$

This equation cannot have pure complex root since multiplier of λ is positive. ■

Proof of Theorem 6

Proof The characteristic equation for $\alpha_x = \alpha_y = \alpha$ is simplified as

$$\lambda^2 + \alpha\lambda - \beta_x \beta_y d_x^k d_y^k + \alpha(\lambda + \alpha) e^{-\lambda \tau_x} = 0.$$

If λ is a multiple root, then it also satisfies equation,

$$2\lambda + \alpha + \alpha e^{-\lambda \tau_x} - \tau_x \alpha(\lambda + \alpha) e^{-\lambda \tau_x} = 0.$$

From the first equation

$$e^{-\lambda \tau_x} = -\lambda + \frac{\beta_x \beta_y d_x^k d_y^k}{\lambda + \alpha}$$

and by substituting it into the second equation, we have

$$2\lambda + \alpha + \left(-\lambda + \frac{\beta_x \beta_y d_x^k d_y^k}{\lambda + \alpha}\right) - \tau_x \left(-\lambda(\lambda + \alpha) + \beta_x \beta_y d_x^k d_y^k\right) = 0$$

which can be written as

$$\lambda^3 \tau_x + \lambda^2 \left(1 + 2\alpha\tau_x\right) + \lambda \left(2\alpha + \alpha^2 \tau_x - \beta_x \beta_y d_x^k d_y^k \tau_x\right) + \left(\alpha^2 + \beta_x \beta_y d_x^k d_y^k (1 - \alpha\tau_x)\right) = 0.$$

If $\lambda = i\omega$, then

$$\omega^2 = \frac{2\alpha + \alpha^2 \tau_x - \beta_x \beta_y d_x^k d_y^k \tau_x}{\tau_x} = \frac{\alpha^2 + \beta_x \beta_y d_x^k d_y^k (1 - \alpha\tau_x)}{1 + 2\alpha\tau_x}$$

This equation can be simplified as follows:

$$2\alpha + 2\tau_x (2\alpha^2 - \beta_x \beta_y d_x^k d_y^k) + \alpha\tau_x^2 \left(2\alpha^2 - \beta_x \beta_y d_x^k d_y^k\right) = 0.$$

If $\beta_x \beta_y \le 0$, then the left hand side is positive, so no solution exists. If $\beta_x \beta_y > 0$, then $\omega_+^2 > 0$ if and only if $\alpha^2 > \beta_x \beta_y d_x^k d_y^k$. In this case the left hand side is positive again showing that no solution exists. ■

Proof of Theorem 7

Proof Select τ_x as the bifurcation parameter and consider λ as the function of τ_x, $\lambda = \lambda(\tau_x)$. Implicitly differentiating the characteristic equation with respect to τ_x gives

$$\left[2\lambda + \alpha + \alpha e^{-\lambda\tau_x} - \alpha\tau_x(\lambda+\alpha)e^{-\lambda\tau_x}\right]\frac{d\lambda}{d\tau_x} - \alpha\lambda(\lambda+\alpha)e^{-\lambda\tau_x} = 0$$

implying that

$$\begin{aligned}\frac{d\lambda}{d\tau_x} &= \frac{\alpha\lambda(\lambda+\alpha)e^{-\lambda\tau_x}}{2\lambda + \alpha + \alpha e^{-\lambda\tau_x} - \alpha\tau_x(\lambda+\alpha)e^{-\lambda\tau_x}} \\ &= \frac{-\lambda^4 - 2\lambda^3\alpha - \lambda^2\alpha^2 + \beta_x\beta_y d_x^k d_y^k \lambda(\lambda+\alpha)}{2\lambda^2 + \alpha\lambda + 2\lambda\alpha + \alpha^2 + \left(1 - \tau_x\lambda - \tau_x\alpha\right)\left(-\lambda^2 - \alpha\lambda + \beta_x\beta_y d_x^k d_y^k\right)}.\end{aligned}$$

Assume that $\lambda = i\omega$, then the numerator becomes

$$\left(-\omega^4 + \omega^2\left(\alpha^2 - \beta_x\beta_y d_x^k d_y^k\right)\right) + i\omega\left(2\omega^2\alpha + \beta_x\beta_y d_x^k d_y^k\alpha\right)$$

and the denominator is simplified as

$$-\omega^2(1+2\alpha\tau_x) + \left(\alpha^2 + \beta_x\beta_y d_x^k d_y^k(1-\alpha\tau_x)\right) + i\left(-\tau_x\omega^3 + \omega\left(2\alpha + \alpha^2\tau_x - \tau_x\beta_x\beta_y d_x^k d_y^k\right)\right).$$

Multiplying the numerator and the denominator by the complex conjugate of the denominator shows that $\mathrm{Re}\left[d\lambda/d\tau_x\right]$ has the same sign as

$$\omega^4 + \omega^2\left(2\alpha^2\right) + \left[\alpha^4 + 2\alpha^2\beta_x\beta_y d_x^k d_y^k - \left(\beta_x\beta_y d_x^k d_y^k\right)^2\right].$$

At $\omega^2 = \omega_+^2 = \alpha^2 - \beta_x\beta_y d_x^k d_y^k$, this expression becomes

$$2\alpha^2\left(2\alpha^2 - \beta_x\beta_y d_x^k d_y^k\right) > 0$$

showing that at the stability switch stability is lost or instability is retained. At $\omega^2 = \omega_-^2 = -\left(\alpha^2 + \beta_x\beta_y d_x^k d_y^k\right)$, $\mathrm{Re}\left[d\lambda/d\tau_x\right]$ has the same sign as

$$2\alpha^2\beta_x\beta_y d_x^k d_y^k$$

which is positive if $\beta_x\beta_y > 0$ and negative if $\beta_x\beta_y < 0$. In the first case stability is lost or instability is retained and in the second case stability is regained or stability is retained. ■

References

Bielczyk, N., Forys, U., & Platkowski, T. (2013). Dynamical models of dyadic interactions with delay. *Journal of Mathematical Sociology*, *37*, 223–249.

Bielczyk, N., Bondnar, M., & Forys, U. (2012). Delay can stabilize: love affairs dynamics. *Applied Mathematics and Computation*, *219*, 3923–3937.

Liao, X., & Ran, J. (2007). Hopf bifurcation in love dynamical models with nonlinear couples and time delays. *Chaos Solitions and Fractals*, *31*, 853–865.

Rinaldi, S., & Gragnani, A. (1998). Love dynamics between secure individuals: A modeling approach. *Nonlinear Dynamics, Psychology, and Life Science*, *2*, 283–301.

Rinaldi, S. (1998a). Love dynamics: The case of linear couples. *Applied Mathematics and Computation*, *95*, 181–192.

Rinaldi, S. (1998b). Laura and Petrach: An intriguing case of cyclical love dynamics. *SIAM Journal on Applied Mathematics*, *58*, 1205–1221.

Son, W.-S., & Park, Y.-J. (2011). *Time Delay Effect on the Love Dynamic Model*, 1–8. arXiv:1108.5786.

Sprott, J. (2004). Dynamical models of love. *Nonlinear Dynamics, Psychology, and Life Sciences*, *8*, 303–314.

Strogatz, S. (1988). Love affairs and differential equations. *Mathematics Magazine*, *61*, 35.

Strogatz, S. (1994). *Nonlinear dynamics and chaos: With applications to physics, biology, chemistry, and engineering*. MA, Addison-Wesley: Reading.

Part III
Applications

Optimizing Baseball and Softball Bats

A. Terry Bahill

Abstract Collisions between baseballs, softballs and bats are complex and therefore their models are complex. One purpose of this paper is to show how complex these collisions can be, while still being modeled using only Newton's principles and the conservation laws of physics. This paper presents models for the speed and spin of balls and bats. These models and equations for bat-ball collisions are intended for use by high school and college physics students, engineering students and most importantly students of the science of baseball. Unlike in previous papers, these models use only simple Newtonian principles to explain simple collision configurations.

1 Précis of This Endeavor

This paper has two primary purposes: first, to help a batter select or create an optimal baseball or softball bat and second, to create models for bat-ball collisions using only fundamental principles of Newtonian mechanics (Table 1). We note that force, velocity, acceleration, impulse and momentum are all vector quantities, although we do not specifically mark them as such.

Newton's principles of motion are idealized as

I. *Inertia*. Every object either remains at rest or continues to move at a constant velocity, unless acted upon by an external force.

$$\sum F = 0 \quad \Leftrightarrow \quad dv/dt = 0$$

A.T. Bahill (✉)
Systems and Industrial Engineering, University of Arizona,
Tucson, AZ 85704, USA
e-mail: terry@sie.arizona.edu

A. Matsumoto (ed.), *Optimization and Dynamics with Their Applications*,
DOI 10.1007/978-981-10-4214-0_10

Table 1 List of variables, inputs, parameters, constants and their abbreviations

Symbol: This table is arranged alphabetically by the symbol	Abbreviation ball = 1 bat = 2 before = b after = a	Description	Typical values for a C243 pro stock wooden bat and a professional major-league baseball player	
			SI units	Baseball units
$\beta_{bat-knob}$	β	Angular velocity of the bat about the knob	rad/s	rpm
CoE		Conservation of energy	Joules	
CoM		Conservation of momentum	kg m/s	
CoAM		Conservation of angular momentum	$kg\,m^2/s$	
CoR		Coefficient of restitution of a bat-ball collision	0.55	0.55
d_{bat}		Length of the bat	0.861 m	34 in.
$d_{bat-cm-ss}$	d_{cm-ss} d	Distance from the center of mass to the sweet spot, which we define as the Center of Percussion	0.134 m	5.3 in.
$d_{bat-knob-cm}$	d_{kcm}	Distance from the center of the knob to the center of mass	0.569 m	22.4 in.
$d_{bat-knob-ss}$	d_{kss}	Distance from the center of the knob to the sweet spot	0.705 m	27.8 in.
$d_{bat-pivot-cm}$		Distance from the pivot point to the center of mass	0.416 m	16.4 in.
$d_{spine-cm}$		Distance from the batter's spine to the center of mass of the bat, an experimentally measured value	1.05 m	41 in.
$d_{bat-ss-end}$		Distance from the sweet spot to the barrel end of the bat	0.149 m	5.9 in.
g		Earth's gravitational constant (at the UofA)	9.718 m/s	
I_{ball}	I_1	Moment of inertia of the ball with respect to its center of mass	0.000079 kg m^2	
I_{bat-cm}	I_2	Moment of inertia of the bat with respect to rotations about its center of mass	0.048 kg m^2	
$I_{bat-knob}$	I_k	Moment of inertia of the bat with respect to rotations about the knob	0.341 kg m^2	
$I_{bat-pivot}$		Moment of inertia of the bat with respect to the pivot point between the hands	0.208 kg m^2	

(continued)

Table 1 (continued)

Symbol: This table is arranged alphabetically by the symbol	Abbreviation ball = 1 bat = 2 before = b after = a	Description	Typical values for a C243 pro stock wooden bat and a professional major-league baseball player	
			SI units	Baseball units
KE_{before}		Kinetic energy of the bat and the ball before the collision	375 J	
KE_{after}		Kinetic energy of the bat and the ball after the collision	216 J	
KE_{lost}		Kinetic energy lost or transformed in the collision	158 J	
m_{ball}	m_1	Mass of the baseball	0.145 kg	5.125 oz
m_{bat}	m_2	Mass of the bat	0.905 kg	32 oz
$\bar{m}$		$\bar{m} = \frac{m_{ball} m_{bat}}{m_{ball} + m_{bat}}$	0.125 kg	4.4 oz
μ_f		Dynamic coefficient of friction for a ball sliding on a wooden bat	0.5	
r_{ball}	r_1	Radius of the baseball	0.037 m	1.45 in.
r_{bat}	r_2	Maximum allowed radius of the bat	0.035 m	1.37 in.
pitch speed		Speed of the ball at the pitcher's release point	−46	−92[a] mph
$v_{ball-before}$	v_{1b}	Velocity of the ball immediately before the collision, 90% of pitch speed	−37 m/s	−83[a] mph
$v_{ball-before-Norm}$	v_{1bN}	Normal component of curveball velocity before collision, $v_{ball-before} \cos 6^{\circ}$	−36.8 m/s	−82.3 mph
$v_{ball-before-Tan}$	v_{1bT}	Tangential component of curveball velocity before collision, $v_{ball-before} \sin 6^{\circ}$	−3.9 m/s	−8.7 mph
$v_{ball-after}$	v_{1a}	Velocity of the ball after the collision, often called the launch speed or the *batted-ball speed.*	41.6 m/s	93 mph
v_{bat}	v_2	Velocity of the bat. If a specific place or time is intended then the subscript may contain cm (center of mass), ss (sweet spot), before (b) or after (a).		

(continued)

Table 1 (continued)

Symbol: This table is arranged alphabetically by the symbol	Abbreviation ball = 1 bat = 2 before = b after = a	Description	Typical values for a C243 pro stock wooden bat and a professional major-league baseball player	
			SI units	Baseball units
$v_{bat-cm-before}$	v_{2cmb}	Velocity of the center of mass of the bat before the bat-ball collision.	23 m/s	51 mph
$v_{bat-cm-after}$	v_{2cma}	Velocity of the center of mass of the bat after the collision.	10.4 m/s	23 mph
$v_{bat-ss-before}$	v_{2ssb}	Velocity of the sweet spot of the bat before the collision.	26 m/s	58[a] mph
$v_{bat-ss-after}$	v_{2ssa}	Velocity of the sweet spot of the bat after the collision.	12 m/s	27 mph
$\omega_{ball-before}$	ω_{1b}	Angular velocity of the ball about its center of mass before the collision. This spin rate depends on the particular type of pitch.	±209 rad/s	±2000 rpm
$\omega_{ball-after}$	ω_{1a}	Angular velocity of the ball about its center of mass after the collision	±209 rad/s	±2000 rpm
$\omega_{bat-before}$	ω_{2b}	Angular velocity of the bat about its center of mass before the collision	Near zero	
$\omega_{bat-after}$	ω_{2a}	Angular velocity of the bat about its center of mass after the collision	−32 rad/s	−303 rpm
$\omega_{spine-before}$		Angular velocity of the batter's arms and the bat about the spine	21 rad/s	201 rpm

[a]The equations of this paper concern variables right before and right after the collision, not at other times. For example, a pitcher could release a fastball with a speed of 92 mph, by the time it got to the collision zone it would have slowed down by 10% to 83 mph. Therefore, in our simulations we used 83 mph for $v_{ball-before}$

II. *Impulse and Momentum.* The rate of change of momentum of a body is directly proportional to the force applied and is in the direction of the applied force.

$$F = \frac{d(mv)}{dt} \Leftrightarrow F = ma$$

Stated differently, the change of momentum of a body is proportional to the impulse applied to the body, and has a direction along the straight line upon which that impulse is applied. An impulse J occurs when a force F acts over an interval of time Δt, and it is given by $J = \int_{\Delta t} F dt$. Since force is the time derivative of momentum, it follows that $J = \Delta p = m\Delta v$. Applying an impulse changes the momentum.

III. *Action/reaction.* For every action there is an equal and opposite reaction.
IV. *Restitution.* The ratio of the relative speeds after and before the collision is defined as the coefficient of restitution (*CoR*). The relative speed of two objects after a collision is a fixed fraction of the relative speed before the collision, regardless of whether one object or the other is initially at rest or the objects are approaching each other. The *CoR* models the energy lost in a collision.

In this paper, we will use these four principles of Newton. We will also use the overarching conservation laws that state, *energy, linear momentum and angular momentum cannot be created or destroyed.* These laws are more general than the principles and apply in all circumstances.

2 Bat-Ball Collisions

In this paper, we are modeling a point in time right before the bat-ball collision and its relationship with another point just after the collision. We are not modeling the behavior (1) during the collision, (2) long before the collision (the pitched ball) or (3) long after the collision (the batted-ball). The flight of the ball has been modeled by Bahill et al. (2009).

My model is for a head-on collision at the sweet spot (ss) of the bat, which I define to be the Center of Percussion (Bahill 2004). Figure 1 is a diagram of such a collision. All figures are drawn for a right-handed batter. This type of analysis was done by Watts and Bahill (1990, 2000). It would produce a “line drive” back to the pitcher.

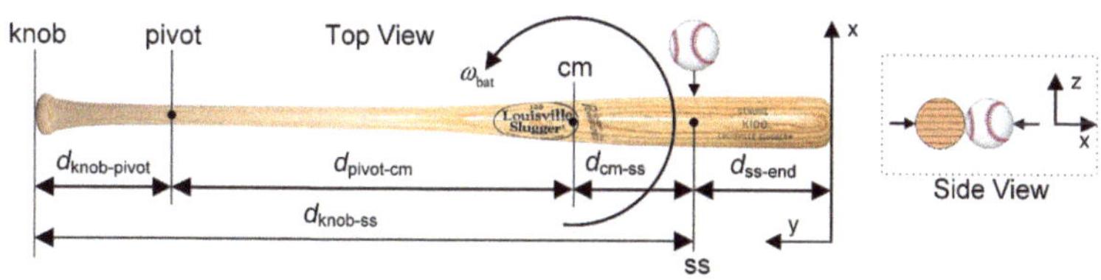

Fig. 1 Model for a collision at the sweet spot (ss) of the bat

3 Equations for Bat-Ball Collisions

3.1 Collisions at the Center of Mass

The literature is abound with linear collisions at the center of mass of an object. In these, kinetic energy is transformed into heat in the ball, vibrations in the bat, acoustic energy in the "crack of the bat" and deformations of the bat or ball. The Coefficient of Restitution (*CoR*) models the energy that is transformed in a frictionless head-on collision between two objects. The equation for the kinetic energy lost in a head-on bat-ball collision at the center of mass (Dadouriam 1913, Eq. (XI), p. 248; Ferreira da Silva 2007, Eq. 23; Brach 2007, Eq. 3.7) is

$$KE_{lost} = \frac{\bar{m}}{2}(\text{collision velocity})^2\left(1 - CoR_{1b}^2\right) \text{ where } \bar{m} = \frac{m_{\text{ball}} m_{\text{bat}}}{m_{\text{ball}} + m_{\text{bat}}}$$
$$KE_{lost} = \frac{\bar{m}}{2}\left(v_{\text{bat}-\text{cm}-\text{before}} - v_{\text{ball}-\text{before}}\right)^2\left(1 - CoR_{1b}^2\right) \quad (1)$$

3.2 Collisions at the Sweet Spot

3.2.1 Coordinate System

We use a right-handed coordinate system with the x-axis pointing from home plate to the pitching rubber, the y-axis points from first base to third base, and the z-axis points straight up. A torque rotating from the x-axis to the y-axis would be positive upward. Over the plate, the ball comes downward at a 10° angle and the bat usually moves upward at about 10°, so later the z-axis will be rotated back 10°.

3.2.2 Assumptions

We made the following assumptions:

A1. We assumed a head-on collision at the sweet spot of the bat.
A2. We neglected permanent deformations of the bat and ball.
A3. We assumed that there were no tangential forces during the collision.
A4. In this paper, we did not model the moment of inertia of the batter's arms.
A5. Collisions at the Center of Percussion produce a rotation about the center of mass, but no translation of the bat.
A6. The collision duration is short, for example, one millisecond.
A7. Because the collision duration is short and the swing is level, we ignored the effects of gravity *during* the collision.
A8. The Coefficient of Restitution (*CoR*) for a baseball wooden-bat collision at major-league speeds is about 0.55.

A9. The dynamic coefficient of friction has been measured by Bahill at $\mu_f = 0.5$. This agrees with measurements by Sawicki et al. (2003) and Cross and Nathan (2006).

A10. Air density affects the flight of the batted-ball. And air density is inversely related to altitude, temperature and humidity, and is directly related to barometric pressure. Of these four, altitude is the most important factor (Bahill et al. 2009). We did not consider these four parameters in this paper, because they are for the flight of the ball, not the collision.

3.2.3 Experimental Validation Data

The experimental data in Table 1 are based on the following assumptions. The batter is using a Louisville Slugger C243 wooden bat and is hitting a regulation major-league baseball. The ball speed at the plate is −83 mph. The velocity of the sweet spot of the bat is 58 mph: this is the average value of the data collected from 28 San Francisco Giants measured by Bahill and Karnavas (1991). These velocities produce a *CoR* of 0.55 and a batted-ball speed of 97 mph, as will be shown in Table 7. Using an ideal launch angle of 31°, we find a batted-ball spin of −2100 rpm (Baldwin and Bahill 2004). With these values, the ball would travel 350 feet, which could produce a home run in all major-league stadiums.

3.2.4 The Model

The model of this paper is for a collision at the sweet spot of the bat with spin on the pitch. The model for the movement of the bat is a translation and a rotation about the center of mass. It has five equations and five unknowns, which are shown in Table 2.

Definition of Variables

To visualize these variables please refer to Fig. 2.

Inputs $v_{ball-before}$, $\omega_{ball-before}$, $v_{bat-cm-before}$, $\omega_{bat-before}$ and *CoR*

$v_{ball-before}$ is the linear velocity of the *ball* in the x-direction before the collision.

$\omega_{ball-before}$ is the angular velocity of the *ball about its center of mass* before the collision.

$v_{bat-cm-before}$ is the linear velocity of the *center of mass of the bat* in the x-direction before the collision.

$\omega_{bat-before}$ is the angular velocity of the *bat about its center of mass* before the collision.

Table 2 The model has five equations and five unknowns

Inputs	$v_{\text{ball-before}}$, $\omega_{\text{ball-before}}$, $v_{\text{bat-cm-before}}$, $\omega_{\text{bat-before}}$ and CoR
Outputs (unknowns)	$v_{\text{ball-after}}$, $\omega_{\text{ball-after}}$, $v_{\text{bat-cm-after}}$, $\omega_{\text{bat-after}}$, and KE_{lost}
Equations	
Conservation of energy, Eq. (2)	$\frac{1}{2}m_{\text{ball}}v^2_{\text{ball-before}} + \frac{1}{2}I_{\text{ball}}\omega^2_{\text{ball-before}} + \frac{1}{2}m_{\text{bat}}v^2_{\text{bat-cm-before}} + \frac{1}{2}I_{\text{bat}}\omega^2_{\text{bat-before}}$ $= \frac{1}{2}m_{\text{ball}}v^2_{\text{ball-after}} + \frac{1}{2}I_{\text{ball}}\omega^2_{\text{ball-after}} + \frac{1}{2}m_{\text{bat}}v^2_{\text{bat-cm-after}} + \frac{1}{2}I_{\text{bat}}\omega^2_{\text{bat-after}} + KE_{\text{lost}}$
Conservation of linear momentum, Eq. (3)	$m_{\text{ball}}v_{\text{ball-before}} + m_{\text{bat}}v_{\text{bat-cm-before}} = m_{\text{ball}}v_{\text{ball-after}} + m_{\text{bat}}v_{\text{bat-cm-after}}$
Definition of *CoR*, Eq. (4)	$CoR_{2b} = -\frac{v_{\text{ball-after}} - v_{\text{bat-cm-after}} - d_{\text{cm-ss}}\omega_{\text{bat-after}}}{v_{\text{ball-before}} - v_{\text{bat-cm-before}} - d_{\text{cm-ss}}\omega_{\text{bat-before}}}$
Newton's second law, Eq. (5)	$d_{\text{cm-ss}}m_{\text{ball}}(v_{\text{ball-after}} - v_{\text{ball-before}}) = -I_{\text{bat}}(\omega_{\text{bat-after}} - \omega_{\text{bat-before}})$
Conservation of angular momentum, Eq. (6s)	$L_{\text{ball-before}} + L_{\text{bat-before}} = L_{\text{ball-after}} + L_{\text{bat-after}}$ $m_1 v_{1b} d + I_1\omega_{1b} + m_1\omega_{1b}d^2 + I_2\omega_{2b} = m_1 v_{1a} d + I_1\omega_{1a} + m_1\omega_{1a}d^2 + I_2\omega_{2a}$

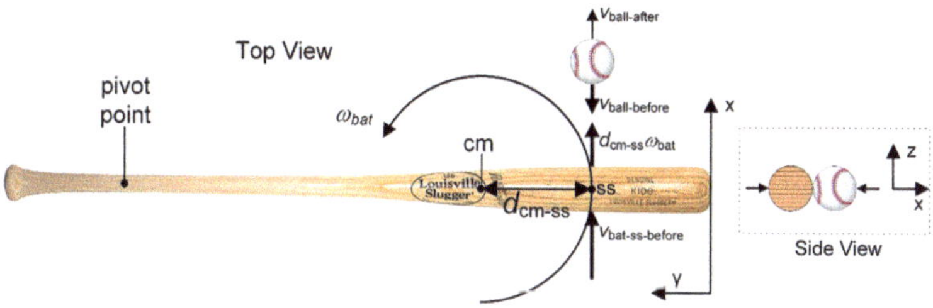

Fig. 2 This figure shows $v_{\text{ball}-\text{before}}, v_{\text{bat}-\text{cm}-\text{before}}, v_{\text{ball}-\text{after}}$ and $d_{\text{cm}-\text{ss}}\omega_{\text{bat}}$, which are used to define the coefficient of restitution

CoR_{2b} is the coefficient of restitution.

Outputs $v_{\text{ball}-\text{after}}$, $\omega_{\text{ball}-\text{after}}$, $v_{\text{bat}-\text{ss}-\text{after}}$, $\omega_{\text{bat}-\text{after}}$ and KE_{lost}

$v_{\text{ball}-\text{after}}$ is the linear velocity of the *batted-ball* in the x-direction after the collision.

$\omega_{\text{ball}-\text{after}}$ is the angular velocity of the *ball about its center of mass* after the collision.

$v_{\text{bat}-\text{ss}-\text{after}}$ is the linear velocity of the *sweet spot* of the bat in the x-direction after the collision.

$\omega_{\text{bat}-\text{after}}$ is the angular velocity of the bat *about its center of mass* after the collision.

KE_{lost} is the kinetic energy lost or transformed in the collision.

We want to solve for $v_{\text{ball}-\text{after}}, \omega_{\text{ball}-\text{after}}, v_{\text{bat}-\text{cm}-\text{after}}, \omega_{\text{bat}-\text{after}}$ and KE_{lost}.

We will use the following fundamental equations of physics: Conservation of Energy, Conservation of Linear Momentum, the Definition of Kinematic *CoR*, Newton's Second Principle and the Conservation of Angular Momentum.

Condensing the Notation for the Equations

First, we want to simplify our notation. We will now make the following substitutions. These abbreviations are contained in Table 1, but for convenience, we repeat them here.

$$d_{\text{cm}-\text{ss}} = d$$
$$I_{\text{bat}} = I_2$$
$$m_{\text{ball}} = m_1$$
$$m_{\text{bat}} = m_2$$
$$v_{\text{ball}-\text{before}} = v_{1b}$$
$$v_{\text{ball}-\text{after}} = v_{1a}$$
$$v_{\text{bat}-\text{cm}-\text{before}} = v_{2b}$$
$$v_{\text{bat}-\text{cm}-\text{after}} = v_{2a}$$
$$\omega_{\text{bat}-\text{before}} = \omega_{2b}$$
$$\omega_{\text{bat}-\text{after}} = \omega_{2a}$$

These substitutions produce the following equations.

Conservation of Energy

The law of conservation of energy states that energy will not be create or destroyed.

$$\frac{1}{2}m_{\text{ball}}v_{\text{ball}-\text{before}}^2+\frac{1}{2}I_{\text{ball}}\omega_{\text{ball}-\text{before}}^2+\frac{1}{2}m_{\text{bat}}v_{\text{bat}-\text{cm}-\text{before}}^2+\frac{1}{2}I_{\text{bat}}\omega_{\text{bat}-\text{before}}^2$$
$$=\frac{1}{2}m_{\text{ball}}v_{\text{ball}-\text{after}}^2+\frac{1}{2}I_{\text{ball}}\omega_{\text{ball}-\text{after}}^2+\frac{1}{2}m_{\text{bat}}v_{\text{bat}-\text{cm}-\text{after}}^2+\frac{1}{2}I_{\text{bat}}\omega_{\text{bat}-\text{after}}^2+KE_{\text{lost}} \quad (2)$$

$$m_1v_{1\text{b}}^2+m_2v_{2\text{b}}^2+I_2\omega_{2\text{b}}^2=+m_1v_{1\text{a}}^2+m_2v_{2\text{a}}^2+I_2\omega_{2\text{a}}^2+2KE_{\text{lost}} \quad (2\text{s})$$

In the label (3s), “s” stands for short.

Conservation of Linear Momentum

The law of conservation of linear momentum states that linear momentum will be conserved in a collision if there are no external forces. We will approximate the bat’s motion before the collision with the tangent to the curve of its arc as shown in Fig. 2. For a collision anywhere on the bat, every point on the bat has the same angular velocity, but the linear velocities will be different, which means that $v_{\text{bat}-\text{before}}$ is a combination of translations and rotations unique for each point on the bat. Conservation of momentum in the direction of the x-axis states that the momentum before plus the external impulse will equal the momentum after the collision. There are no external impulses during the bat-ball collision: therefore, this is the equation for Conservation of Linear Momentum

$$m_{\text{ball}}v_{\text{ball}-\text{before}}+m_{\text{bat}}v_{\text{bat}-\text{cm}-\text{before}}=m_{\text{ball}}v_{\text{ball}-\text{after}}+m_{\text{bat}}v_{\text{bat}-\text{cm}-\text{after}} \quad (3)$$

$$m_1v_{1\text{b}}+m_2v_{2\text{b}}=m_1v_{1\text{a}}+m_2v_{2\text{a}} \quad (3\text{s})$$

Definition of the Coefficient of Restitution

The kinematic Coefficient of Restitution (*CoR*) was defined by Sir Isaac Newton as the ratio of the relative velocity of the two objects after the collision to the relative velocity before the collision.

In our models, for a collision at the sweet spot (ss) of the bat we have

$$CoR_{2\text{b}}=-\frac{v_{\text{ball}-\text{after}}-v_{\text{bat}-\text{cm}-\text{after}}-d_{\text{cm}-\text{ss}}\omega_{\text{bat}-\text{after}}}{v_{\text{ball}-\text{before}}-v_{\text{bat}-\text{cm}-\text{before}}-d_{\text{cm}-\text{ss}}\omega_{\text{bat}-\text{before}}} \quad (4)$$

$$CoR_{2\text{b}}=-\frac{v_{1\text{a}}-v_{2\text{a}}-d\omega_{2a}}{v_{1\text{b}}-v_{2\text{b}}-d\omega_{2\text{b}}} \quad (4\text{s})$$

These variables are illustrated in Fig. 2. A note on notation: $\omega_{\text{bat}-\text{after}}$ is the angular velocity of the bat *about its center of mass* after the collision and

$v_{\text{bat}-\text{cm}-\text{before}}$ is the linear velocity *of the center of mass* of the bat in the x-direction before the collision: this is a combination of translation and rotation.

Newton's Second Principle

Watts and Bahill (1990) derived the following equation from Newton's second principle that states that a force acting on an object produces acceleration in accordance with the equation $F = ma$. If an object is accelerating, then its velocity and momentum is increasing. This principle is often stated as; applying an impulsive force to an object will change its momentum. According to Newton's third principle, when a ball hits a bat at the sweet spot there will be a force on the bat in the direction of the negative x-axis, let us call this $-F_1$, and an equal but opposite force on the ball, called F_1. This force will be applied during the duration of the collision. When a force is applied for a short period of time, it is called an impulse. According to Newton's second principle, an impulse will change momentum. The force on the bat will create a torque of $-d_{cm-ss}F_1$ around the center of mass of the bat. An impulsive torque will produce a change in angular momentum of the *bat*.

$$-d_{cm-ss}F_1 t_c = I_{bat}(\omega_{\text{bat}-\text{after}} - \omega_{\text{bat}-\text{before}})$$

Now this impulse will also change the linear momentum of the *ball*.

$$F_1 t_c = m_{\text{ball}}(v_{\text{ball}-\text{after}} - v_{\text{ball}-\text{before}})$$

Multiply both sides of this equation by $d_{\text{cm}-\text{ss}}$ and add these two equations to get

$$d_{\text{cm}-\text{ss}} m_{\text{ball}}(v_{\text{ball}-\text{after}} - v_{\text{ball}-\text{before}}) = -I_{\text{bat}}(\omega_{\text{bat}-\text{after}} - \omega_{\text{bat}-\text{before}}) \quad (5)$$

$$dm_1(v_{1a} - v_{1b}) = -I_2(\omega_{2a} - \omega_{2b}) \quad (5s)$$

For now, we have ignored ω_{ball}. We will reconsider this later.

Conservation of Angular Momentum

The initial and final angular momenta comprise ball translation, ball rotation, bat translation and bat rotation about its center of mass.

$$L_{\text{initial}} = L_{\text{final}}$$

$$\begin{aligned} & m_1 v_{1b} d + (I_1 + m_1 d^2)\omega_{1b} + I_2\omega_{2b} \\ & = +m_1 v_{1a} d + (I_1 + m_1 d^2)\omega_{1a} + I_2\omega_{2a} \end{aligned} \quad (6s)$$

Summary of abbreviations that will be used in the following sections, with units:

$$C = v_{1b} - v_{2b} - d\omega_{2b} \quad \text{m/s}$$

$$D = \frac{m_1 d^2}{I_2} \quad \text{unit less}$$

$$K = (m_1 I_2 + m_2 I_2 + m_1 m_2 d^2) \quad \text{kg}^2\,\text{m}^2$$

$$\mathrm{L} = v_{2b} m_2 I_2 (1 + CoR_{2b}) + \omega_{2b} m_2 d I_2 (1 + CoR_{2b}) \quad \text{kg}^2\,\text{m}^3/\text{s}$$

$$\bar{m} = \frac{m_1 m_2}{m_1 + m_2} \quad \text{kg}$$

Note that none of these abbreviations contains the outputs $v_{\mathrm{ball-after}}$, $\omega_{\mathrm{ball-after}}$, $v_{\mathrm{bat-cm-after}}$, $\omega_{\mathrm{bat-after}}$ and KE_{lost}. The most useful abbreviations are the ones that are constants independent of velocities after the collision. These abbreviations are only used during the derivations. They are removed from the output equations. We will now use the Newtonian principles in Eqs. (3)–(5) to find $v_{\mathrm{ball-after}}$, $v_{\mathrm{bat-cm-after}}$, and $\omega_{\mathrm{bat-after}}$.

Finding Ball Velocity After the Collision

First, we solve for $v_{\mathrm{ball-after}}$.

Start with Eq. (5) and solve for ω_{2a}

$$d m_1 (v_{1a} - v_{1b}) = -I_2 (\omega_{2a} - \omega_{2b})$$

$$\boxed{\omega_{2a} = \omega_{2b} - \frac{d m_1}{I_2}(v_{1a} - v_{1b})}$$

This equation was derived from Eq. (5). We will use this expression repeatedly. We know that for baseball and softball ω_{2b} is close to zero, but for generality, we will leave it in for as long as we can.

Next, we use Eq. (4) and solve for v_{2a}

$$CoR_{2b} = -\frac{v_{1a} - v_{2a} - d\omega_{2a}}{v_{1b} - v_{2b} - d\omega_{2b}}$$

$$CoR_{2b}(v_{1b} - v_{2b} - d\omega_{2b}) = -v_{1a} + v_{2a} + d\omega_{2a}$$

$$\boxed{v_{2a} = v_{1a} + CoR_{2b}(v_{1b} - v_{2b} - d\omega_{2b}) - d\omega_{2a}}$$

This equation was derived from Eq. (4). We will use this expression repeatedly. Next, substitute ω_{2a} into this v_{2a} equation. We put substitutions in squiggly braces {} to make it obvious what has been inserted.

$$v_{2a} = v_{1a} + CoR_{2b}\ (v_{1b} - v_{2b} - d\omega_{2b}) - d\left\{\omega_{2b} - \frac{dm_1}{I_2}(v_{1a} - v_{1b})\right\}$$

Let $D = \frac{m_1 d^2}{I_2}$ and $C = v_{1b} - v_{2b} - d\omega_{2b}$

$$v_{2a} = v_{1a} + \{D\}(v_{1a} - v_{1b}) + CoR_{2b}\{C\} - d\omega_{2b}$$
$$v_{2a} = v_{1a}(1 + D) - v_{1b}D + CoR_{2b}C\ - d\omega_{2b}$$

Now substitute this $m_2 v_{2a}$ into Eq. (3)

$$m_1 v_{1b} + m_2 v_{2b} = m_1 v_{1a} + \{m_2 v_{1a}(1 + D) - m_2 D v_{1b} + m_2 CoR_{2b}\ C - m_2 d\omega_{2b}\}$$

Replace the dummy variables C and D and

$$v_{1a}\left[m_1 + m_2 + \frac{m_1 m_2 d^2}{I_2}\right] = v_{1b}\left[m_1 + \frac{m_1 m_2 d^2}{I_2} - m_2 CoR_{2b}\right] + m_2 v_{2b}$$
$$+ m_2 CoR_{2b} v_{2b}\ + \omega_{2b} m_2 d(1 + CoR_{2b})$$

Multiply by I_2.

$$v_{1a}\left[m_1 I_2 + m_2 I_2 + m_1 m_2 d^2\right] = v_{1b}\left[m_1 I_2 + m_1 m_2 d^2 - m_2 CoR_{2b} I_2\right] + m_2 v_{2b} I_2$$
$$+ m_2 CoR v_{2b}\ I_2 + \omega_{2b} m_2 d I_2(1 + CoR_{2b})$$

Rearrange

$$\boxed{v_{1a} = \frac{v_{1b}(m_1 I_2 - m_2 I_2\ CoR_{2b} + m_1 m_2 d^2) + v_{2b} m_2 I_2(1 + CoR_{2b}) + \omega_{2b} m_2 d I_2(1 + CoR_{2b})}{m_1 I_2 + m_2 I_2 + m_1 m_2 d^2}} \tag{7}$$

This equation was derived from Eqs. (3)–(5).
Now we want to rearrange this normal form equation into its canonical form.

$$\text{Let } K = \left(m_1 I_2 + m_2 I_2 + m_1 m_2 d^2\right)$$
$$\mathrm{L} = v_{2b} m_2 I_2(1 + CoR_{2b}) + \omega_{2b} m_2 d I_2(1 + CoR_{2b})$$
$$v_{1a} = \frac{v_{1b}(m_1 I_2 - m_2 I_2 CoR_{2b} + m_1 m_2 d^2)}{K} + \frac{L}{K}$$

add $\left(v_{1b} - \frac{v_{1b}K}{K}\right)$ to the right side

$$v_{1a} = v_{1b} + \frac{v_{1b}(m_1 I_2 - m_2 I_2 CoR_{2b} + m_1 m_2 d^2) - v_{1b}(m_1 I_2 + m_2 I_2 + m_1 m_2 d^2)}{K} + \frac{L}{K}$$

$$v_{1a} = v_{1b} + \frac{-v_{1b} m_2 I_2 (1 + CoR_{2b}) + L}{K}$$

Finally, we get the canonical form:

$$\boxed{v_{1a} = v_{1b} - \frac{(v_{1b} - v_{2b}) m_2 I_2 (1 + CoR_{2b}) - \omega_{2b} m_2 d I_2 (1 + CoR_{2b})}{m_1 I_2 + m_2 I_2 + m_1 m_2 d^2}} \tag{8}$$

This equation was derived from Eqs. (3)–(5).

$$v_{\text{ball-after}} = v_{\text{ball-before}} - \frac{(v_{\text{ball-before}} - v_{\text{bat-cm-before}}) m_{\text{bat}} I_{\text{bat}} (1 + CoR_{2b}) - \omega_{bat-before} m_{bat} d I_{bat} (1 + CoR_{2b})}{m_{\text{ball}} I_{\text{bat}} + m_{\text{bat}} I_{\text{bat}} + m_{\text{ball}} m_{\text{bat}} d^2_{\text{cm-ss}}}$$

Finding Bat Velocity After the Collision

We solve for $v_{\text{ball-after}}$. As before, we start with Eq. (5) and solve for ω_{2a}

$$\omega_{2a} = \omega_{2b} - \frac{d m_1}{I_2}(v_{1a} - v_{1b})$$

Next use Eq. (4) and solve for v_{2a}

$$CoR_{2b} = -\frac{v_{1a} - v_{2a} - d\omega_{2a}}{v_{1b} - v_{2b} - d\omega_{2b}}$$

$$v_{2a} = v_{1a} + CoR_{2b}(v_{1b} - v_{2b} - d\omega_{2b}) - d\omega_{2a}$$

We will use this expression repeatedly. Substitute ω_{2a} into this v_{2a} equation. I put the substitution in squiggly braces {} to make it obvious what has been inserted.

$$v_{2a} = v_{1a} + CoR_{2b}\ (v_{1b} - v_{2b} - d\omega_{2b}) - d\left\{\omega_{2b} - \frac{d m_1}{I_2}(v_{1a} - v_{1b})\right\}$$

Let $C = v_{1b} - v_{2b} - d\omega_{2b}$

$$v_{2a} = v_{1a} + \frac{m_1 d^2}{I_2}(v_{1a} - v_{1b}) + CoR_{2b}\{C\} - \omega_{2b} d$$

Equation (7) in the previous section is

$$v_{1a} = \left\{ v_{1b} - \frac{(v_{1b} - v_{2b})m_2 I_2(1 + CoR_{2b}) - \omega_{2b} m_2 d I_2(1 + CoR_{2b})}{K} \right\}$$

Put this into both places for v_{1a} in the v_{2a} equation above.

$$\begin{aligned} v_{2a} = & \left\{ v_{1b} - \frac{(v_{1b} - v_{2b})m_2 I_2(1 + CoR_{2b}) - \omega_{2b} m_2 d I_2(1 + CoR_{2b})}{K} \right\} \\ & + \frac{m_1 d^2}{I_2} \left(\left\{ v_{1b} - \frac{(v_{1b} - v_{2b})m_2 I_2(1 + CoR_{2b}) - \omega_{2b} m_2 d I_2(1 + CoR_{2b})}{K} \right\} - v_{1b} \right) \\ & + CoR_{2b} C - \omega_{2b} d \end{aligned}$$

Now multiply by K

$$\begin{aligned} v_{2a}K = & v_{1b}K - (v_{1b} - v_{2b}) m_2 I_2 (1 + CoR_{2b}) + \omega_{2b} m_2 d I_2 (1 + CoR_{2b}) \\ & + \frac{m_1 d^2}{I_2} \left[v_{1b}K - (v_{1b} - v_{2b}) m_2 I_2 (1 + CoR_{2b}) + \omega_{2b} m_2 d I_2 (1 + CoR_{2b}) - v_{1b}K \right] \\ & + CoR_{2b}\, CK - \omega_{2b} dK \end{aligned}$$

Cancel the terms in color

$$\begin{aligned} v_{2a}K = & v_{1b}K - (v_{1b} - v_{2b}) m_2 I_2 (1 + CoR_{2b}) + \omega_{2b} m_2 d I_2 (1 + CoR_{2b}) \\ & - (v_{1b} - v_{2b}) m_1 m_2 d^2 (1 + CoR_{2b}) + \omega_{2b} m_1 m_2 d^3 (1 + CoR_{2b}) \\ & + CCoR_{2b}\, K - \omega_{2b} dK \end{aligned}$$

Let us break up the $(v_{1b} - v_{2b})$ terms and substitute $C = v_{1b} - v_{2b} - d\omega_{2b}$.

$$\begin{aligned} v_{2a}K = & v_{1b}K - v_{1b} m_2 I_2 (1 + CoR_{2b}) + v_{2b} m_2 I_2 (1 + CoR_{2b}) + \omega_{2b} m_2 d I_2 (1 + CoR_{2b}) \\ & - v_{1b} m_1 m_2 d^2 (1 + CoR_{2b}) + v_{2b} m_1 m_2 d^2 (1 + CoR_{2b}) + \omega_{2b} m_1 m_2 d^3 (1 + CoR_{2b}) \\ & + v_{1b} CoR_{2b}\, K - 2\mathrm{b}v CoR_{2b}\, K - \omega_{2b} dK(1 + CoR) \end{aligned}$$

Rearrange

$$\begin{aligned} v_{2a}K = & v_{1b}K - v_{1b} m_2 I_2 (1 + CoR_{2b}) - v_{1b} m_1 m_2 d^2 (1 + CoR_{2b}) + v_{1b} CoR_{2b}\, K \\ & + v_{2b} m_2 I_2 (1 + CoR_{2b}) + v_{2b} m_1 m_2 d^2 (1 + CoR_{2b}) - v_{2b} CoR_{2b}\, K \\ & + \omega_{2b} m_2 d I_2 (1 + CoR_{2b}) + \omega_{2b} m_1 m_2 d^3 (1 + CoR_{2b}) - \omega_{2b} dK(1 + CoR) \end{aligned}$$

Now let us break up the $(1 + CoR_{2b})$ terms.

$$\begin{aligned} v_{2a}K = & v_{1b}K - v_{1b} m_2 I_2 - v_{1b} m_2 I_2 CoR_{2b} - v_{1b} m_1 m_2 d^2 - v_{1b} m_1 m_2 d^2 CoR_{2b} + v_{1b} CoR_{2b}\, K \\ & + v_{2b} m_2 I_2 + v_{2b} m_2 I CoR_{2b} + v_{2b} m_1 m_2 d^2 + v_{2b} m_1 m_2 d^2 CoR_{2b} - v_{2b} CoR_{2b} K \\ & + \omega_{2b} m_2 d I_2 + \omega_{2b} m_2 d I_2 CoR_{2b} + \omega_{2b} m_1 m_2 d^3 + \omega_{2b} m_1 m_2 d^3 CoR_{2b} - \omega_{2b} dK - \omega_{2b} dK CoR_{2b} \end{aligned}$$

Are any of these terms the same? No. OK, now let's substitute

$K = (m_1 I_2 + m_2 I_2 + m_1 m_2 d^2)$ and hope for cancellations.

$$\begin{aligned}
v_{2a}K = {} & v_{1b}\left(m_1 I_2 + m_2 I_2 + m_1 m_2 d^2\right) - v_{1b} m_2 I_2 - v_{1b} m_2 I_2 CoR_{2b} \\
& - v_{1b} m_1 m_2 d^2 - v_{1b} m_1 m_2 d^2 CoR_{2b} + v_{1b} CoR_{2b}\left(m_1 I_2 + m_2 I_2 + m_1 m_2 d^2\right) \\
& + v_{2b} m_2 I_2 + v_{2b} m_2 I_2 CoR_{2b} + v_{2b} m_1 m_2 d^2 + v_{2b} m_1 m_2 d^2 CoR_{2b} \\
& - v_{2b} CoR_{2b}\left(m_1 I_2 + m_2 I_2 + m_1 m_2 d^2\right) \\
& + \omega_{2b} m_2 d I_2 + \omega_{2b} m_2 d I_2 CoR_{2b} + \omega_{2b} m_1 m_2 d^3 + \omega_{2b} m_1 m_2 d^3 CoR_{2b} \\
& - \omega_{2b}\left(m_1 d I_2 + m_2 d I_2 + m_1 m_2 d^3\right) - \omega_{2b}\left(m_1 d I_2 + m_2 d I_2 + m_1 m_2 d^3\right) CoR_{2b}
\end{aligned}$$

The terms in color cancel, leaving

$$v_{2a}K = v_{1b} m_1 I_2 (1 + CoR_{2b}) + v_{2b}(-m_1 I_2 CoR_{2b} + m_2 I_2 + m_1 m_2 d^2) - \omega_{2b} m_1 d I_2 (1 + CoR_{2b})$$

Continuing

$$v_{2a}K = +v_{1b} m_1 I_2 (1 + CoR_{2b}) + v_{2b}\left(-m_1 I_2 CoR_{2b} + m_2 I_2 + m_1 m_2 d^2\right) - \omega_{2b} m_1 d I_2 (1 + CoR_{2b})$$

distribute the second term and add $- v_{2b} m_1 I_2 + v_{2b} m_1 I_2$

$$\begin{aligned}
v_{2a}K &= +v_{1b} m_1 I_2 (1 + CoR_{2b}) - v_{2b} m_1 I_2 CoR_{2b} - v_{2b} m_1 I_2 + v_{2b} m_1 I_2 + v_{2b} m_2 I_2 + v_{2b} m_1 m_2 d^2 - \omega_{2b} m_1 d I_2 (1 + CoR_{2b}) \\
&= \left(v_{1b} - v_{2b}\right) m_1 I_2 (1 + CoR_{2b}) + v_{2b} m_1 I_2 + v_{2b} m_2 I_2 + v_{2b} m_1 m_2 d^2 - \omega_{2b} m_1 d I_2 (1 + CoR_{2b}) \\
&= (v_{1b} - v_{2b}) m_1 I_2 (1 + CoR_{2b}) + v_{2b} K - \omega_{2b} m_1 d I_2 (1 + CoR_{2b}) \\
&= v_{2b} K + (v_{1b} - v_{2b}) m_1 I_2 (1 + CoR_{2b}) - \omega_{2b} m_1 d I_2 (1 + CoR_{2b})
\end{aligned}$$

Finally divide by K

$$\boxed{v_{2a} = v_{2b} + \frac{(v_{1b} - v_{2b}) m_1 I_2 (1 + CoR_{2b}) - \omega_{2b} m_1 d I_2 (1 + CoR_{2b})}{(m_1 I_2 + m_2 I_2 + m_1 m_2 d^2)}}$$

This equation was derived from Eqs. (3)–(5) and (7). We can change this into our normal form by first combining the two terms over one common denominator.

$$\begin{aligned}
v_{2a} &= v_{2b}\frac{(m_1 I_2 + m_2 I_2 + m_1 m_2 d^2)}{(m_1 I_2 + m_2 I_2 + m_1 m_2 d^2)} + \frac{(v_{1b} - v_{2b}) m_1 I_2 (1 + CoR_{2b}) - \omega_{2b} m_1 d I_2 (1 + CoR_{2b})}{(m_1 I_2 + m_2 I_2 + m_1 m_2 d^2)} \\
&= \frac{v_{2b}(m_1 I_2 + m_2 I_2 + m_1 m_2 d^2) + (v_{1b} - v_{2b}) m_1 I_2 (1 + CoR_{2b}) - \omega_{2b} m_1 d I_2 (1 + CoR_{2b})}{(m_1 I_2 + m_2 I_2 + m_1 m_2 d^2)}
\end{aligned}$$

and then simplifying

$$\boxed{v_{2a} = \frac{v_{2b}(-m_1 I_2 CoR_{2b} + m_2 I_2 + m_1 m_2 d^2) + v_{1b} m_1 I_2 (1 + CoR_{2b}) - \omega_{2b} m_1 d I_2 (1 + CoR_{2b})}{(m_1 I_2 + m_2 I_2 + m_1 m_2 d^2)}}$$

or we can write this more compactly as

$$v_{2a}K = v_{2b}\left(-m_1I_2CoR_{2b} + m_2I_2 + m_1m_2d^2\right) + v_{1b}m_1I_2(1 + CoR_{2b}) - \omega_{2b}m_1dI_2(1 + CoR_{2b})$$

Finding the Bat Angular Velocity After the Collision

Now we want to find ω_{2a} (the angular velocity of the bat after the collision) in terms of the input parameters. We know that ω_{2b} is about zero, but for generality, we will leave it in for now.

This is v_{1a} from the canonical form of Eq. (7).

$$v_{1a} = \left\{ v_{1b} - \frac{(v_{1b} - v_{2b})m_2I_2(1 + CoR_{2b}) - \omega_{2b}m_2dI_2(1 + CoR_{2b})}{m_1I_2 + m_2I_2 + m_1m_2d^2} \right\}$$

From Eq. (5) solve for ω_{2a}

$$\omega_{2a} = \omega_{2b} - \frac{m_1d}{I_2}(v_{1a} - v_{1b})$$

Substitute v_{1a} into this equation for ω_{2a}

$$\omega_{2a} = \omega_{2b} - \frac{m_1d}{I_2}\left\{ v_{1b} - \frac{(v_{1b} - v_{2b})m_2I_2(1 + CoR_{2b}) - m_2d\omega_{2b}I_2(1 + CoR_{2b})}{m_1I_2 + m_2I_2 + m_1m_2d^2} \right\} + \frac{m_1d}{I_2}v_{1b}$$

Finally

$$\boxed{\omega_{2a} = \omega_{2b} + \frac{(v_{1b} - v_{2b})m_1m_2d(1 + CoR) - \omega_{2b}m_1m_2d^2(1 + CoR_{2b})}{m_1I_2 + m_2I_2 + m_1m_2d^2}}$$

This equation was derived from Eqs. (5) and (7). We can change this into our normal form by first combining the two terms over one common denominator.

$$\omega_{2a} = \omega_{2b}\frac{m_1I_2 + m_2I_2 + m_1m_2d^2}{m_1I_2 + m_2I_2 + m_1m_2d^2} + \frac{(v_{1b} - v_{2b})m_1m_2d(1 + CoR) - m_1m_2d^2\omega_{2b}(1 + CoR_{2b})}{m_1I_2 + m_2I_2 + m_1m_2d^2}$$

$$= \frac{\omega_{2b}\left(m_1I_2 + m_2I_2 + m_1m_2d^2\right) + (v_{1b} - v_{2b})m_1m_2d(1 + CoR) - m_1m_2d^2\omega_{2b}(1 + CoR_{2b})}{m_1I_2 + m_2I_2 + m_1m_2d^2}$$

Cancel duplicate terms and we get the normal form

$$\boxed{\omega_{2a} = \frac{\omega_{2b}(m_1 I_2 + m_2 I_2 - m_1 m_2 d^2 CoR_{2b}) + (v_{1b} - v_{2b}) m_1 m_2 d (1 + CoR_{2b})}{m_1 I_2 + m_2 I_2 + m_1 m_2 d^2}}$$

Three Output Equations in Three Formats

We will now summarize by giving equations for $v_{\text{ball-after}}$, $v_{\text{bat-cm-after}}$ and $\omega_{\text{bat-after}}$ in three formats. First normal form

$$v_{1a} = \frac{v_{1b}(m_1 I_2 - m_2 I_2\ CoR_{2b} + m_1 m_2 d^2) + v_{2b} m_2 I_2 (1 + CoR_{2b}) + \omega_{2b} m_2 d I_2 (1 + CoR_{2b})}{m_1 I_2 + m_2 I_2 + m_1 m_2 d^2}$$

$$v_{2a} = \frac{v_{2b}(-m_1 I_2 CoR_{2b} + m_2 I_2 + m_1 m_2 d^2) + v_{1b} m_1 I_2 (1 + CoR_{2b}) - \omega_{2b} d m_1 I_2 (1 + CoR_{2b})}{(m_1 I_2 + m_2 I_2 + m_1 m_2 d^2)}$$

$$\omega_{2a} = \frac{\omega_{2b}(m_1 I_2 + m_2 I_2 - m_1 m_2 d^2 CoR_{2b}) + (v_{1b} - v_{2b}) m_1 m_2 d (1 + CoR_{2b})}{m_1 I_2 + m_2 I_2 + m_1 m_2 d^2}$$

Second canonical form

$$v_{1a} = v_{1b} - \frac{(v_{1b} - v_{2b}) m_2 I_2 (1 + CoR_{2b}) - \omega_{2b} m_2 d I_2 (1 + CoR_{2b})}{m_1 I_2 + m_2 I_2 + m_1 m_2 d^2}$$

$$v_{2a} = v_{2b} + \frac{(v_{1b} - v_{2b}) m_1 I_2 (1 + CoR_{2b}) - \omega_{2b} d m_1 I_2 (1 + CoR_{2b})}{(m_1 I_2 + m_2 I_2 + m_1 m_2 d^2)}$$

$$\omega_{2a} = \omega_{2b} + \frac{(v_{1b} - v_{2b}) m_1 m_2 d (1 + CoR_{2b}) - \omega_{2b} m_1 m_2 d^2 (1 + CoR_{2b})}{m_1 I_2 + m_2 I_2 + m_1 m_2 d^2}$$

Now let

$$A = \left\{ \frac{[(v_{1b} - v_{2b}) - \omega_{2b} d](1 + CoR_{2b})}{m_1 I_2 + m_2 I_2 + m_1 m_2 d^2} \right\}$$

and we get our reduced canonical form:

$$v_{1a} = v_{1b} - A m_2 I_2$$
$$v_{2a} = v_{2b} + A m_1 I_2$$
$$\omega_{2a} = \omega_{2b} + A m_1 m_2 d$$

Please note that A is not a constant. It depends on the inputs v_{1b}, v_{2b} and ω_{2b}. Also, notice that ω_{ball} does not appear in these output equations. It will appear later. We now want to add the equation for conservation of energy, Eq. (2).

Adding Conservation of Energy and Finding KE_{lost}

This approach, of adding conservation of energy to the bat-ball collision equations, is unique in the science of baseball literature. For a head-on collision at the center of mass of the bat, we had that

$$KE_{\text{lost}-\text{config}-\text{cm}} = \frac{\bar{m}}{2}\left(v_{\text{bat}-\text{cm}-\text{before}} - v_{\text{ball}-\text{before}}\right)^2\left(1 - CoR_{1\text{h}}^2\right) \tag{9}$$

However, for a collision at the sweet spot this equation for kinetic energy lost is not valid, because we now also have angular kinetic energy in the rotation of the bat. There are no springs in the system and the bat swing is level, therefore there is no change in potential energy. Before the collision, there is kinetic energy in the bat created by rotation of the batter's body and arms plus the translational kinetic energy of the ball. In Fig. 2, the sweet spot is the distance $d_{\text{cm}-\text{ss}}$ from the center of mass.

$$KE_{\text{before}} = \frac{1}{2}m_{\text{ball}}v_{\text{ball}-\text{before}}^2 + \frac{1}{2}m_{\text{bat}}v_{\text{bat}-\text{cm}-\text{before}}^2 + \frac{1}{2}I_{\text{ball}}\omega_{\text{ball}-\text{before}}^2 + \frac{1}{2}I_{\text{bat}}\omega_{\text{bat}-\text{before}}^2$$

As always, ω means rotation about the center of mass of the object. The collision will make the bat spin about its center of mass. If the collision is at the Center of Percussion for the pivot point, it will produce a rotation about the center of mass, but no translation.

$$KE_{\text{after}} = \frac{1}{2}m_{\text{ball}}v_{\text{ball}-\text{after}}^2 + \frac{1}{2}m_{\text{bat}}v_{\text{bat}-\text{cm}-\text{after}}^2 + \frac{1}{2}I_{\text{ball}}\omega_{\text{ball}-\text{after}}^2 + \frac{1}{2}I_{\text{bat}}\omega_{\text{bat}-\text{after}}^2$$

We now add kinetic energy of the rotating curveball. We will add two terms with ball spin $\left(\frac{1}{2}I_{\text{ball}}\omega_{\text{ball}-\text{before}}^2\right.$ and $\left.\frac{1}{2}I_{\text{ball}}\omega_{\text{ball}-\text{after}}^2\right)$ to the Conservation of Energy equation, to create

$$\frac{1}{2}m_{\text{ball}}v_{\text{ball}-\text{before}}^2 + \frac{1}{2}m_{\text{bat}}v_{\text{bat}-\text{cm}-\text{before}}^2 + \frac{1}{2}I_{\text{ball}}\omega_{\text{ball}-\text{before}}^2 + \frac{1}{2}I_{\text{bat}}\omega_{\text{bat}-\text{before}}^2$$
$$= \frac{1}{2}m_{\text{ball}}v_{\text{ball}-\text{after}}^2 + \frac{1}{2}m_{\text{bat}}v_{\text{bat}-\text{cm}-\text{after}}^2 + \frac{1}{2}I_{\text{ball}}\omega_{\text{ball}-\text{after}}^2 + \frac{1}{2}I_{\text{bat}}\omega_{\text{bat}-\text{after}}^2 + KE_{lost}$$
$$KE_{\text{before}} = KE_{\text{after}} + KE_{\text{lost}}$$

The KE_{before} and the $KE_{\text{after}} >$ are easy to find. It is the KE_{lost} that is hard to find.

In the next section on "Adding Conservation of Angular Momentum," we will prove that for head-on collisions without friction $\omega_{ball-before} = \omega_{ball-after}$. Therefore, the ball spin terms in these conservation of energy equations cancel resulting in

$$0 = m_1v_{1\text{b}}^2 + m_2v_{2\text{b}}^2 + I_2\omega_{2\text{b}}^2 - m_1v_{1\text{a}}^2 - m_2v_{2\text{a}}^2 - I_2\omega_{2\text{a}}^2 - 2KE_{\text{lost}}$$

From before, we have

$$A = \left[\frac{(v_{1b} - v_{2b})(1 + CoR_{2b}) - d\omega_{2b}}{m_1 I_2 + m_2 I_2 + m_1 m_2 d^2}\right]$$
$$v_{1a} = v_{1b} - Am_2 I_2$$
$$v_{1a} = v_{1b} - Am_2 I_2$$
$$v_{2a} = v_{2b} + Am_1 I_2$$
$$\omega_{2a} = \omega_{2b} + Am_1 m_2 d$$
$$\omega_{1a} = \omega_{1b}$$

Substituting v_{1a}, v_{2a} and ω_{2a} into the new conservation of energy equation yields

$$KE_{\text{lost}} = \frac{1}{2}\left\{\begin{array}{l} m_1 v_{1b}^2 + m_2 v_{2b}^2 + I_2\omega_{2b}^2 - m_1(v_{1b} - Am_2 I_2)^2 \\ - m_2(v_{2b} + Am_1 I_2)^2 - I_2(\omega_{2b} + Am_1 m_2 d)^2 \end{array}\right\}$$

Now we want to put this into the form that we had for Eq. (1). The following derivation is original. First, expand the squared terms.

$$2KE_{\text{lost}} = m_1 v_{1b}^2 + m_2 v_{2b}^2 + I_2\omega_{2b}^2 - m_1(v_{1b}^2 - 2v_{1b}Am_2 I_2 + A^2 m_2^2 I_2^2)$$
$$-m_2(v_{2b}^2 + 2v_{2b}Am_1 I_2 + A^2 m_1^2 I_2^2) - I_2\left(\omega_{2b}^2 + 2\omega_{2b}Am_1 m_2 d + A^2 m_1^2 m_2^2 d^2\right)$$

cancel terms in the same color and distribute the leading term

$$2KE_{\text{lost}} = 2v_{1b}Am_1 m_2 I_2 - A^2 m_1 m_2^2 I_2^2$$
$$-2v_{2b}Am_1 m_2 I_2 - A^2 m_1^2 m_2 I_2^2 - I_2\left(2\omega_{2b}Am_1 m_2 d + A^2 m_1^2 m_2^2 d^2\right)$$

Rearrange

$$2KE_{\text{lost}} = 2v_{1b}Am_1 m_2 I_2 - 2v_{2b}Am_1 m_2 I_2 - A^2 m_1^2 m_2 I_2^2 - A^2 m_1 m_2^2 I_2^2 - \left(2\omega_{2b}Am_1 m_2 d + A^2 m_1^2 m_2^2 d^2\right) I_2$$

factor

$$2KE_{\text{lost}} = Am_1 m_2 I_2\left[2(v_{1b} - v_{2b}) - A(m_1 I_2 + m_2 I_2 + m_1 m_2 d^2) - 2\omega_{2b} d\right]$$

Substitute A

$$2KE_{\text{lost}} = Am_1 m_2 I_2\left[2(v_{1b} - v_{2b}) - \left\{\frac{(v_{1b} - v_{2b})(1 + CoR_{2b}) - d\omega_{2b}}{m_1 I_2 + m_2 I_2 + m_1 m_2 d^2}\right\}(m_1 I_2 + m_2 I_2 + m_1 m_2 d^2) - 2\omega_{2b} d\right]$$
$$2KE_{\text{lost}} = Am_1 m_2 I_2[2(v_{1b} - v_{2b}) - (v_{1b} - v_{2b})(1 + CoR_{2b}) + d\omega_{2b} - 2\omega_{2b} d]$$

factor $(v_{1b} - v_{2b})$ out of the first two terms

$$2KE_{lost} = Am_1m_2I_2[(v_{1b} - v_{2b})(1 + CoR_{2b}) - d\omega_{2b}]$$

substitute A

$$2KE_{lost} = \left\{\frac{(v_{1b} - v_{2b})(1 + CoR_{2b}) - d\omega_{2b}}{m_1I_2 + m_2I_2 + m_1m_2d^2}\right\}m_1m_2I_2[(v_{1b} - v_{2b})(1 + CoR_{2b}) - d\omega_{2b}]$$

$$2KE_{lost} = \frac{m_1m_2I_2}{m_1I_2 + m_2I_2 + m_1m_2d^2}\{(v_{1b} - v_{2b})(1 + CoR_{2b}) - d\omega_{2b}\}[(v_{1b} - v_{2b})(1 + CoR_{2b}) - d\omega_{2b}]$$

After a little bit of algebra we get

$$2KE_{lost} = \frac{m_1m_2I_2}{m_1I_2 + m_2I_2 + m_1m_2d^2}\left[(v_{1b} - v_{2b})^2(1 - CoR_{2b}^2) - 2(v_{1b} - v_{2b})\omega_{2b}d + \omega_{2b}^2d^2\right]$$

$$\boxed{KE_{lost} = \frac{1}{2}\frac{m_1m_2I_2}{m_1I_2 + m_2I_2 + m_1m_2d^2}\left[(v_{1b} - v_{2b})^2(1 - CoR_{2b}^2) - 2(v_{1b} - v_{2b})\omega_{2b}d + \omega_{2b}^2d^2\right]}$$

This is a general result. It is original and unique. For a collision at the center of mass, d $= 0$. Therefore,

$$KE_{lost} = \frac{1}{2}\frac{m_1m_2}{m_1 + m_2}(v_{1b} - v_{2b})^2(1 - CoR^2)$$

When we substitute, $\bar{m} = \frac{m_1m_2}{m_1 + m_2}$ we get

$$KE_{lost} = \frac{\bar{m}}{2}(v_{1b} - v_{2b})^2(1 - CoR^2) \tag{10}$$

Which is the same as the following equation that has been derived in the literature.

$$KE_{lost} = \frac{\bar{m}}{2}(v_{bat-cm-before} - v_{ball-before})^2(1 - CoR^2)$$

Adding Conservation of Angular Momentum

In this section, we will prove that for a head-on collision without considering friction for a pitch of any spin there will be no change in the spin of the ball. To do this we will use the law of conservation of angular momentum about the center of mass of the bat. When the ball contacts the bat, as shown in Fig. 3, the ball has linear momentum of $m_{ball}v_{ball-before}$. However, the ball does not know if it is

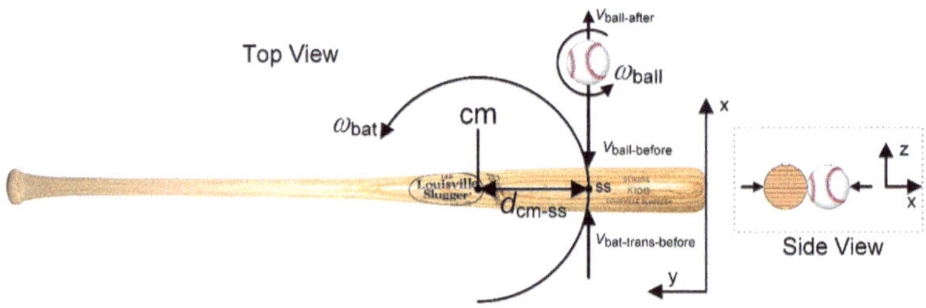

Fig. 3 This figure shows $v_{\text{ball}-\text{before}}$, $v_{\text{ball}-\text{before}}$, ω_{ball}, $d_{\text{cm}-\text{ss}}$ and ω_{bat}, which are used in the conservation of angular momentum equation

translating or if it is tied on a string and rotating about the center of mass of the bat. Following conventional physics, we will model the ball as rotating about the bat's center of mass at a distance $d = d_{\text{cm}-\text{ss}}$. Therefore, the ball has an initial angular momentum of $m_{\text{ball}} d_{\text{cm}-\text{ss}} v_{\text{ball}-\text{before}}$ about the bat's center of mass. In addition, it is possible to throw a curveball so that it spins about the vertical, z-axis, as also shown in Fig. 3. We call this a purely horizontal curveball (although it will still drop due to gravity, more than it will curve horizontally). The curveball will have angular momentum of $I_{\text{ball}} \omega_{\text{ball}-\text{before}}$ where $I_{\text{ball}} = 0.4 m_{\text{ball}} r_{\text{ball}}^2$ about an axis parallel to the z-axis. However, this is its momentum about *its* center of mass and we want the momentum about the center of mass of the *bat*. Therefore, we use the parallel axis theorem producing $(I_{\text{ball}} + m_{ball} d^2)\omega_{\text{ball}-\text{before}}$.

The bat has an initial angular momentum of $I_{\text{bat}} \omega_{\text{bat}-\text{before}}$. It also has an angular momentum about the bat's center of mass of due to the bat translation momentum $m_{\text{bat}} d v_{\text{bat}-\text{before}}$, however, in this case $d = 0$ because the center of mass of the bat is passing through its center of mass. L is the symbol used for angular momentum. I guess all the cool letters (like *F, m, a, v, I, ω, d*, etc.) were already taken, so they were stuck with the blah symbol *L*. Therefore, the initial angular momentum about the center of mass of the bat is

$$L_{\text{initial}} = m_1 v_{1\text{b}} d + \left(I_1 + m_1 d^2\right)\omega_{1b} + I_2 \omega_{2\text{b}}$$

All of these momenta are positive, pointing out of the page.

For the final angular momentum, we will treat the ball, as before, as an object rotating around the axis of the center of mass of the bat with angular momentum, $m_{\text{ball}} v_{\text{ball}-\text{after}} d_{\text{cm}-\text{ss}}$. Now we could treat the bat as a long slender rod with a moment of inertia of $m_{\text{bat}} d_{\text{bat}}^2/12$, where d_{bat} is the bat length. However, this is only an approximation and we have actual experimental data for the bat moment of inertia. Therefore, the bat angular momentum is $I_{\text{bat}} \omega_{\text{bat}-\text{after}}$. Thus, our final angular momentum about the center of mass of the bat is

$$L_{\text{final}} = m_1 v_{1\text{a}} d + \left(I_1 + m_1 d^2\right)\omega_{1\text{a}} + I_2 \omega_{2\text{a}}$$

The law of conservation of angular momentum states that the initial angular momentum about some axis equals the final angular momentum about that axis.

$$\boxed{\begin{array}{l} L_{\text{initial}} = L_{\text{final}} \\ m_1 v_{1b} d + (I_1 + m_1 d^2)\omega_{1b} + I_2\omega_{2b} = m_1 v_{1a} d + (I_1 + m_1 d^2)\omega_{1a} + I_2\omega_{2a} \end{array}}$$

Previously we used Eq. (5), Newton's second principle and solved for ω_{2a}.

$$dm_1(v_{1a} - v_{1b}) = -I_2(\omega_{2a} - \omega_{2b}) \tag{11}$$

$$\omega_{2a} = \omega_{2b} - \frac{dm_1}{I_2}(v_{1a} - v_{1b})$$

So let us substitute this into our conservation of angular momentum equation above.

$$m_1 v_{1b} d + I_1\omega_{1b} + m_1\omega_{1b} d^2 + I_2\omega_{2b} = m_1 v_{1a} d + I_1\omega_{1a} + m_1\omega_{1a} d^2 + I_2\left\{\omega_{2b} + \frac{dm_1}{I_2}(v_{1b} - v_{1a})\right\}$$

We want to solve this for ω_{1a}

$$-I_1\omega_{1a} - m_1\omega_{1a} d^2 = -m_1 v_{1b} d - I_1\omega_{1b} - I_2\omega_{2b} - m_1\omega_{1b} d^2 + m_1 v_{1a} d + I_2\omega_{2b} + dm_1(v_{1b} - v_{1a})$$

Cancel the terms in color and rearrange

$$\omega_{1a}(I_1 + m_1 d^2) = \omega_{1b}(I_1 + m_1 d^2)$$

$$\boxed{\omega_{1a} = \omega_{1b}}$$

We have now proven that for a pitch with any spin about the z-axis, the spins before and after are the same. What about a pitch that has spin about the z-axis and also about the y-axis, like most pitches? The collision will not change ball rotation. As shown above, it will not change the spin about the z-axis. We could write another set of equations for angular momentum about the y-axis. However, the bat has no angular momentum about the y-axis, so there is nothing to affect the ball spin about the y-axis. In conclusion, a head-on collision between a bat and a ball will not change the spin on the ball. Some papers have shown a relationship between ball spin before and ball spin after, but they were using oblique collisions (Nathan et al. 2012; Kensrud et al. 2016) (Table 3).

The numbers in the Excel simulation satisfy the following checks: (1) Conservation of linear momentum, (2) Conservation of angular momentum, (3) Coefficient of restitution, (4) Newton's second principle, an impulse changes momentum, (5) Conservation of energy and (6) Kinetic energy lost. Table 4 shows the kinetic energies for the same simulation.

The first purpose of this paper is to model bat-ball collisions using only Newton's principles and the conservation equations. We did it. Our equations are complete, consistent and correct.

Table 3 Simulation values for bat-ball collisions at the sweet spot

	SI units (m/s, rad/s, or J)	Baseball units
Inputs		
$v_{\text{ball}-\text{before}}$	−37	−83 mph
$\omega_{\text{ball}-\text{before}}$	209	2000 rpm
$v_{\text{bat}-\text{cm}-\text{before}}$	26	58 mph
$\omega_{\text{bat}-\text{before}}$	0.1	1 rpm
CoR_{2b}	0.55	
Outputs		
$v_{\text{ball}-\text{after}}$	43	97 mph
$\omega_{\text{ball}-\text{after}}$	$=\omega_{\text{ball}-\text{before}}$	
$v_{\text{bat}-\text{cm}-\text{after}}$	13	29 mph
$\omega_{\text{bat}-\text{after}}$	−32	−310 rpm
KE_{lost}	165	

Table 4 Kinetic energies

KE ball linear velocity before	99.3
KE bat linear velocity before	304.2
KE ball angular velocity before	1.7
KE bat angular velocity before	0.0
KE before total	405.2
KE ball linear velocity after	136.1
KE bat linear velocity after	77.2
KE ball angular velocity after	1.7
KE bat angular velocity after	25.2
KE after	240.2
KE loss	165.0
KE after + KE loss	405.2

3.2.5 Analytic Sensitivity Analysis

The second purpose of this paper is to show how the batter can select and tailor an optimal baseball or softball bat. From the viewpoint of the batter, the only model output that is important is the speed of the batted-ball. Therefore, we will now find the sensitivity of the batted-ball speed, $v_{\text{ball}-\text{after}}$, with respect to the system parameters. The eight system parameters are $v_{\text{ball}-\text{before}}$, m_{ball}, I_{bat}, m_{bat}, CoR_{2b}, $d_{\text{cm}-\text{ss}}$, $v_{\text{bat}-\text{cm}-\text{before}}$ and $\omega_{\text{bat}-\text{before}}$. For baseball and softball, the batted – ball speed, v_{1a}, is the most important output. The larger it is the more likely the batter will get on base safely (Baldwin and Bahill 2004). Therefore, let us start with v_{1a} from Eq. (7).

$$v_{1a} = v_{1b} - \frac{(v_{1b} - v_{2b})m_2 I_2 (1 + CoR_{2b}) - \omega_{2b} m_2 d I_2 (1 + CoR_{2b})}{m_1 I_2 + m_2 I_2 + m_1 m_2 d^2}$$

In order to perform an analytic sensitivity analysis we first need the partial derivatives of v_{1a} with respect to the eight parameters. These partial derivatives are often called the absolute sensitivity functions. Move the minus sign and simplify the numerator.

$$v_{1a} = v_{1b} + \frac{(1 + CoR_{2b})[(-v_{1b} + v_{2b})m_2 I_2 + \omega_{2b} m_2 d I_2]}{(m_1 I_2 + m_2 I_2 + m_1 m_2 d^2)}$$

Let $K = (m_1 I_2 + m_2 I_2 + m_1 m_2 d^2)$

$$H = (1 + CoR_{2b})[(-v_{1b} + v_{2b})m_2 I_2 + \omega_{2b} m_2 d I_2]$$

$$v_{1a} = v_{1b} + \frac{H}{K}$$

$$\frac{\partial v_{1a}}{\partial v_{1b}} = 1 - \frac{m_2 I_2 (1 + CoR_{2b})}{K} \quad \text{unitless}$$

$$\frac{\partial v_{1a}}{\partial \omega_{2b}} = \frac{m_2 d I_2 (1 + CoR_{2b})}{K} \quad \text{m}$$

$$\frac{\partial v_{1a}}{\partial CoR_{2b}} = \frac{(-v_{1b} + v_{2b})m_2 I_2 + \omega_{2b} m_2 d I_2}{K} \quad \text{m/s}$$

$$\frac{\partial v_{1a}}{\partial v_{2b}} = \frac{m_2 I_2 (1 + CoR_{2b})}{K} \quad \text{unitless}$$

Alternatively, we could start with

$$v_{1a} = v_{1b} - A m_2 I_2$$

$$A = \left\{ \frac{[(v_{1b} - v_{2b}) - \omega_{2b} d](1 + CoR_{2b})}{m_1 I_2 + m_2 I_2 + m_1 m_2 d^2} \right\}$$

$$v_{1a} = v_{1b} - \left\{ \frac{[(v_{1b} - v_{2b}) - \omega_{2b} d](1 + CoR_{2b})}{m_1 I_2 + m_2 I_2 + m_1 m_2 d^2} \right\} m_2 I_2$$

$$\frac{\partial v_{1a}}{\partial v_{2b}} = \frac{m_2 I_2 (1 + CoR_{2b})}{K}$$

This gives the same result. For the following partial derivatives, we need the derivative of a quotient.

$$\left[\frac{f(x)}{g(x)}\right]' = \frac{g(x) f' f'(x) - f(x) g'(x)}{[g(x)]^2}$$

$$\frac{\partial v_{1a}}{\partial d} = \frac{(1 + CoR_{2b}) K m_2 \omega_{2b} I_2 - 2 H m_1 m_2 d}{K^2} \quad 1/\text{s}$$

$$\frac{\partial v_{1a}}{\partial m_2} = \frac{K(1+CoR_{2b})\{(-v_{1b}+v_{2b})I_2+\omega_{2b}dI_2\}-H(I_2+m_1d^2)}{K^2} \quad \text{m/kg s}$$

$$\frac{\partial v_{1a}}{\partial m_1} = -\frac{(I_2+m_2d^2)H}{K^2} \quad \text{m/kg s}$$

$$\frac{\partial v_{1a}}{\partial I_2} = \frac{K(1+CoR_{2b})[(-v_{1b}+v_{2b})m_2+\omega_{2b}m_2d]-H(m_1+m_2)}{K^2} \quad \text{1/kg m s}$$

$$\frac{\partial^2 v_{1a}}{\partial v_{2b}\partial m_2} = \frac{I_2(1+CoR_{2b})[K-m_2(I_2+m_1d^2)]}{K^2} \quad \text{1/kg}$$

In the above partial derivatives, units on the left and right sides of the equations are the same. This is a simple, but important accuracy check. We perform such a dimensional analysis on all of our equations.

We did not show the derivations of all of the second-order partial derivatives. We choose the interaction of the bat mass and the bat speed, above, because it was expected to be large based on principles of physiology. Additionally, the forthcoming discussion on optimizing the bat suggests an interaction between the bat mass and moment of inertia. Therefore, we will now derive one more interaction term, the interaction between bat mass and moment of inertia, I_{bat} and m_{bat}.

Given

$$\frac{\partial v_{1a}}{\partial m_2} = \frac{K(1+CoR_{2b})\{(-v_{1b}+v_{2b})I_2+\omega_{2b}dI_2\}-H(I_2+m_1d^2)}{K^2}$$

Find $\frac{\partial^2 v_{1a}}{\partial I_2 \partial m_2}$

We will be dealing with I_2, so let us isolate it. First replace K and H, $\frac{\partial v_{1a}}{\partial m_2}$ becomes

$$=\left[(m_1+m_2)I_2+m_1m_2d^2\right](1+CoR_{2b})(-v_{1b}+v_{2b}+\omega_{2b}d)I_2$$

$$-(1+CoR_{2b})(-v_{1b}+v_{2b}+\omega_{2b}d)m_2I_2(I_2+m_1d^2)$$

$$=(1+CoR_{2b})(-v_{1b}+v_{2b}+\omega_{2b}d)(m_1I_2^2+m_2I_2^2+m_1m_2d^2I_2-m_2I_2^2-m_1m_2d^2I_2)$$

Cancel the terms in color and consolodate the terms without I_2 by letting

$$E=(1+CoR_{2b})(-v_{1b}+v_{2b}+\omega_{2b}d)$$

The numerator $\frac{\partial v_{1a}}{\partial m_2}$ of becomes

$$= Em_1 I_2^2$$

Therefore,

$$\frac{\partial^2 v_{1a}}{\partial I_2 \partial m_2} = \frac{2Em_1 I_2 K^2 - Em_1 I_2^2 2K(m_1 + m_2)}{K^4}$$

$$= \frac{2Em_1 I_2 K[K - I_2(m_1 + m_2)]}{K^4}$$

substitute for the second K in the numerator

$$= \frac{2Em_1 I_2 K[\{m_1 I_2 + m_2 I_2 + m_1 m_2 d^2\} - I_2(m_1 + m_2)]}{K^4}$$

$$= \frac{2Em_1 I_2 K[m_1 I_2 + m_2 I_2 + m_1 m_2 d^2 - m_1 I_2 - m_2 I_2]}{K^4}$$

cancel the terms in color

$$= \frac{2Em_1^2 m_2 d^2 I_2 K}{K^4}$$

substitute E and K

$$\frac{\partial^2 v_{1a}}{\partial I_2 \partial m_2} = \frac{2\{(1 + CoR_{2b})(-v_{1b} + v_{2b} + \omega_{2b} d)\} m_1^2 m_2 d^2 I_2}{\{m_1 I_2 + m_2 I_2 + m_1 m_2 d^2\}^3} \quad 1/\text{kg}^2 \bullet \text{m} \bullet \text{s}$$

Now we want to form the *semirelative-sensitivity functions,* which are defined as

$$\tilde{S}_\alpha^F = \left.\frac{\partial F}{\partial \alpha}\right|_{\text{NOP}} \alpha_0$$

where NOP and the subscript 0 mean that all functions, inputs and parameters assume their nominal operating point values (Smith et al. 2008).

$$\tilde{S}_\alpha^F = \left.\frac{\partial F}{\partial \alpha}\right|_{\text{NOP}} \alpha_0$$

$$\tilde{S}_{v_{1b}}^{v_{1a}} = 1 - \left.\frac{m_2 I_2 (1 + CoR_{2b})}{K}\right|_{\text{NOP}} v_{1b_0}$$

$$\tilde{S}_{v_{2b}}^{v_{1a}} = \left.\frac{m_2 I_2 (1 + CoR_{2b})}{K}\right|_{\text{NOP}} v_{2b_0}$$

$$\tilde{S}_{\omega_{2b}}^{v_{1a}} = \left.\frac{m_2 d I_2 (1 + CoR_{2b})}{K}\right|_{\text{NOP}} \omega_{2b_0}$$

$$\tilde{S}_{CoR}^{v_{1a}} = \left.\frac{(-v_{1b} + v_{2b}) m_2 I_2 + \omega_{2b} m_2 d I_2}{K}\right|_{\text{NOP}} CoR_{2b_0}$$

Table 5 Typical values and sensitivities

Variable	Nominal values		Range of realistic values		$\tilde{S}_{\alpha}^{F} = \frac{\partial F}{\partial \alpha}\big\|_{NOP}\alpha_0$ semirelative sensitivity values
	SI units	Baseball units	SI units	Baseball units	
Inputs					
$v_{ball-before}$	−37 m/s	−83 mph	26.8–40.2 m/s	60–90 mph	10
$v_{bat-cm-before}$	26 m/s	58 mph	24.6–27.2 m/s	58 ± 10 mph	31
$\omega_{ball-before}$	209 rad/s	2000 rpm	188–230 rad/s	2000 ± 100 rpm	0
$\omega_{bat-before}$	0.1 rad/s	1 rpm	−0.1–0.1 rad/s	±2 rpm	0.02
Parameters					
CoR_{2b}	0.55		0.45–0.65	0.55 ± 0.1	28
d_{cm-ss}	0.134 m	5.3 in.	0.13–0.14 m	5.3 ± 2 in.	−7
m_{ball}	0.145 kg	5.125 oz	0.142–0.156 kg	5.125 ± 0.125 oz	−14
m_{bat}	0.905 kg	32 oz	0.709–0.964 kg	25–34 oz	10
I_{bat-cm}	0.048 kg m^2	2624 oz in^2	0.036–0.06 kg m^2	1968–3280 oz in^2	4
$v_{bat-before}$ interacting with m_{bat}					4
I_{bat} interacting with m_{bat}					1

$$\tilde{S}_{d}^{v_{1a}} = \frac{(1+CoR_{2b})K\omega_{2b}m_2 I_2 - 2Hm_1 m_2 d}{K^2}\bigg|_{NOP} d_0$$

$$\tilde{S}_{m_2}^{v_{1a}} = \frac{K(1+CoR_{2b})[(-v_{1b}+v_{2b})I_2 + d\omega_{2b}I_2] - H(I_2 + m_1 d^2)}{K^2}\bigg|_{NOP} m_{2_0}$$

$$\tilde{S}_{m_1}^{v_{1a}} = -\frac{(I_2 + m_2 d^2)H}{K^2}\bigg|_{NOP} m_{1_0}$$

$$\tilde{S}_{I_2}^{v_{1a}} = \frac{K(1+CoR_{2b})[(-v_{1b}+v_{2b})m_2 + m_2 d\omega_{2b}] - H(m_1 + m_2)}{K^2}\bigg|_{NOP} I_{2_0}$$

$$\tilde{S}_{v_{2b}-m_2}^{v_{1a}} = \frac{I_2(1+CoR_{2b})[K - m_2(I_2 + m_1 d^2)]}{K^2}\bigg|_{NOP} v_{2b_0} m_{2_0}$$

$$\tilde{S}_{I_2-m_2}^{v_{1a}} = \frac{2[(1+CoR_{2b})(-v_{1b}+v_{2b}+\omega_{2b}d)]m_1^2 m_2 d^2 I_2}{[m_1 I_2 + m_2 I_2 + m_1 m_2 d^2]^3}\bigg|_{NOP} I_{2_0} m_{2_0}$$

Table 5 gives the nominal input and parameter values, along with a range of physically realistic values for collegiate and professional batters and the semirelative sensitivity values. The bigger the sensitivity is, the more important the variable is for maximizing batted-ball speed.

The right column of Table 5 shows that the most important variable, in terms of maximizing batted-ball speed, is the speed of the bat before the collision. This is certainly no surprise. The second most important variable is the coefficient of restitution, CoR_{2b}. The least important variables are the angular velocities, $\omega_{ball-before}$ and $\omega_{bat-before}$. The sensitivities to distance between the center of mass and the sweet spot of the bat, d_{cm-ss}, and the mass of the ball, m_{ball}, are negative, which merely means that as they increase the batted-ball speed decreases. Cross (2011) wrote that in his model the most sensitive variables were also the bat speed followed by the *CoR*. His sensitivity to the mass of the ball was also negative. The bottom two rows of Table 5 show that the interaction terms are small, which means that the model is well behaved. For example, the interaction of the mass of the bat with the bat speed is smaller than either the influence of the mass of the bat by itself or the bat speed by itself. The interaction of bat mass and moment of inertia is surprisingly small.

3.2.6 Optimizing with Commercial Software

We applied *What's Best!*, a subset of the LINGO solvers, to our model. We constrained each variable to stay within physically realistic limits under natural conditions. Such values are shown in Table 5. We have previously gotten good results using this technique when doing empirical sensitivity analyses (Bahill et al. 2009). Then we asked the optimizer to give us the set of values that would maximize batted-ball speed. The optimizer applied a nonlinear optimization program. The results were the same as in Table 5! That is, for variables with positive sensitivities, the optimizer choose the maximum values. For variables with negative sensitivities, the optimizer choose the minimum values. Using all of the optimal values at the same time increased the batted-ball speed from 43 to 56 m/s (96–125 mph). Using this optimal set of values only changed the sensitivities slightly.

1. The numerical sensitivity values mostly increased. This is a direct result of the definition of the semirelative sensitivity function where the partial derivative is multiplied by the parameter value. If parameter values increase, then the sensitivities increase.
2. However, and most importantly, the rank order stayed the same except that the batted-ball speed became more sensitive to $v_{ball-before}$ than to m_{bat}. In the optimal set, both of these sensitivities increased, but because the value of $v_{ball-before}$ changed from 37 to 40 m/s whereas the value of m_{bat} only changed from 0.90 to 0.96 kg, the change in the sensitivity to $v_{ball-before}$ was bigger.

This all means that the sensitivity analysis is robust. Its results remain basically the same after big changes in the variables.

We then tried a different optimization technique. Instead of constraining each variable to stay within realistic physical limits, we allowed the optimizer to change each variable by at most ±10% and then give us the set of values that maximizes batted-ball speed. The numerical values changed but the rank order stayed the same, except for $v_{ball-before}$ and m_{bat} just as it did with the realistic values technique.

Both empirical sensitivity analyses and optimization can constrain each variable to stay within specified realistic physical limits or change each variable by a certain percentage. Both techniques gave the same results. However, we prefer the former technique (Bahill et al. 2009).

We found an interesting relationship between the sensitivity analyses and optimization: they gave the same results! For variables with positive sensitivities, the optimizer chooses the maximum values. For variables with negative sensitivities, the optimizer chooses the minimum values. But of course, this finding is not original. Sensitivity analyses are commonly used in optimization studies (Choi and Kim 2005). These studies typically apply sensitivity analysis after optimization. They try to find values or limits for the objective function or the right-hand sides of the constraints that would change the decisions. However, in our study, we applied optimization after the sensitivity analysis and we had only one variable in our objective function. Therefore, our problem was much simpler than sensitivity analyses in the optimization literature.

3.2.7 Optimizing the Bat

The second purpose of this paper is to help the batter acquire an optimal baseball or softball bat. Therefore, we ask, How can the batter use these sensitivity and optimization results to select or customize an optimal bat? First, it is no surprise that bat speed, $v_{bat-cm-before}$, is the most important variable in Table 5. Its effect is shown

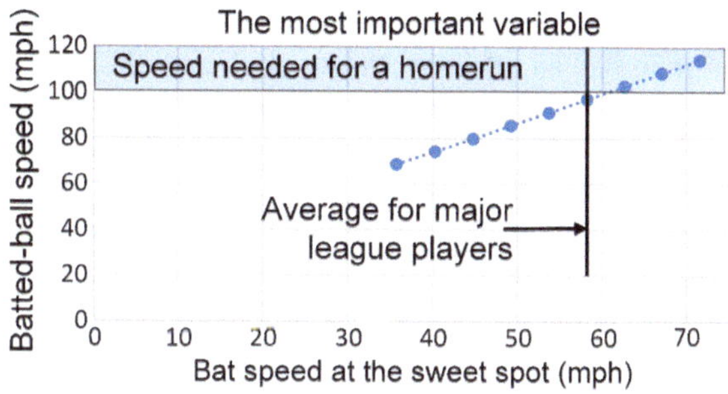

Fig. 4 The most important variable in our model is the bat speed at the sweet spot before the collision. For this figure, first we computed the batted-ball speed with $v_{\text{ball-after}} = v_{\text{ball-before}} - Am_{\text{bat}}I_{\text{bat}}$ and then we plotted the batted-ball speed as a function of the bat sweet spot speed before the collision. Remember that A is not a constant, it depends on the velocity of the ball and bat before the collision and on the angular velocity of the bat before the collision

in Fig. 4, where the slope of the line is the absolute sensitivity. For decades, Little League coaches have taught their boys to practice and gain strength so that they could increase their bat speeds. They also said that it is very important to reduce the variability in the bat swings: Every swing should be the same. "Don't try to kill the ball." Given our new information, we now recommend that Little League coaches continue to give the same advice: increase bat speed and reduce variability. Practice is the key. Baldwin (2007), a major-league pitcher with a career 3.08 ERA, sagaciously wrote that if you lose a game, don't blame the umpire or your teammates; just go home and practice harder.

Our measurements of over 300 batters showed that variability in the speed of the swing decreases with level of performance from Little League to Major League Baseball. For major leaguers the bat speed standard deviations were typically around ±5% (Bahill and Karnavas 1989), which is a very small value for physiological data.

The variable with the second largest sensitivity is the coefficient of restitution (*CoR*). The *CoR* of a bat-ball collision depends on where the ball hits the bat. It is difficult, but absolutely essential, for the batter to control this. He or she must hit the ball with the sweet spot of the bat. The *CoR* also depends on the manufacturing process. The NCAA now measures the Bat-ball Coefficient of Restitution (BBCOR) for sample lots coming off the manufacturing line. Therefore, amateurs are all going to get similar BBCORs. However, a lot can still be done with the *CoR* for aluminum and composite bats. For example, the performance of composite bats typically improves with age because of the break-in process; repeatedly striking the bat eventually breaks down the bat's composite fibers and resinous glues. 'Rolling' the bat also increases its flexibility. Rolling the bat stretches the composite fibers and accelerates the natural break-in process simulating a break-in period of hitting, say, 500 balls.

For wooden bats, the batter could try to influence the *CoR* by choosing the type of wood that the bat is made of. Throughout history, the most popular woods have been white ash, sugar maple and hickory. However, hickory is heavy, so most professionals now use ash or maple. A new finding about bat manufacturing is that the slope of the grain has an effect on the strength and elasticity of the bat. As a result, the wood with the straightest gain is reserved for professionals and wood with the grain up to 5° off from the long-axis of the bat is used for amateurs. Furthermore, the manufacturer's emblem is stamped on the flat grain side of ash bats so that balls collide with edge grain as shown in Fig. 1, whereas the emblem is stamped on the edge grain side of maple bats because they are stronger when the collision is on the flat grain side.

The next largest sensitivities are for the mass of the ball and its speed before the collision, m_{ball} and $v_{\mathrm{ball-before}}$. However, the batter can do nothing to influence the mass of the ball or the ball speed before the collision, so we will not concern ourselves with them. Likewise, the batter has no control over the ball spin, $\omega_{\mathrm{ball-before}}$, so we will ignore it when selecting bats. Now if this discussion were being written from the perspective of the pitcher (Baldwin 2007), then these three parameters would be very important.

The next most important variable in Table 5 is the mass of the bat. Therefore, we will now consider the mass and other related properties of the bat. The sensitivity of the batted-ball speed with respect to the mass of the bat is positive, meaning (if everything else is held constant) as the mass goes up so does the batted-ball speed. However, everything else cannot be held constant, because the heavier bat cannot be swung as fast (Bahill and Karnavas 1989) due to the force-velocity relationship of human muscle. This physiological relationship was not included in the equations of this paper because in this paper we only modeled the *physics* of the collision, notwithstanding physiology trumping physics in this case. The net result of physics *in conjunction with physiology* is that lighter bats are better for almost all batters (Bahill 2004).

Perhaps due to this general feeling, back in the 1960s and 70s, it was popular for professionals to 'cork' the bat. This reduced the mass of the bat, but because it also reduced the moment of inertia, it did not improve performance significantly (Nathan et al. 2011). However, it is now legal to make a one to two-inch diameter hole 1.25 in. deep into the barrel end of the bat. Most batters do this because it makes the bat lighter with few adverse effects. Other bat parameters that are being studied include the type of wood (density, strength, elasticity and straightness of the grain) and the type of materials (density, strength, break-in period and vibrational frequency).

For an aluminum bat, some batters reduce the thickness of the barrel wall by shaving the inside of the barrel. This reduces the bat mass, which according to physics *and* physiology, increases batted-ball speed.

The distance between the center of mass of the bat and the sweet spot, $d_{\mathrm{cm-ss}}$, is the next most important parameter. We presumed that the sweet spot of the bat was the center of percussion (CoP) of the bat. All batters try to hit the ball on the sweet spot of the bat. To help the batter, manufacturers of aluminum bats have been moving the CoP by moving the internal weight from the end of the bat toward the knob http://www.acs.psu.edu/drussell/bats/cop.html. It is now an annual game of cat and mouse. The manufacturers move the CoP, then the rule makers change their rules, then the manufacturers move … etc.

Finally, we come to the moment of inertia of the bat, I_{bat}, with respect to its center of mass. The physics, revealed with the sensitivity analysis, states that although the moment of inertia is one of the least important variables, it would help to increase its value. More importantly, physiology showed that all batters would profit from using end-loaded bats (Bahill 2004). There are many ways to change the moment of inertia of a bat. Most aluminum bats start with a common shell and then the manufacturer adds a weight inside to bring the bat up to its stated weight. The important question then becomes, *where* should the weight be added? It has been suggested that they add weight in the knob because this would comply with the regulations and would not decrease bat speed. However, the results of Bahill (2004) show that they should add the weight in the barrel end of the bat making it *end loaded*. This will increase the batted-ball speed. For a wooden bat, the moment of inertia can be changed by cupping out the barrel end, adding weight to the knob or tapering the barrel end. Assume that the end of the barrel of a bat is only used to "protect" the outside edge of the plate: no one hits home runs on the end of the bat. Therefore, a professional could use a bat where the last 3 in. (7 cm) was tapered from 2½ inches (6.4 cm) down to 1¾ of an inch (4.4 cm). This would decrease the weight, decrease the moment of inertia about the center of mass and would move the sweet spot 2% closer to the knob: these changes would probably benefit some players. However, such modifications would have to be individually designed for each player.

Most people can feel the difference between bats with different moments of inertia. In 1985, a coach with the San Francisco Giants showed us a legal custom-made bat with a large moment of inertia created by leaving it with a huge knob. He presumed that his players already understood the influence of bat weight on bat speed so he was trying to expand their understanding to the influence of bat moment of inertia on the speed of the swing. One of our University of Arizona softball players described our biggest moment of inertia bat, "That's the one that pulls your arms out."

The bat moment of inertia is the only parameter under the control of the batter for which a consensus does not exist in the science of baseball literature. The bat moment of inertia is an enigma because for most, but not all, batters as the bat moment of inertia goes up the *bat* speed goes down, and at the same time the *batted-ball* speed goes up (Bahill 2004; Smith and Kensrud 2014). For Bahill's (2004) batters, 20% had positive slopes for bat speed versus moment of inertia, for moments of inertia in the range of 0.03–0.09 kg m^2. Therefore, he showed the actual data for all players rather than averaging them, because averaging graphical data is usually meaningless. Perhaps more physiological studies would help clear up this issue. Our best generalization is that all batters would profit from using end-loaded bats. Smith and Kensrud (2014) concluded their paper with "Batter swing speed decreased with increasing bat inertia, while … the hit-ball speed increases with bat inertia."

Summarizing, these are the most important factors for understanding bat performance: bat weight, the coefficient of restitution, the moment of inertia and characteristics of humans swinging the bats.

In the future, it will be possible to see how the coefficient of friction μ_f affects the batted-ball speed. Then we will be able to decide if the varnish or paint on the bat should be made rough-textured or smooth, or if bats should be rubbed or oiled in order to improve bat performance.

To improve bat performance manufacturers could reduce the variability of bat and ball parameters. Major-league bats were custom made for us by Hillerich and Bradsby Co. The manufacturing instructions were "Professional Baseball Bat, R161, Clear Lacquer, 34 in., 32 oz, make as close to exact as possible, end brand—genuine model R161 pro stock, watch weights" emphasis added. The result was six bats with an average weight of 32.1 oz and a standard deviation of 0.5! This large standard deviation surprised us. We assume there is the same variability in bats used by major-league players.

There is also variability in the ball. We assume that the center of mass of the ball is coincident with the geometric center of the ball. However, put a baseball or softball in a bowl of water. Let the movement subside. Then put an X on the top the ball. Now spin it and let the motion subside again. The X will be on top again. This shows that for most baseballs and softballs the center of mass is not coincident with the geometric center of the ball.

3.2.8 Summary of Bat Selection

These sensitivity and optimality analyses show that the most important variable, in terms of increasing batted-ball speed, is bat speed before the collision. This is in concert with ages of baseball folklore and principles of physiology. Therefore, batters should develop strength, increase coordination and practice so that their swings are fast and with low variability.

These analyses show that the next most important parameter is the coefficient of restitution, the *CoR*. Engineers and bat regulators are free to play their annual cat and mouse game of increasing *CoR* then writing rules and making tests that prohibit these changes. Indeed, most recent bat research has gone into increasing the *CoR* of bat-ball collisions.

Pitch speed, ball spin and the mass of the ball are important. However, the batter cannot control them. Therefore, they cannot help the batter to choose or modify a bat.

The next most important parameter is the bat mass, m_{bat}. However, physics recommends heavy bats, whereas the force-velocity relationship of muscle recommends light bats. In this case, physiology trumps physics. Each person's preferred bat should be as light as possible while still fitting within baseball needs, regulations and availability.

The last interesting parameter from the sensitivity analysis and the optimization study is the bat moment of inertia, I_{bat}. These studies suggest that a larger bat moment of inertia would be better. However, a lot of the physics literature recommends smaller moments of inertia. Conversely, an experimental physiology study stated that all players would profit from using end-loaded bats (Bahill 2004). Therefore, this is the only parameter under the control of the batter for which a consensus does not exist in the science of baseball literature.

The second purpose of this paper is to show what the batter can do to achieve optimal bat performance. The most important thing is practice. Next, batters should select lightweight bats. They should then select bats that increase the *CoR* by all legal means. Finally, they should choose bats with a larger moment of inertia, bats that are often called end-loaded.

3.2.9 The Ideal Bat Weight™

So far, the equations in this paper were equations of physics. However, we repeatedly mentioned physiology. Now is the time to step back and look at physiology. This section is based on Bahill and Karnavas (1991).

Our instrument for measuring bat speed, the[1] Bat Chooser™, has two vertical laser beams, each with associated light detectors. Our batters swung the bats through the laser beams. A computer recorded the time between interruptions of the light beams. Knowing the distance between the light beams and the time required for the bat to travel that distance, the computer calculated the speed of the sweet spot, which we defined as the center of percussion. We told the batters to swing each bat as fast as they could while still maintaining control. We said, "Pretend you are trying to hit a Nolan Ryan fastball."

In our experiments, each batter swung six bats through the light beams. The bats ran the gamut from super-light to super heavy; yet they had similar lengths and weight distributions. In our developmental experiments, we tried about four dozen bats. We used aluminum bats, wooden bats, plastic bats, heavy metal warm-up bats, bats with holes in them, bats with lead in them, major-league bats, college bats, softball bats, Little League bats, brand-new bats and bats made in the 1950s.

In one set of experiments, we used six bats of significantly different weights but similar lengths of about 34 in. (89 cm), with centers of mass about 23 in. from the end of the handle (see Table 6).

In a 20-min interval, each subject swung each bat through the instrument five times. The order of presentation was randomized. The selected bat was announced by a speech synthesizer, for example: "Please swing bat Hank **A**aron, that is, bat **A**." (We named our bats after famous baseball players who had names starting with the letter assigned to the bat.)

For each swing, we recorded the bat weight and the speed of the center of mass, which we converted to the speed of the center of percussion. However, that was as far as physics could take us; we then had to look to the principles of physiology.

Physiologists have long known that muscle speed decreases with increasing load. This is why bicycles have gears; gears enable riders to maintain the muscle speed that imparts maximum power through the pedals, while the load, as reflected by the bicycle speed, varies greatly. To discover how the muscle properties of individual baseball players affect their best bat weights, for each player, we plotted

[1]Bat Chooser and Ideal Bat Weight are trademarks of Bahill Intelligent Computer Systems.

Table 6 Test bats used by major-league players

Name	Weight (oz)	Weight (kg)	Distance from knob to center of mass (in.)	Distance from knob to center of mass (m)	Average sweet spot speed (mph) from Fig. 5	Description
D	49.0	1.39	22.5	0.57	88	Aluminum bat filled with water
C	42.8	1.21	24.7	0.63	74	Wooded bat, filled with lead
A	33.0	0.94	23.6	0.60	65	Wooded bat
B	30.6	0.87	23.3	0.59	65	Wooden bat
E	23.6	0.67	23.6	0.60	61	Wooden bat
F	17.9	0.51	21.7	0.55	60	Wooden handle mounted on a light steel pipe with a 6 oz weight at the end

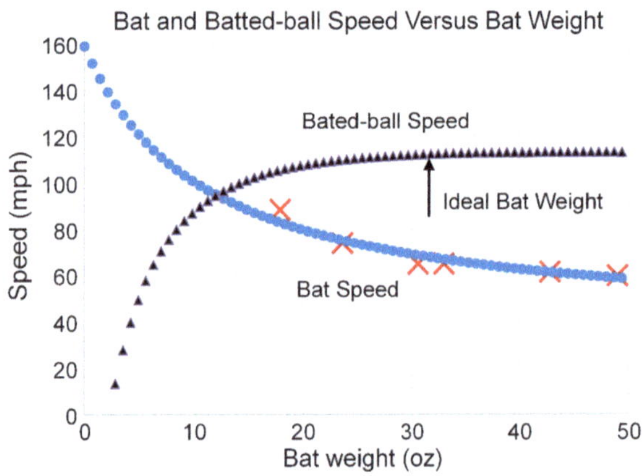

Fig. 5 Measured bat speed (*red Xs*), a hyperbola fit to this data (*blue dots*) and the calculated batted-ball speed (*black triangles*) for a 90 mph pitch to one of the fastest San Francisco Giants

bat speeds as a function of bat weight to produce graphical numerical models known as the muscle force-velocity relationships (see Fig. 5). The red Xs represent the average of the five swings of each bat; the standard deviations were small for physiological data.

Over the past 75 years, physiologists have used three equations to describe the force-velocity relationship of muscles: straight lines, hyperbolas and exponentials. Each of these equations has produced the best fit for some experimenters, under certain conditions and with certain muscles. However, usually the hyperbola fits the data best. In our experiments, we tried all three equations and chose the one that had the best fit to the data of each subject's 30 swings. For the data of the force-velocity relationships illustrated in Fig. 5, we found that a hyperbola provided the best fit.

These curves indicate how bat speed varies with bat weight. We now want to find the bat weight that will make the ball leave the bat with the highest speed and thus have the greatest chance of eluding the fielders. We call this the maximum-batted-ball-speed bat weight. To calculate this bat weight we must couple the muscle force-velocity relationships to the equations of physics.

For the major-league player whose data are shown in Fig. 5, the best fit for his force-velocity data was the hyperbola, $(m_{\text{bat}} + 11) \times (v_{\text{bat - before}} - 36) = 1350$ units are ounces and mph, bluc dots. This batter had some of the fastest swing speeds on the team. When we substituted this equation into the batted-ball speed equation, Eq. (7), we were able to plot the ball speed after the collision as a function of bat weight, black triangles in Fig. 5.

$$v_{1a} = \frac{v_{1b}(m_1 I_2 - m_2 I_2\ CoR_{2b} + m_1 m_2 d^2) + v_{2b} m_2 I_2 (1 + CoR_{2b}) + \omega_{2b} m_2 d I_2 (1 + CoR_{2b})}{m_1 I_2 + m_2 I_2 + m_1 m_2 d^2}$$

$$(m_{\text{bat}} + 11) \times (v_{2b} - 36) = 1350$$

$$v_{2b} = \left\{ \frac{36 m_2 + 1746}{m_2 + 11} \right\}$$

$$v_{1a} = v_{1b} \frac{(m_1 I_2 - m_2 I_2\ CoR_{2b} + m_1 m_2 d^2)}{K} + \left\{ \frac{36 m_2 + 1746}{m_2 + 11} \right\} \frac{m_2 I_2 (1 + CoR_{2b})}{K} + \omega_{2b} \frac{m_2 d I_2 (1 + CoR_{2b})}{K}$$

In this equation, I_2 is also a function of m_2. This curve shows that the maximum-batted-ball-speed bat weight for this subject is about 45 oz, which is much heavier than that used by any batters. However, this batted-ball speed curve is almost flat between 30 and 49 oz. This player normally used a 32-oz bat. Evidently the greater control permitted by the 32-oz bat outweighed the one per cent increase in speed that could be achieved with the 45-oz bat.

However, the maximum-batted-ball-speed bat weight is not the best bat weight for any player. Because a lighter bat will give a batter better control, more accuracy and more time to compute the ball's impact point. Obviously, a trade-off must be made between batted-ball speed and control. Because the batted-ball speed curve is so flat around the point of the maximum-batter-ball-speed, we believe there is little advantage in using a bat as heavy as the maximum-batter-ball-speed bat weight. Therefore, we have defined the [1]*ideal bat weight*™ to be the weight where the ball speed curve drops 1 per cent below the maximum-batter-ball speed. Using this criterion, the ideal bat weight for this batter is 31.75 oz. We believe this gives a good trade-off between distance and accuracy.

As can be seen from the batted-ball speed equation, both v_{1a} and the ideal bat weight increase with pitch speed. However, we do not recommend that a batter use a heavier bat against a fire-baller, because heavier bats increase the swing time and decrease the prediction time.

The ideal bat weight is specific to each individual; it is not correlated with height, weight, age, circumference of the upper arm, or any combination of these factors, nor is it correlated with any other obvious physical factors. Although, Bahill and Morna Freitas (1995) mined our database of 163 subjects and 36 factors and determined some rules of thumb that could make suggestions.

3.2.10 Bat Speed

Throughout this paper we have used a before collision bat speed of 58 mph (26 m/s). This is the average sweet spot speed that we measured for 28 members of the San Francisco Giants baseball team. However, our subjects were not paid and therefore they were not highly motivated: furthermore, they did not actually hit a ball: both of these circumstances increase the variance of swing speeds. Some studies in the literature filtered their data and only included selected batters, usually the fastest. Internet sites that are trying to sell their equipment and services cite sizzling bat speeds between 70 and 90 mph (31–40 m/s). We think that these numbers are bogus. The big web sites such as mlb.com, espn.com/mlb/and hit-trackeronline.com give the leaders in many categories, meaning that they a have selected the 20 fastest players out of 750. This would be misleading if the reader thought that these statistics were *representative* of major-league batters, which they do.

Table 7 gives average sweet spot speeds for six studies of male college and professional batters. When multiple bats were used, we chose the wooden bats closest to that described in Table 1.

Figure 4 shows that the average major-league batter has a high enough bat speed to occasionally hit a home run, when the batted-ball has the ideal spin and launch angle. However, over half of major-league batters seldom hit homeruns. Indeed, of the 2200 active players listed by MLB.com half of them have never hit a home run in their major-league careers. Our equations show that a ball velocity before the collision, v_{1b}, of 83 mph (37 m/s) and a bat sweet spot speed, v_{2b}, of 58 mph

Table 7 Bat sweet spot speed before the collision

Average speed of the sweet spot (m/s)	Average speed of the sweet spot (mph)	Subjects	References
32	71	Unknown	King et al. (2012)
31	69	7 selected male professional baseball players	Welch et al. (1995)
30	68	19 male baseball players	Crisco et al. (2002)
26	58	28 San Francisco Giants	Database of Bahill and Karnavas (1989)
26	58	7 male college baseball players	Koenig et al. (2004)
26	58	17 male college baseball players	Fleisig et al. (2002)

(26 m/s) would produce a batted-ball speed, v_{1a}, of 97 mph (43 m/s), which would be almost enough for a home run in any major-league stadium. Our rule of thumb is that it takes a batted-ball speed of 100 mph (45 m/s) to produce a homerun. The following is Eq. (7).

$$v_{1a} = \frac{v_{1b}(m_1 I_2 - m_2 I_2\ CoR_{2b} + m_1 m_2 d^2) + v_{2b} m_2 I_2 (1 + CoR_{2b}) + \omega_{2b} m_2 d I_2 (1 + CoR_{2b})}{m_1 I_2 + m_2 I_2 + m_1 m_2 d^2}$$

For a major league wooden bat, as described in Table 1,

$$v_{1a} = -0.28 v_{1b} + 1.28 v_{2b} + 0.17 \omega_{2b}$$

where the units are either mph and rpm or m/s and rad/s. Remember that v_{1b} is a negative number. So far, we have made no approximations; everything has been exactly according to Newton's principles. But now we will create our *rule of thumb* by rounding, substituting $\omega_{2b} = 0$ and using pitch speed instead the speed of the ball at the beginning of the collision.

$$\boxed{v_{\text{batted}-\text{ball}} = -0.25 v_{\text{pitch}-\text{speed}} + 1.3 v_{\text{bat}-\text{before}}}$$

For oblique collisions, the batted-ball speed would be less, but backspin on the ball in flight would keep it up in the air longer, so those two effects partially cancel out (Kensrud et al. 2016).

Most recent studies of bat speed have used video cameras and commercial prepackaged software to measure and compute bat speed. There are no calibration tests. Most of these systems report higher bat speeds than other methods of measuring bat speed. On television, the batted-ball speed is often called the exit speed or the exit velocity.

3.2.11 Seeing the Collision

When a baseball bat moving at 58 mph (26 m/s) hits a baseball traveling in the opposite direction at 83 mph (37 m/s) there is a violent collision, which was shown in figure 5.3. Table 5.3 shows that during the collision the kinetic energy in the motion of the bat changes by 81 Joules (J): a loss of 106 J in linear translational kinetic energy, a gain of 25 J in angular kinetic energy. Notably, 81 J is equivalent to dropping a bowling ball from your waist onto your toe or having a dove fly into your windshield while you are driving down a highway at 80 mph (130 km/hr).

Frame by frame analysis of high-speed video of a major-league batter showed that at the beginning of the collision there was (1) a big abrupt change in the ball velocity as it swung from negative to positive, (2) a sudden drop in the linear velocity of the sweet spot of the bat and (3) a sharp change in the angle of the bat.

Now, imagine a film of Ted Williams hitting a baseball. His swing is smooth and graceful although the kinetic energy of his bat changes by 202 Joules during a collision. The reason his swing seems so smooth is that we mainly visualize the movement of his body, arms, hands and the bat. We model this movement with the bat's angular rotation about the knob, β. The change in this angular motion is not visually obvious because it is just a short small jerk (a few degrees) in the middle of a big swinging motion. Hence, what we see does not change much. On the other hand, the bat's linear translational motion, β, decreases from 26 to 13 m/s. However, we do not visualize this translational motion well, because his swing looks like a big rotation: it does not look like a translation. As a result, the movement that we visualize well, does not change much. Whereas, the movement that changes a lot, β, is not visualized well. This explains why people do not perceive an abrupt jerk when the bat and ball collide.

What about the batter? Would he be able to see the effects of this violent collision? Probably not. Bahill and LaRitz (1984) showed that no batter can keep his eye on the ball from the pitcher's release point to the bat-ball collision. Their graduate students fell behind when the ball was 9 ft (2.7 m) in front of the plate. Comparatively, their major-league baseball player was able to keep his position error below 2° until the ball was 5.5 ft (1.7 m) from the plate. Then he fell behind. This finding runs contrary to baseball's hoary urban legend that Ted Williams could see the ball hit his bat. However, in reality, Ted Williams could not see the ball hit his bat. In a letter that he sent to Bahill dated January 23, 1984 he wrote,

> Received your letter and have also had a chance to read your research, and I fully agree with your findings.
>
> I always said I couldn't see a ball hit the bat except on very, very rare occasions and that was a slow pitch that I swung on at shoulder height. I cam[e] very close to seeing the ball hit the bat on those occasions.

In summary, the bat-ball collision is violent. But nobody perceives it, because (1) even in slow motion, the spectator only sees the smooth movement of the batters body, arms, hands, and bat, which glide continuously, (2) movements that change abruptly, such as the bat's linear translational velocity, are difficult to visualize because they are so quick, (3) batters are not able to see the bat-ball collision at all and (4) the bat-ball collision only lasts one millisecond. This explains why nobody sees an abrupt jerk when the bat hits the ball, not even Ted Williams.

4 Summary

One purpose of this paper was to show how complicated bat-ball collisions could be while still being modeled using only Newton's principles and the conservation laws. The model of this paper is the most complex configuration for which our model is valid. Our model was explained with Figs. 1 and 3. The five equations that we used were listed in Table 2.

The following canonical form equations comprise our model for bat-ball collisions.

$$KE_{\text{lost}} = \frac{1}{2}\frac{m_1 m_2 I_2}{m_1 I_2 + m_2 I_2 + m_1 m_2 d^2}\left[(v_{1\text{b}} - v_{2\text{b}})^2(1 - CoR_{2\text{b}}^2) - 2(v_{1\text{b}} - v_{2\text{b}})\omega_{2\text{b}} d + d^2\omega_{2\text{b}}^2\right]$$

$$v_{1\text{a}} = v_{1\text{b}} - \frac{(v_{1\text{b}} - v_{2\text{b}})m_2 I_2(1 + CoR_{2\text{b}}) - \omega_{2\text{b}} m_2 d I_2(1 + CoR_{2\text{b}})}{m_1 I_2 + m_2 I_2 + m_1 m_2 d^2}$$

where $v_{2\text{b}} = v_{\text{bat-trans-before}} + d_{\text{cm-ss}}\omega_{\text{bat-before}}$

$$v_{2\text{a}} = v_{2\text{b}} + \frac{(v_{1\text{b}} - v_{2\text{b}})m_1 I_2(1 + CoR_{2\text{b}}) - \omega_{2\text{b}} d m_1 I_2(1 + CoR_{2\text{b}})}{(m_1 I_2 + m_2 I_2 + m_1 m_2 d^2)}$$

$$\omega_{2\text{a}} = \omega_{2\text{b}} + \frac{(v_{1\text{b}} - v_{2\text{b}})m_1 m_2 d(1 + CoR_{2\text{b}}) - \omega_{2\text{b}} m_1 m_2 d^2(1 + CoR_{2\text{b}})}{m_1 I_2 + m_2 I_2 + m_1 m_2 d^2}$$

$$\omega_{1a} = \omega_{1b}$$

If we let

$$A = \left\{\frac{[(v_{1\text{b}} - v_{2\text{b}}) - \omega_{2\text{b}} d](1 + CoR_{2\text{b}})}{m_1 I_2 + m_2 I_2 + m_1 m_2 d^2}\right\}$$

then we get

$$v_{1\text{a}} = v_{1\text{b}} - A m_2 I_2$$
$$v_{2\text{a}} = v_{2\text{b}} + A m_1 I_2$$
$$\omega_{2\text{a}} = \omega_{2\text{b}} + A m_1 m_2 d$$
$$\omega_{1a} = \omega_{1b}$$

A second purpose of this paper was to show how the individual batter can find and customize an optimal baseball or softball bat for him or herself. The sensitivity analysis and optimization study of this paper showed that the most important variable, in terms of increasing batted-ball speed, is bat speed before the collision. However, in today's world, the coefficient of restitution and the bat mass are experiencing the most experimentation trying to improve bat performance. Although, the bat moment of inertia provides more room for future improvement. Above all, future studies must include physics in conjunction with physiology in order to improve bat performance.

Acknowledgements I thank Lynden Mahrt for comments on the manuscript. I acknowledge brilliant mathematical derivations from the Great Szidarovszky.

References

Bahill, A. T. (2004). The ideal moment of inertia for a baseball or softball bat. *IEEE Transactions on Systems, Man, and Cybernetics Part A: Systems and Humans, 34*(2), 197–204.

Bahill, A. T., Baldwin, D. G., & Ramberg, J. S. (2009). Effects of altitude and atmospheric conditions on the flight of a baseball. *International Journal of Sports Science and Engineering, 3*(2), 109–128. http://www.worldacademicunion.com/journal/SSCI/online.htm. ISSN 1750-9823 (print).

Bahill, A. T., & Karnavas, W. J. (1989). Determining ideal baseball bat weights using muscle force-velocity relationships. *Biological Cybernetics, 62,* 89–97.

Bahill, A. T., & Karnavas, W. J. (1991). The ideal baseball bat. *New Scientist, 130*(1763), 26–31.

Bahill, A. T., & LaRitz, T. (1984). Why can't batters keep their eyes on the ball. *American Scientist, 72*, 249–253.

Bahill, A. T., & Morna Freitas, M. (1995). Two methods for recommending bat weights. *Annals of Biomedical Engineering, 23*(4), 436–444.

Baldwin, D. (2007). *Snake Jazz*, Xlibris Corp. www.Xlibris.com.

Baldwin, D. G., & Bahill, A. T. (2004). A model of the bat's vertical sweetness gradient. In M. Hubbard, R. D. Mehta, & J. M. Pallis (Eds.), *The engineering of sport 5. Proceedings of the 5th International Engineering of Sport Conference, September 13–16, 2004, Davis, CA, International Sports Engineering Association (ISEA), Sheffield, England* (Vol. 2, pp. 305–311).

Brach, R. M. (2007). *Mechanical impact dynamics: Rigid body collisions.* Wiley.

Choi, K. K., & Kim, M. H. (2005). *Structural sensitivity analysis and optimization 1, linear systems.* New York, NY: Springer Science + Bussiness Media Inc.

Crisco, J. J., Greenwald, R. M., Blume, J. D., & Penna, L. H. (2002). Batting performance of wood and metal baseball bats. *Medicine & Science in Sports & Exercise,* 1675–1684. doi:10.1249/01.MSS.0000031320.62025.57.

Cross, R. (2011). *Physics of baseball and softball.* Springer.

Cross, R., & Nathan, A. M. (2006). Scattering of a baseball by a bat. *American Journal of Physics, 74*(1), 896–904.

Dadouriam, H. M. (1913). *Analytic mechanics for students of physics and engineering* (p. 248). New York: D. Van Nostrand Co.

Ferreira da Silva, M. F. (2007). Meaning and usefulness of the coefficient of restitution. *European Journal of Physics, 28,* 1219–1232. doi:10.1088/0143-0807/28/6/019.

Fleisig, G. S., Zheng, N., Stodden, D. F., & Andrews, J. R. (2002). Relationship between bat mass properties and bat velocity. *Sports Engineering, 5,* 1–8. doi:10.1046/j.1460-2687.2002.00096.x.

Kensrud, J. R., Nathan, A. M., & Smith, L. V. (2016). Oblique collisions of baseballs and softballs with a bat. *American Journal of Physics.*

King, K., Hough, J., & McGinnis, R. (2012). A new technology for resolving the dynamics of a swinging bat. *Sports Engineering, 15,* 41–52.

Koenig, K., Mitchell, N. D., Hannigan, T. E., & Cluter, J. K. (2004). The influence of moment of inertia on baseball/softball bat swing speed. *Sports Engineering, 7,* 105–117.

Nathan, A. M., Crisco, J. J., Greenwald, R. M., Russell, D. A., & Smith, L. V. (2011a). A Comparative study of baseball bat performance. *Sports Engineering, 13,* 153–162.

Nathan, A. M., Smith, L. V., Faber, W. L., & Russell, D. A. (2011b). Corked bats, juiced balls, and humidors: The physics of cheating in baseball. *American Journal of Physics, 79*(6), 575–580.

Nathan, A. M., Cantakos, J., Kesman, R., Mathew, B., & Lukash, W. (2012). Spin of a batted baseball, 9th Conference of the International Sports Engineering Association (ISEA). *Procedia Engineering, 34,* 182–187. doi:10.1016/j.proeng.2012.04.032.

Sawicki, G. S., Hubbard, M., & Stronge, W. J. (2003). How to hit home runs: Optimum baseball bat swing parameters for maximum range trajectories. *American Journal of Physics, 71*(11), 1152–1162.

Smith, E. D., Szidarovszky, F., Karnavas, W. J., & Bahill, A. T. (2008). Sensitivity analysis, a powerful system validation technique. *Open Cybernetics & Systemics Journal, 2,* 39–56.

Smith, L., & Kensrud, J. (2014). Field and laboratory measurements of softball player swing speed and bat performance. *Sports Engineering, 17,* 75–82. doi:10.1007/s12283-013-0126-y.
Watts, R. G., & Bahill, A. T. (1990). *Keep your eye on the ball: Curveballs, knuckleballs, and fallacies of baseball* (1st ed.). New York: W. H. Freeman and Co.
Watts, R. G., & Bahill, A. T. (2000). *Keep your eye on the ball: Curveballs, knuckleballs, and fallacies of baseball* (2nd ed.). New York: W. H. Freeman and Co.
Welch, C. M., Banks, S. A., Cook, F. F., & Draovitch, P. (1995). Hitting a baseball: A biomechanical description. *Journal of Orthopaedic and Sports Physical Therapy, 22,* 193–201.

Author Biography

Terry Bahill is an Emeritus Professor of Systems Engineering and of Biomedical Engineering at the University of Arizona in Tucson. He received his Ph.D. in electrical engineering and computer science from the University of California, Berkeley. He is the author of seven engineering books and over two hundred and fifty papers, over one hundred of them in peer-reviewed scientific journals. Bahill has worked with dozens of high-tech companies presenting seminars on Systems Engineering, working on system development teams and helping them to describe their Systems Engineering processes. He holds a U.S. patent for the Bat Chooser™, a system that computes the Ideal Bat Weight™ for individual baseball and softball batters. He was elected to the Omega Alpha Association, the systems engineering honor society. He received the Sandia National Laboratories Gold President's Quality Award. He is a Fellow of the Institute of Electrical and Electronics Engineers (IEEE), of Raytheon Missile Systems, of the International Council on Systems Engineering (INCOSE) and of the American Association for the Advancement of Science (AAAS). He is the Founding Chair Emeritus of the INCOSE Fellows Committee. His picture is in the Baseball Hall of Fame's exhibition "Baseball as America." You can view this picture at http://sysengr.engr.arizona.edu/.

Reverse Logistic Network Design for End-of-Life Wind Turbines

Suna Cinar and Mehmet Bayram Yildirim

Abstract Energy generation from wind turbines shows an increasing trend for the last two decades. As the amount of wind generation increases, wind turbine (WT) operators face challenges with finding alternative disposal options for WTs over their useful life. Wind farm operator (decision makers) can benefit from a well-designed reverse logistics network to determine the best disposal alternative for WT end-of-life use (EOL). This chapter is an example of the recovery of valuable material that can be recycled/recovered or remanufactured at the end of WTs useful life by designing an effective reverse logistics network. Here, a mixed integer linear programming (MILP) model is proposed to determine a long-term strategy for WT EOL. The objective in this model is to minimize the transportation and operating cost as well as finding the best locations for recycling and remanufacturing centers. The results of this study show that due to the high operating cost at remanufacturing centers, sending most WTs to them is costlier than sending them to recycling centers. In addition, it was found that transportation cost depends on the amount of flow that has been sent to the recycling or remanufacturing center.

Keywords Mixed-integer linear optimization · Reverse logistic · End-of-life use · Wind energy · Wind turbine

Abbreviations

WT	Wind turbine
EOL	End-of-life
MILP	Mixed-integer linear programming
WTRLN	Wind turbine reverse logistic network
RLN	Reverse logistic network

S. Cinar (✉) · M.B. Yildirim
Department of Industrial and Manufacturing Engineering,
Wichita State University, 1845 Fairmount Street, Box 35
Wichita, KS 67260-0035, USA
e-mail: sxcinar@wichita.edu; cinarsuna@yahoo.com

M.B. Yildirim
e-mail: bayram.yildirim@wichita.edu

A. Matsumoto (ed.), *Optimization and Dynamics with Their Applications*,
DOI 10.1007/978-981-10-4214-0_11

1 Introduction

Due to the increased awareness of environmental issues and more restrictive environmental regulations, renewable energy sources such as wind, solar, hydro, and geothermal are becoming more popular. In addition, due to the increase in total energy consumption, attaining a sustainable energy supply will be a challenge in the near future for the world. Using renewable energy sources effectively is one of the options to help overcome this problem. The main driver for interest in wind turbines is to produce electrical power with very low CO_2 emissions, which is one of the largest contributors of greenhouse gas emissions, the insidious cause of climate change (Ghenai 2012).

Life expectancy for WTs is about 20 years (Post 2013; Haapala and Prempreeda 2014). Due to increasing demand of using wind energy as a renewable energy source, at some point, many WTs will reach the end of their service life. Thus, a sustainable process that can be used when WTs reach the end of their service life is needed in order to maximize the environmental and economic benefits of wind energy and to minimize the environmental impact.

In the literature, a significant number of studies focus on the reverse logistics network for different EOL products, such as electric/electronic products and vehicles. Reverse logistics strategies and different application methods for various products are already being reviewed by others. However, no study providing an optimization model and a detailed analysis of WTs EOL using reverse logistics has been found. Therefore, this study has attempted to cover this lack of knowledge through the development of an effective reverse logistic network (RLN) design with a mathematical programming model for WTs EOL use. Unlike previous research, the work here modeled generation points as collection points. Installing numerous collection centers in near proximity to generation points (and inspection centers) may make sense when collecting small assets, like plastic bottles for curbside collection. However, this is not a rational approach for large assets like WTs, where generation points are typically far apart, which would render a network of collection centers complicated and economically infeasible. The improved model formulation provides a much more realistic representation of real-world economics for large assets, and therefore yields a more accurate optimization of long-term costs. In summary, the reverse logistics network considered in this chapter is different than networks in the reverse logistics literature. While RLN design for EOL has been studied by many researchers under various settings, there is still need for further research that examines the modeling of different recycling ratios in order to overcome quality uncertainty of WT component.

This study presents an MILP formulation to solve the wind turbine reverse logistics network (WTRLN) problem. This optimization model is applicable to all kinds of WTs, and the model is applied to study the different recycling and remanufacturing ratios over a finite horizon, in order to overcome the uncertainty associated with RLN design. The investigation of different disposal options (such as recycling and remanufacturing) for EOL WTs in terms of cost components could be part of a decision support framework.

Before providing details of the proposed model, it is best to understand the WT supply chain, which along with the role of the reverse logistics network, is discussed in Sect. 2. The remainder of this chapter is organized as follows. The mathematical model, the calculation of input parameters and application of the model are presented in detail in Sect. 3. All computational results for the base-case scenario and different scenarios are given in Sect. 4. Finally, some concluding remarks with future directions are provided in Sect. 5.

2 Wind Turbine Supply Chain

Over the last decade, wind power has grown at around 7% a year, increasing by a factor of 10. The U.S. Department of Energy aims for 20% of U.S. electricity to be wind produced by 2030 (Centers of Excellence 2009).

To reach this goal, there must be enough raw materials to manufacture the WTs. Every megawatt capacity of WT requires 200 kg of neodymium. In addition, the heavy rare earth metal dysprosium, which is used to increase the longevity of magnets in WT, is becoming more difficult to find. To overcome the difficulty of supplying rare earth metals, mines could be developed, but the increased mining of rare earth metals could create more environmental degradation and human health hazards (Cho 2012). In addition, the refinement process for rare earth metals uses toxic acids and results in polluted wastewater that must be properly disposed of. Although recycling cannot satisfy the rapidly growing demand for rare earth metals, it is one way to help decrease the shortage. Recycling and reusing materials also saves energy that is used in mining and processing, conserves resources, and reduces pollution and greenhouse gas emissions. As the use of these materials in WTs increases, recycling would make more economic sense. Recycled content could become valuable as a secondary source on the market, which can ease periods of tight supply (Bauer et al. 2010; Hahn and Gilman 2014). Figure 1 represents a generic WT supply chain. As can be seen, the last step of WT supply chain management is end of life.

A framework for EOL options for WTs is provided in Fig. 2. An integrated WT reverse logistics network should address the following significant factors: plans for collection of WT components, estimation of recyclable and recovered quantities of WT parts, and remanufacturing and remarketing of recovered WTs items. Reverse logistics activities for WT EOL use can be grouped into three stages: (1) product/part collection, (2) inspection, separation, and sorting, and (3) recovery and disposition.

Because of increasing demand for renewable energy resources, the main manufacturers of WTs are struggling to keep up with the increasing demand for new units (Ghenai 2012). Due to the recent growth of demand for wind energy, there is a

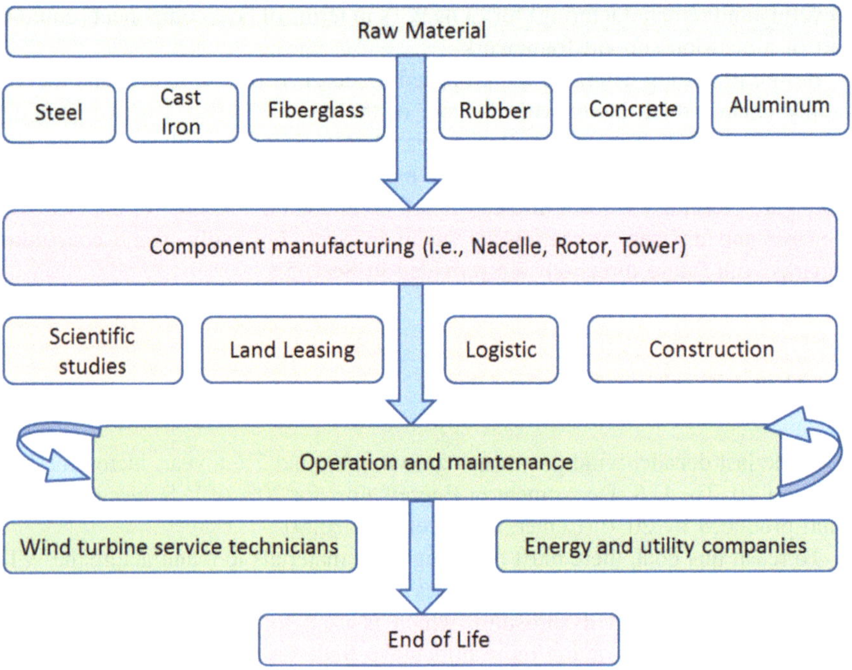

Fig. 1 Wind turbine supply chain (adapted from U.S. Department of Labor 2010)

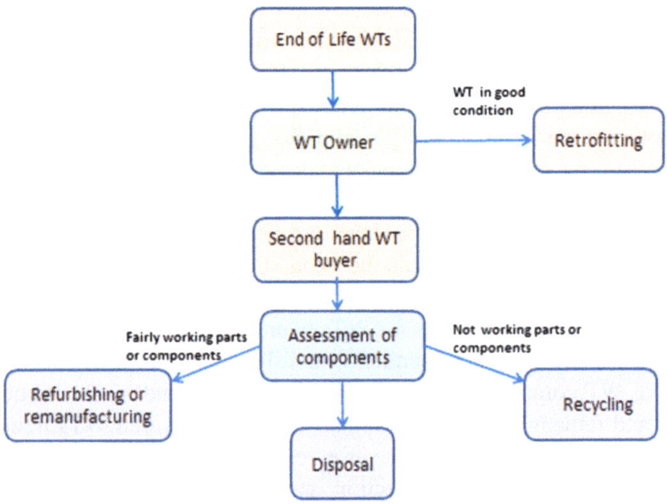

Fig. 2 End-of-life options for wind turbines (adapted from U.S. Department of Labor 2010)

Table 1 Global shortage of wind turbine components

Component	Shortage risk
Blades	High
Bearing	High
Gearbox	High
Controls	Low
Generators	Low
Castings	Low
Tower	Low

Adapted from Lehner and Roastogi (2012)

global shortage affecting some of the components. Table 1 summarizes the components of WTs and their shortage risk (Lehner and Roastogi 2012).

A longer waiting period for most parts of the major turbine makes remanufacturing very attractive, since traditional remanufacturing activities are capable of returning most WT components to "as-new" condition (Walton and Parker 2008). Therefore, remanufacturing the valuable components of WTs at the EOL may be an effective way to meet their increasing demand.

In addition, because of the growing demand for energy in developing countries, and the interest in renewable energy sources, i.e., wind energy, which provides a sustainable and environmentally friendly power supply, remanufacturing of EOL WTs could be helpful to satisfy this growing need for power. Most of these developing countries may not be able to afford brand new WTs as a source of renewable energy. Therefore, providing used refurbished WTs in these locations offers several benefits, such as lower capital investment, shorter project duration, reduction of CO_2 emissions, and a contribution to sustainable development (Hulshorst 2008).

Several studies discussed the reduction in carbon dioxide by comparing different alternatives for treatment and replacement of old WTs. The highest amount of CO_2 emissions for energy generation from WTs was found to be in the material production phase, which is 60–64% of total emissions, and the next was in wind turbine production. Transportation, disassembly, and renovation/maintenance contributes only 2–3% of CO_2 emissions (Rydh et al. 2004). Sosa Skrainka (2012) analyzed the environmental impact of remanufacturing WTs and concluded that remanufacturing of the component inside the nacelle has a smaller impact on the environment than manufacturing new components. Arvesen and Hertwich (2012) assessed the life-cycle environmental impacts of wind power and estimated that the EOL phase of WTs reduces emissions, decreasing greenhouse gas emissions by 19%.

Table 2 shows the benefits of recycling parts of the WTs at the end life of their useful life, producing less CO_2 than the landfilling process. It can be seen that the dominant phase that is consuming more energy and producing more CO_2 emissions is the material phase and primary material production of the WT parts. More energy is consumed and high amount of CO_2 is released in the atmosphere during these two phases. Results also show the benefits of recycling materials at the end life of

Table 2 Energy and CO_2 footprint summary—wind turbine

Life cycle of wind turbine with landfilling			Life cycle of wind turbine with recycling		
Phase	Energy (J)	CO_2 (kg)	Phase	Energy (J)	CO_2 (kg)
Material	1.759E+013	1.2546E+006	Material	1.759E+013	1.2546E+006
Manufacture	1.3593E+012	107669.7209	Manufacture	1.3593E+012	107669.7209
Transport	2.4336E+011	17278.6954	Transport	2.4336E+011	17278.6954
Use	1.6778E+011	11912.557	Use	1.6778E+011	11912.557
EOL Landfilling	2.1826E+011	13095.7080	EOL recycling	–6.8512E+011	–495917.2797
Total	1.9583E+013	1.4054E+006	Total	1.2513E+013	895503.8906

Adapted from Ghenai (2012)

the WT. If all materials are sent to the landfill at the WT end of life, then 2.18 E +011 J of energy (1.1% of total energy) is needed to process these materials, and 13,095.71 kg of CO_2 (0.9% increase of total CO_2) are released to the atmosphere. If WT material is recycled at the EOL, then a total energy of 6.85E+012 J (54.8% of total energy) is recovered. A net reduction of CO_2 emissions by 495,917.28 kg (55.4% of total CO_2) is obtained by recycling the WT material (Ghenai 2012).

Based on the environmental and economic factors, the supply chain would greatly benefit if a reverse logistics network was integrated into the whole supply chain process. By doing so, the WT supply chain would become environmentally more responsible by recycling, reusing, or remanufacturing the WT. In addition, there is a possibility of economic gain from recycling and remanufacturing. Recovery of products and parts can be good alternatives to manufacturing new products and parts and virgin resources (Krikke et al. 1999; Geyer and Jackson 2004). It is clear that an effective reverse logistics for WTs can generate direct gains by reducing the use of raw materials, adding value with recovery, reducing disposal costs, recycling to save landfill space and energy, and reducing CO_2 emissions, in turn providing a more sustainable supply chain. In Sect. 3, the proposed reverse logistics supply chain for WTs is proposed, and the proposed model notations, parameters, and formulations are introduced.

3 Wind Turbine Reverse Logistics Network Mathematical Model

This section presents a mathematical model of the WTRLN problem. First, the reverse logistics network considered in this work is described. Then the variables and parameters of the model for the proposed RLN design model are given. Last, a formulation of the model is proposed.

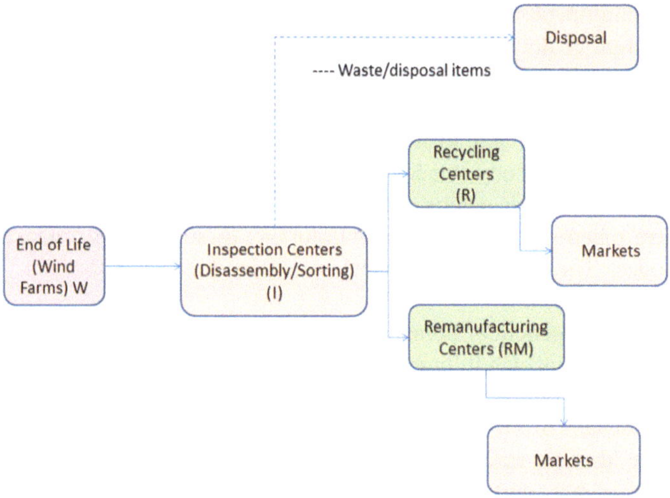

Fig. 3 Reverse logistics of wind turbines

The main objective of this model is to minimize the cost associated with logistics and operating cost of different disposal options (i.e., recycling or remanufacturing) for EOL WTs. The proposed model considers the design of a multi-echelon reverse logistics network that consists of wind farms, inspection centers, remanufacturing centers, recycling centers, and secondary market. The general structure of an RLN for WTs is shown in Fig. 3.

As can be seen, reverse flow starts at the wind farms. It is assumed that the major components of WTs (i.e., nacelles, blades, and tower, etc.) are dismantled and transported to inspection centers, which sort the materials and components by identifying quality of the parts. At inspection centers, better conditioned WT parts are transported to remanufacturing centers, while WT parts that are in bad condition are sent to recycling centers. It is difficult to predict the physical condition of WTs at the end of their useful life. Therefore, constraints are introduced to provide flexibility to run the model with different recycling ratios. The following assumptions are made:

- Locations are known.
  - Potential inspection, recycling, and remanufacturing centers
  - Markets for recycling and remanufacturing
  - Disposal centers
- There is no storage in the inspection/recycling and remanufacturing centers, therefore no holding cost.

- Dismantling operations take place at wind farms, and WT parts are transported to inspection centers for testing and cleaning.
- A fixed cost is associated with opening inspection, recycling, disposal, and remanufacturing centers.
- Even though a WT has many components, only three main components are considered in this study (blades, nacelle (gearbox and generator), and tower).
- Transportation cost is determined per mile, and total transportation costs in the objective function are obtained by multiplying these costs by distances between two nodes. These distances are calculated by haversine formula (Longitude Store.com 2014).
- Wind farms as generator points or collection centers are used interchangeably.
- For the initial runs, landfilling (disposal) cost is not considered in this model. It is assumed that only a small percent of WT components are going to be sent to disposal centers from the inspection, recycling, and remanufacturing centers. Therefore, the cost associated with disposal activities such as transportation cost will be minimal, and this cost is already included in the operating cost of inspection, recycling, and remanufacturing centers.
- Only one type of WT is considered in this model.

3.1 Model Notation

In order to propose our model for the problem, the sets, indexes, parameters, cost, and decision variables used in the model are given as follows:

c wind turbine components, $c \in C = \{1, \ldots, |C|\}$
j all possible locations $j \in J = \{1, \ldots, |J|\}$
t time periods, $t \in T = \{1, \ldots, |T|\}$
w location of wind farms, $w \in W = \{1, \ldots, |J|\}$
i potential inspection centers, $\mathrm{i} \in \mathrm{I} = \{1, \ldots, |\mathrm{I}|\} \subseteq J$
m potential remanufacturing centers, $m \in M = \{1, \ldots, |M|\} \subseteq J$
r potential recycling centers, $r \in R = \{1, \ldots, |R|\} \subseteq J$
s potential markets, $s \in S = \{1, \ldots, |S|\} \subseteq J$
ds potential disposal centers, $ds \in DS = \{1, \ldots, |DS|\} \subseteq J$

Parameters

Q_{wct} supply of WT component c at wind farm w in period t
DSM_{sct} demand of WT component c at market s in period t
DR_{rct} demand of WT component c at recycling center r in period t
DM_{mct} demand of WT component c at remanufacturing center m in period t
DL_{dsct} demand of WT component c at disposal center ds in period t

$CAPI_{it}$ capacity of inspection center i in period t
$CAPR_{rt}$ capacity of recycling center r in period t
$CAPM_{mt}$ capacity of remanufacturing center m in period t
$CAPD_{dst}$ capacity of disposal center ds in period t
α % of WT component c sent from inspection center to recycling center
β % of WT component c sent from inspection center to remanufacturing center
γ % of WT component c sent from inspection center to disposal center

Costs

FCI_{it} fixed cost opening inspection center i in period t ($)
FCM_{mt} fixed cost opening remanufacturing center m in period t ($)
FCR_{rt} fixed cost of opening recycling center r in period t ($)
OPI_{cit} cost of processing one unit of WT component c at inspection center i in period t ($)
OPR_{crt} cost of processing one unit of WT component c at recycling center r in period t ($)
OPM_{cmt} cost of processing one unit of WT component c at remanufacturing plant m in period t ($)
T_{wict}, T_{irct}, $T_{imct,}$ T_{msct}, T_{idsct} transportation distance of one unit of WT component c at time period t from w to i, i to r, i to m, m to s, or i to ds (mile)
θ unit transpiration cost factor ($/mile)
dr inflation rate

Decision Variables

$X1_{wict}$ number of WT components c shipped from wind farm w to inspection center i in period t
$X2_{imct}$ number of WT components c shipped from inspection center i to remanufacturing center m in period t
$X3_{irct}$ number of WT components c shipped from inspection center i to recycling center r in period t
$X4_{msct}$ number of WT components c shipped from remanufacturing center m to market s in period t
$X5_{idsct}$ number of WT components c shipped from inspection center i to disposal center ds in period t

Binary Variables

$$Y_{it} = \begin{cases} 1 & \text{if an inspection center } i \in I \text{ is operating in period } t \in T, \\ 0 & \text{otherwise} \end{cases}$$

$$Z_{mt} = \begin{cases} 1 & \text{if a remanufacturing center } m \in M \text{ is operating in period } t \in T, \\ 0 & \text{otherwise} \end{cases}$$

$$U_{rt} = \begin{cases} 1 & \text{if a remanufacturing center } r \in R \text{ is operating in period } t \in T, \\ 0 & \text{otherwise} \end{cases}$$

$$A_{dst} = \begin{cases} 1 & \text{if a disposal center } ds \in DS \text{ is operating in period } t \in T, \\ 0 & \text{otherwise} \end{cases}$$

3.2 Mathematical Model

The proposed model decision variables are to determine the location and number of inspection, recycling, remanufacturing, and disposal centers to open in each time period and the flow (amount components send to each center) between these centers. This model aims to minimize the costs of EOL WT recovery, including transportation costs; operating costs of inspection, recycling, and remanufacturing centers; and capital cost of opening inspection, recycling, and remanufacturing centers.

$$\begin{aligned} \text{Min} & \sum_i \sum_t FCI_{it} * (Y_{it} - Y_{i,t-1}) * (1 + dr)^{-t} + \sum_m \sum_t FCM_{mt} * (Z_{mt} - Z_{m,t-1}) * (1 + dr)^{-t} \\ & + \sum_r \sum_t FCR_{rt} * (U_{rt} - U_{r,t-1}) * (1 + dr)^{-t} + \sum_t \sum_c \sum_w \sum_i T_{wict} * \theta * X1_{wict} * (1 + dr)^{-t} \\ & + \sum_t \sum_c \sum_i \sum_r T_{irct} * \theta * X3_{irct} * (1 + dr)^{-t} + \sum_t \sum_c \sum_i \sum_m T_{imct} * \theta * X2_{imct} * (1 + dr)^{-t} \\ & + \sum_t \sum_c \sum_m \sum_s T_{msct} * \theta * X4_{msct} * (1 + dr)^{-t} + \sum_t \sum_c \sum_i \sum_m OPM_{cmt} * X2_{imct} * (1 + dr)^{-t} \\ & + \sum_t \sum_c \sum_i \sum_r OPR_{crt} * X3_{irct} * (1 + dr)^{-t} + \sum_t \sum_c \sum_w \sum_i OPI_{cit} * X1_{wict} * (1 + dr)^{-t} \end{aligned} \quad (1)$$

Constraint (2) is a flow balance constraint which is the number of disassemble WTs parts at wind farms (generation points) equal to the number of WT parts sent to inspection centers.

$$Q_{wct} = \sum\nolimits_{i \in I} X1_{wict} \quad w \in W, \ c \in C, \ t \in T \quad (2)$$

Constraints (3–4) model the flow balance between inspection centers, and recycling and remanufacturing centers, i.e., the total number of WT components at the inspection centers is equal to number of WT components shipped to recycling and remanufacturing center.

$$\sum_{w \in W} \alpha * X1_{wict} = \sum_{m \in M} X2_{imct} \quad i \in I,\ c \in C,\ t \in T \tag{3}$$

$$\sum_{w \in W} \beta * X1_{wict} = \sum_{r \in R} X3_{irct} \quad i \in I,\ c \in C,\ t \in T \tag{4}$$

Constraint (5) shows the total inflow component coming from remanufacturing centers is equal to the outflow of WTs sold to secondary market.

$$\sum_{s \in S} X4_{msct} = \sum_{i \in I} X2_{imct} \quad m \in M,\ c \in C,\ t \in T \tag{5}$$

Constraint (6) formulates the number of WTs sold to the secondary market are no more than the demand for the remanufactured WTs at each time period.

$$\sum_{s \in S} X4_{msct} \leq \sum_{s \in S} DSM_{sct} \quad m \in M,\ c \in C,\ t \in T \tag{6}$$

Constraint (7) assures that the number of WT components sent to a recycling center is no more than the demand of component at each time period.

$$\sum_{r \in R} X3_{irct} \leq \sum_{r \in R} DR_{rct} \quad i \in I,\ c \in C,\ t \in T \tag{7}$$

Constraint (8) ensures that the amount of WT component sent to remanufacturing center is no more than the demand of component at each time period.

$$\sum_{m \in M} X2_{imct} \leq \sum_{m \in M} DM_{mct} \quad i \in I,\ c \in C,\ t \in T \tag{8}$$

Constraint (9) is the capacity constraint for production in the inspection center.

$$\sum_{w \in W} \sum_{c \in C} X1_{wict} \leq CAPI_{it} * Y_{it} \quad i \in I,\ t \in T \tag{9}$$

Constraint (10) is the capacity constraint for production in the recycling center.

$$\sum_{i \in I} \sum_{c \in C} X3_{irct} \leq CAPR_{rt} * U_{rt} \quad r \in r,\ t \in T \tag{10}$$

Constraint (11) is the capacity constraint for production in the remanufacturing center.

$$\sum_{i \in I} \sum_{c \in C} X2_{imct} \leq CAPM_{mt} * Z_{mt} \quad m \in M,\ t \in T \tag{11}$$

Constraints (12–14) ensure that once a center is installed, it remains operating until the end of the planning horizon.

$$Y_{it} \leq Y_{i,t+1} \quad i \in I,\ t \in T \tag{12}$$

$$U_{rt} \leq U_{r,t+1} \quad r \in R,\ t \in T \tag{13}$$

$$Z_{mt} \leq Z_{m,t+1} \quad m \in M,\ t \in T \tag{14}$$

Constraint (15) is the non-negativity constraint, and constraint (16) is the integrality constraint.

$$X1_{wict}, X2_{imct}, X3_{irct}, X4_{msct}, X5_{idsct} \geq 0 \tag{15}$$

$$Y_{it} \in \{0,1\},\ Z_{mt} \in \{0,1\},\ U_{rt} \in \{0,1\},\ i \in I,\ r \in R,\ M \in M,\ t \in T \tag{16}$$

In the following sections, we present an application of the model considering EOL WTs.

4 Case Study

The WTRLN model explained in Sect. 3 has been applied to the case of a RLN design for EOL WTs. A five-echelon network consisting of ten wind farms (generating plants) was considered for the model implementation. A simple illustration of the model—a single WT type with three components—is considered. The other necessary input parameters used in the model with detailed explanations are provided next.

4.1 Input Parameters

Here, the data collected from various resources to formulate the case study is presented. In this study, it is assumed that WTs are to be collected from wind farms. For each wind farm location, the number of WTs is determined randomly, and the distance matrix is created between ten wind farm locations. It is assumed that three types of WT components are sent to each center. Based on expert opinion and literature data, the following section provides the detailed cost data and the base of each cost selected for this study.

The transportation cost between wind farms, potential recycling and remanufacturing centers, and potential markets are based on the transportation cost provided for transporting blades from the manufacturing facility to wind farms. Five cost categories for transporting blades from the manufacturing facility to the wind farms are summarized in Table 3. As a conservative, approach, it is assumed that for each WT component, the transportation cost is more or less similar to the transportation cost for the blades. A WT has several main components. This model

Table 3 Transportation cost of wind turbine blades

Transportation cost category	Cost factor (per mile)
Freight	$1.55
Overdimension charge	$1.25
Escort charge (per escort)	$1.40
Total unit cost	$4.2

Adapted from Sandia National Laboratories (2003)

Table 4 Dismantling cost of wind turbine

Procedure	Cost factor ($)
Dismantling hubs	1,200
Dismantling blades	1,200
Dismantling nacelle	1,200
Dismantling tower	5,500
Hiring crane and Demobilizing	22,500
Other additional procedures	5,000
Dismantling total	36,600

Adapted from Repowering Solutions (2011)

considers the transportation of only three large components (blades, tower, and hub or nacelle). These components will be transported to recycling or remanufacturing centers, and transportation unit cost ($unit/mile) for each component will be the same. (Sandia National Laboratories 2003).

The dismantling cost for WTs at wind farms can be considered a fixed cost that can be added to the operating cost of the inspection center for each WT. Based on the literature review, the total dismantling cost per WT is estimated to be $36,600. A cost breakdown is given in Table 4. Labor cost and foundation disassembly or site remediation activities costs are not taken into consideration in this study (Repowering Solutions 2011). The typical price of replacement components (set of rotor blades, gearbox, and generator) is 15–20% of the price of new components. A new turbine costs approximately $1,400–$1,600 per kilowatt-hour of generating capacity, and the remanufactured cost is in the range of $700–$800 per kilowatt-hour (McDermott 2009).

The price for a refurbished/remanufactured WT is claimed to be up to 50% lower than new turbines. Based on this cost data, it is assumed that 50% of the refurbished WT cost is due to repairing or buying new parts and installing them during remanufacturing (Walton and Parker 2008). Research by Tegen et al. (2012) indicates that the gearbox price was 137 $/kW for a 1.5 MW baseline project in 2010. The price for a new gearbox for a 2 MW turbine can hover between $184,000 and $218,500. A gearbox with a standard refurbish, in which bearings are replaced, the gear teeth are overhauled and reground, and the components measured costs from $103,500 to more than $115,000 (Knight 2011).

A typical remanufacturing center for WTs includes a dirt room with a service bay for receiving, staging, initial washing, and tearing down the gearbox/main shaft assemblies; a temperature-controlled clean environment room; testing room; oil conditioning system; and condition monitoring equipment (vibration, oil, temperature analysis). The typical size of the warehouse can be 33,500 square feet, with a storage room for complete kW and MW gearbox parts and assemblies. Based on this, it is clear that remanufacturing operation and installation costs would be higher than operation and installation costs of recycling centers. A summary of the cost data used in this study is provided in Table 5. To be able to determine the total profit of remanufacturing and recycling operations, the total material composition of a 1.65-MW WT and total recycling cost of each material are provided in Tables 6 and 7, respectively.

Table 7 shows the breakdown of salvage and disposal costs for a typical 1.65-MW WT. Based on the composition of a 1.65-MW WT, it is assumed that the composition of a 1.5-MW WT would be close to that of a 1.65-MW WT composition, and the total disposal cost and salvage value of a 1.5-MW WT is calculated based on that assumption. One should note that the cost of metal fluctuates daily. Therefore, the cost data provided at the time of this study may not be accurate in the future.

The generator replacement was selected over replacing other nacelle components, because it contained a large amount of copper, e.g., the generator consisted of around 35% copper and 65% steel, compared to around 1% copper, 1% aluminum, and 98% steel in the gearbox (Ancona and McVeigh 2001). All cost values are assumed to increase by the yearly inflation rate of 1.7%, as published by the U.S. government (U.S. Inflation Calculator 2008–2015).

4.2 Wind Turbine Reverse Logistics Network Model Run

The WTRLN was coded in the General Algebraic Modeling System (GAMS). A set of data was prepared to reflect the real case situation. Fifty (50) time periods, each representing 1 year, were used during each model run. The impact of key parameters was evaluated based on the results. The effect of different disposal options of EOL was investigated in order to determine the cost of each option.

For the initial base run, it was assumed that 40% of the total supply would be remanufactured and the remaining 60% would be recycled. It was assumed that this is not the case for all WTs, since several factors may affect their remaining life and that some of the WTs may still be in good conditions, or vise versa. Therefore, several other scenarios with different increases or decreases in recycling ratios were run. Each of these scenarios was modelled using ratio (α) values between 0.1 and 0.9, in increments of 0.1, to evaluate the effects of recycling/remanufacturing costs during the decision-making process. During model runs, the other parameters were kept the same. For the first three scenarios, considering that only small percentages of WT parts would be sent to the disposal center from the inspection center, the disposal center-related cost was not added to these scenarios.

The sensitivity analysis involved the investigation of the impact of high and low transportation costs and high and low operating costs of recycling and

Table 5 Summary of cost data

Item		Cost	Reference
Transportation cost		$4.2 per mile	Sandia National Laboratories (2003)
New wind turbine cost (GE 1.5 XLE 1.5 MW		$1,400,000	Cost analysis of material composition of the wind turbine blades for Wobben Windpower/ENERCON GmbH Model E-82, Wagner Sousa de Oliveira and Antonio Jorge Fernandes, Cyber Journals: Multidisciplinary Journals in Science and Technology, Journal of Selected Areas in Renewable Energy (JRSE), January Edition (2012) and Repowering Solutions (2011)
Remanufactured turbine cost (GE 1.5 SL)		$500,000	Cost analysis of material composition of the wind turbine blades for Wobben Windpower/ENERCON GmbH Model E-82, Wagner Sousa de Oliveira and Antonio Jorge Fernandes, Cyber Journals: Multidisciplinary Journals in Science and Technology, Journal of Selected Areas in Renewable Energy (JRSE), January Edition (2012) and Repowering Solutions (2011)
Operating cost	Operating cost at remanufacturing center	[$10,000–$50,000]	Estimated based on expert opinion (gearbox, generator, towers or blades) Renew Energy Maintenance (2012)
	Operating cost at inspection center plus added dismantling cost	[$1,000–$5,000] [$35,000 added dismantling cost]	Estimated based on expert opinion (gearbox, generator, towers or blades) Renew Energy Maintenance (2012)
	Operating cost at recycling center	[1,000–5,000]	Estimated based on expert opinion (gearbox, generator, towers or blades) Renew Energy Maintenance (2012)
Installation cost of centers	Inspection, remanufacturing, and recycling centers	[15,000–70,000]	Estimated based on expert opinion Renew Energy Maintenance (2012)

(continued)

Table 5 (continued)

Item		Cost	Reference
Remanufactured turbine component cost	Gearbox 10–15% total cost of WT	\$50,000–\$75,000	Cost analysis of material composition of the wind turbine blades for Wobben Windpower/ENERCON GmbH Model E-82, Wagner Sousa de Oliveira and Antonio Jorge Fernandes, Cyber Journals: Multidisciplinary Journals in Science and Technology, Journal of Selected Areas in Renewable Energy (JRSE), January Edition (2012) and Repowering Solutions (2011)
	Generator 5–10% total cost of WT	\$25,000–\$50,000	
	Tower cost 10–25% total cost of WT	\$50,000–\$125,000	
	Blades 10–15% total cost of WT	\$50,000–\$75,000	
Recycling cost profit	Generator	\$12,500	Estimated based on typical materials and quantities required for Vestas V82 1.65-MW turbine.
	Gear box	\$7,000	
	Tower	\$75,600	

remanufacturing centers. Opening either a low- or high-capacity remanufacturing center is another parameter that is expected to affect the model results. Therefore, different scenarios with different supply data were run to see how the model reacts under different supply conditions.

4.3 Model Results

We carried out analyses on various scenarios to understand the possible changes in the network with variation in recycling and remanufacturing quantities. Table 8 shows the solution to Scenario 1. For the first scenario, different recycling ratios were used to observe the effect of flow on the total network cost. It was observed that with increasing recycling ratio, the total network cost decreased from \$1,091,403,000 to \$765,191,900, which represents a 30% decrease. This is because the operating cost of remanufacturing centers is higher that of recycling centers. Therefore, it can be concluded that in the assumed case study, in addition to transportation and other logistics costs, operating cost is also an important factor in the design of a reverse logistics network. Table 9 shows the number of inspection, recycling, and remanufacturing centers opened for Scenario 1. For different recycling ratios, the network requires four inspection centers, a maximum of two remanufacturing centers, and two recycling centers. With a decreasing recycling ratio, the number of recycling centers decreased, respectively. This is because fewer WT components are sent to recycling centers.

Table 6 Typical materials and quantities required for Vestas V82 1.65-MW wind turbine

Turbine component	Materials	Tonnes per turbine*	Percent recycled
Tower		**135.2**	
	Steel	126.1	90
	Aluminium	2.6	90
	Electronics	2.2	
	Plastic	2.0	
	Copper	1.3	90
	Oil	1.0	
Nacelle		**50.6**	
	Cast iron	18.0	90
	Steel, engineering	13.0	90
	Stainless steel	7.8	90
	Steel	6.3	90
	Fibreglass	1.8	
	Copper	1.6	90
	Plastic	1.0	
	Aluminium	0.5	90
	Electronics	0.3	90
	Oil	0.3	
Rotor		**42.1**	
	Cast iron	11.3	90
	Steel	4.2	90
	Steel, engineering	1.5	90
Blades	Epoxy, fiber glass, birchwood, balsa wood, etc.	25.2	
Foundation		**832**	
	Concrete	805	90
	Steel	27	90
Internal cables		**0.82**	
	Aluminum	0.35	90
	Plastic	0.30	
	Copper	0.17	90
Transformer station		**0.95**	
	Steel	0.50	90
	Copper	0.13	90
	Transformer oil	0.21	
	Other: insulation, paint, wood, porcelain etc.	0.11	
External cables		**14.9**	
	Plastic	8.35	
	Aluminum	5.24	90
	Copper	1.31	90

*_Source_ Life cycle assessment of electricity produced from onshore sited wind power plants based on Vestas V82-1.65 MW turbines (2006) https://www.vestas.com/~/media/vestas/about/sustainability/pdfs/lca%20v82165%20mw%20onshore2007.pdf

Table 7 Salvage value and disposal cost of material and mass for 1.5-MW wind turbine

Salvage value		Ton[3]	Cost ($/ton)	($)
Recycling	Steel	186.4	522[1]	97,039
	Copper	4.51	4,550[1]	3,000
	Aluminum	8.69	1,482[1]	2,443
Total revenue				102,482
Disposal cost		Ton	Cost ($/ton)	($)
Disposal	Epoxy, plastic, fiber	38.65	33.35[2]	1,289
	Concrete	805	85.59[2]	68,905
	Lubricant	300[3]		
Disposal cost				70,208

[1]Atlantic County Utilities Authority (2016)
[2]Milanese (2009)
[3]Given away free to used oil collectors

In addition, Fig. 4 shows the percent distribution of operating and transportation cost of recycling and remanufacturing centers in comparison to overall network cost. It is clear that when the recycling ratio decreases, the operation and maintenance cost increases up to 14% due to the high cost of processing at the remanufacturing center. The percent contribution of the transportation cost for recycling and remanufacturing centers stays between 3% and 4%. Figure 5a and b depict the recycling and remanufacturing ratios versus total network cost.

In the initial run, it was assumed that all WTs reaching their EOL would be dismantled and either recycled or remanufactured. But this may not always be the case. It can be assumed that some of the WTs that reach their useful life may work several more years with proper maintenance. If this would be the case, then consideration may be given to maintaining these WTs on site and keeping them in place and using them for several more years. Therefore, instead of installing only one size of remanufacturing plant, two different sizes of remanufacturing centers are considered—one with a low operating capacity and one with a high operating capacity. In this case, if the supply of the WTs is less than predicted, then instead of opening large remanufacturing centers, the model can have flexibility to decide to open a remanufacturing center with a low capacity. This would decrease the fixed cost of a remanufacturing center and, overall, reduce the reverse logistics network cost. Therefore, for Scenario 2, the constraint number (11) was modified, a binary variable defining the "if then constraint" for opening either capacity remanufacturing center based on supply was added.

$$\sum_{i \in I} \sum_{c \in C} X2_{imct} \leq CAPML_{mt} * V_{mt} + CAPMH_{mt} * K_{mt} \quad m \in M, \ t \in T \tag{17}$$

where $CAPML_{mt}$ is a lower capacity of remanufacturing center m, and $CAPMH_{mt}$ is a higher capacity of remanufacturing center m. Decision variables include the following:

Table 8 Solution to scenario 1

Scenario	R ratio	RM ratio	Network cost ($)	Operating cost R ($)	Operating cost RM ($)	Transportation cost to R ($)	Transportation cost to RM ($)
0 (base case)	60	40	3,498,145,000	38,556,000	311,760,000	105,046,600	70,031,040
1	50	50	3,796,639,000	32,130,000	389,700,000	87,538,810	87,538,810
2	40	60	4,095,133,000	25,704,000	467,640,000	70,031,040	105,046,600
3	30	70	4,393,687,000	19,278,000	545,580,000	52,523,280	122,554,300
4	20	80	4,692,181,000	12,852,000	623,520,000	35,015,520	140,062,100
5	10	90	4,990,675,000	6,426,000	701,460,000	17,507,760	157,569,900

RM Remanufacturing Center, *R* Recycling Center

Table 9 Numbers of inspection, recycling, and remanufacturing centers opened for scenario 1

Recycling	Inspection centers	Recycling centers	Remanufacutring centers
0 (base case)	4	2	1
1	4	2	1
2	4	2	1
3	4	1	2
4	4	1	2
5	4	1	2

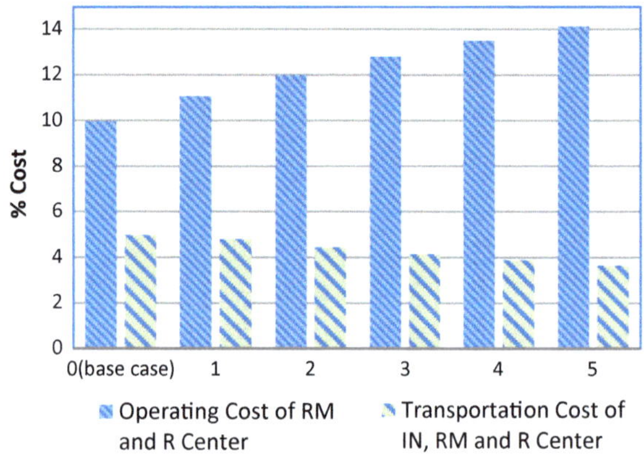

Fig. 4 Recycling and Remanufacturing operating and transportation cost percent distribution

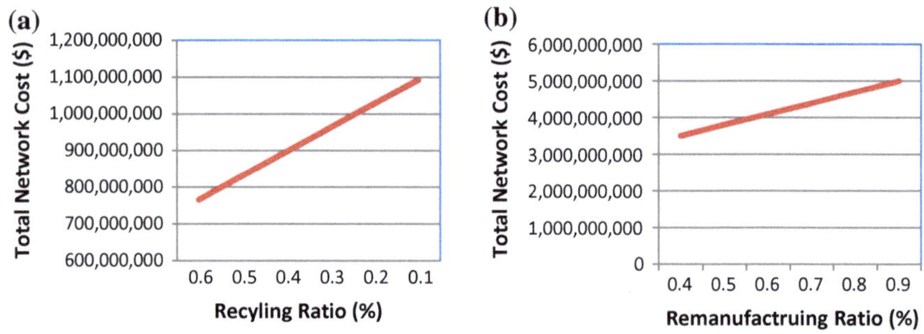

Fig. 5 **a** Recycling ratios versus total network cost. **b** Remanufacturing ratios versus total network cost

$$V_{mt} = \begin{cases} 1 & \text{if a lower capacity remanufacturing center } m \in M \text{ is operating in period } t \in T, \\ 0 & \text{otherwise} \end{cases}$$

$$K_{mt} = \begin{cases} 1 & \text{if a higher capacity remanufacturing center } m \in M \text{ is operating in period } t \in T, \\ 0 & \text{otherwise} \end{cases}$$

For Scenario 2, additional constraints for opening either low- or high-capacity remanufacturing centers were introduced. Because each type of remanufacturing center has different installation and operation costs, it is obvious that opening a low-capacity remanufacturing center would also be less costly. For this strategy, the model was modified and run for different recycling ratios. The results of Scenario 2 are given in Tables 10 and 11. Figure 6 shows the cost difference between Scenarios 1 and 2, which is less than 3%. Even though this percentage number looks very minimal, in terms of dollar amount, it is roughly \$900,000–\$1,000,000, which can be a substantial savings under a tight budget constraint.

Scenario 3 makes the assumption that due to catastrophic events, such as a tornado or any other natural disaster, most of the WTs that did not reach their useful life, suddenly become available as supply. Based on this assumption, the supply of WTs along with demand were dramatically increased. Thus, it was possible to see that increasing supply and demand will definitely increase the reverse logistics network cost. As indicated previously, forecasting cost under catastrophic events will provide flexibility to the decision maker to allocate the budget to either the recycling or remanufacturing option for EOL WT use. Results of Scenario 3 are given in Tables 12 and 13.

Scenario 4 considers that some of the WTs cannot be recycled or remanufactured due to their present condition, and only 10% of WT parts are assumed to be sent to the disposal center from the inspection center. This is a very conservative assumption. In reality, compared to the total weight of a 1.65-MW WT, which is roughly 1,631 tons, the components that need to be sent to the disposal center (such as fiberglass, oil, plastic, and rubber) only comprise about 2% of the total weight (Haapala and Prempreeda 2014). To account for the disposal center costs, in addition to revising the objective function and budget constraints, the following constraints were added to the original model:

Constraint (18), the total ratio of components that are sent to recycling, remanufacturing, and disposal centers, is equal to one.

$$\alpha + \beta + \gamma = 1 \tag{18}$$

Constraint (19) models the flow balance between inspection centers and disposal centers.

Table 10 Solution to scenario 2

Scenario	R ratio	RM ratio	Network cost ($)	Operating cost R ($)	Operating cost RM ($)	Transportation cost to R ($)	Transportation cost to RM ($)
0 (base case)	60	40	3,113,349,050	38,556,000	311,760,000	121,854,056	70,031,040
1	50	50	3,379,008,710	32,130,000	389,700,000	101,545,020	87,538,810
2	40	60	3,644,668,370	25,704,000	467,640,000	81,236,006	105,046,600
3	30	70	3,910,381,430	19,278,000	545,580,000	60,927,005	122,554,300
4	20	80	4,176,041,090	12,852,000	623,520,000	40,618,003	140,062,100
5	10	90	4,441,700,750	6,426,000	701,460,000	20,309,002	157,569,900

RM Remanufacturing Center, *R* Recycling Center

Table 11 Numbers of inspection, recycling, and remanufacturing centers opened for extended model

Recycling	Inspection centers	Recycling centers	Remanufacutring centers
0 (base case)	4	2	1-0
1	4	2	1-0
2	4	2	0-1
3	4	1	1-1
4	4	1	1-1
5	4	1	1-1

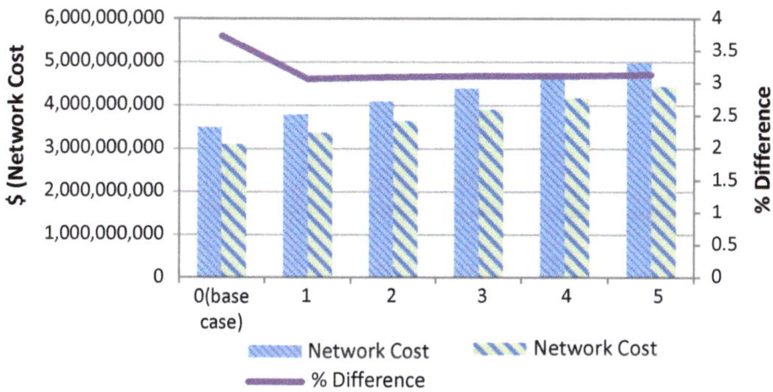

Fig. 6 Percent cost difference between base and extended model

$$\sum_{w \in M} \gamma * X1_{wict} \sum_{ds \in DS} X5_{idsct} \quad i \in I,\ c \in C,\ t \in T \tag{19}$$

Constraint (20) assures that the number of WT components sent to a disposal center is no more than the demand of component at each time period.

$$\sum_{ds \in DS} X5_{idsct} \leq \sum_{ds \in DS} DL_{dsct} \quad i \in I, c \in C, t \in T \tag{20}$$

Constraint (21) is the capacity constraint for production in the disposal center, which assures that the amount of components sent to disposal centers are not more than the total capacity of disposal center.

$$\sum_{i \in I} \sum_{c \in C} X5_{idsct} \leq CAPD_{dst} * A_{dst} \quad ds \in DS,\ t \in T \tag{21}$$

These results were compared with the model that has only one type of remanufacturing center capacity (Scenario 1). Comparing results with the previous run, it can be seen that the network cost decreased from 4 to 3% for Scenario 4, due to the fact that fewer WTs were sent to the remanufacturing and recycling centers, which have higher processing costs than disposal centers. Table 14 shows the results for

Table 12 Solution to scenario 3

Scenario	R ratio	RM ratio	Network cost ($)	Operating cost R ($)	Operating cost RM ($)	Transportation cost to R ($)	Transportation cost to RM ($)
0 (base case)	60	40	4,022,866,750	45,496,080	367,876,800	123,954,988	82,636,627
1	50	50	4,366,134,850	37,913,400	459,846,000	103,295,796	103,295,796
2	40	60	4,709,402,950	30,330,720	551,815,200	82,636,627	123,954,988
3	30	70	5,052,740,050	22,748,040	643,784,400	61,977,470	144,614,074
4	20	80	5,396,008,150	15,165,360	735,753,600	41,318,314	165,273,278
5	10	90	5,739,276,250	7,582,680	827,722,800	20,659,157	185,932,482

RM Remanufacturing Center, *R* Recycling Center

Table 13 Numbers of inspection, recycling, and remanufacturing centers opened for scenario 3

Recycling	Inspection centers	Recycling centers	Remanufacutring centers (low-high)
0 (base case)	5	3	1-1
1	5	3	1-1
2	5	2	0-2
3	5	2	0-3
4	5	1	0-3
5	5	1	0-3

Scenario 4. The cost difference between Scenarios 1 and 4 are shown in Fig. 7. In addition, the cost of opening a disposal center is not taken into account for this scenario. It is assumed that the waste will be shipped to existing municipal landfills.

4.4 *Sensitivity Analysis*

It is clear that there is a relationship between network cost, operation cost, and transportation cost. Therefore, a sensitivity analysis was performed to see the effect of operating cost and transportation cost on the total network cost. The transportation and operating costs (i.e., 10, 20, and 30%) at remanufacturing and recycling centers were increased. Sensitivity analysis showed (Tables 15, 16 and 17) that even if the transportation cost increased by 10%, the optimal solution increases by less than 1%. This analysis indicates that in the assumed case, it is not the logistics but probably the operating costs that have more impact on the reverse logistics network decision. Thus, by increasing the operating cost for remanufacturing and recycling centers by 10%, the optimal solution increased by more than 3%, which proves that operating cost has more impact on the WTs reverse logistics network decision.

By analyzing each scenario, it is clear that in addition to transportation cost, operating cost is also one of the main cost contributors to overall reverse logistics cost for this case study. Increasing the recycling ratio increases the operating cost at recycling centers, and decreasing the recycling ratio increases the operating cost at remanufacturing centers. Therefore, wind farm decision makers should pay attention to the operating cost of each disposal alternative for their end-of-life wind turbines.

By analyzing the results of four scenarios, it is clear that the model indicates that the total overall costs for the third scenario with different recycling ratios are higher than costs of the first and second scenarios. This is expected, since the demand and supply data were modified, and there are more inspection centers and remanufacturing centers opened. As summarized throughout this chapter, by increasing the recycling ratio, the transportation cost of sending the WT components to a recycling

Table 14 Solution to scenario 4 (adding disposal center to scenario 1)

Scenario	R ratio	RM ratio	Network cost ($)	Operating cost R ($)	Operating cost RM ($)	Transportation cost to R ($)	Transportation cost to RM ($)	Transportation cost to IN ($)
0 (base case)	60	40	3,351,139,523	45,496,080	233,820,000	105,046,600	52,606,070	7,945,372
1	50	50	3,649,622,489	37,913,400	311,760,000	87,538,810	70,024,408	7,945,372
2	40	60	3,948,107,062	30,330,720	389,700,000	70,031,040	87,530,499	7,945,372
3	30	70	4,246,400,777	22,748,040	467,640,000	52,523,280	105,036,561	7,945,372
4	20	80	4,549,777,972	15,165,360	545,580,000	35,015,520	122,542,707	7,945,372
5	10	90	4,843,653,063	7,582,680	623,520,000	17,507,760	140,048,808	7,945,372

IN Inspection Center, *RM* Remanufacturing Center, *R* Recycling Center

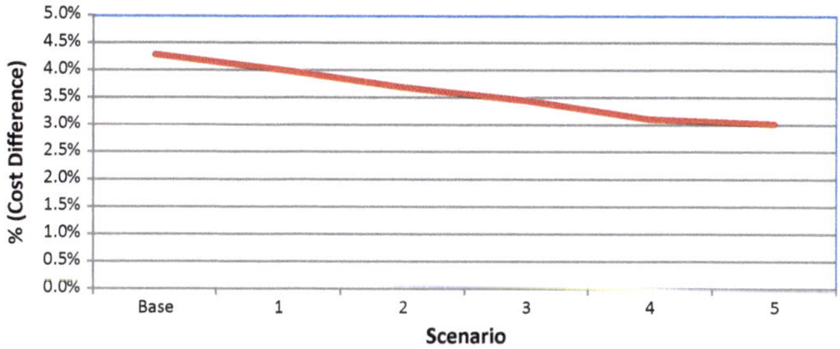

Fig. 7 Percent cost difference between scenarios 1 and 4

Table 15 Sensitivity analysis results for transportation cost

Scenario	Transportation cost increase %	Network cost ($)	Transportation cost to R ($)	Transportation cost to RM ($)
0 (base case)		3,498,145,000	38,556,000	311,760,000
1	10	3,522,461,798	38,784,000	311,850,000
2	20	3,546,252,761	39,056,000	312,070,000
3	30	3,570,204,407	39,225,000	312,260,000

RM Remanufacturing Center, *R* Recycling Center

Table 16 Sensitivity analysis results for operating costs at remanufacturing center

Scenario	Operating cost increase RM (%)	Network cost ($)	Operating cost R ($)	Operating cost RM ($)
0 (base case)		3,498,145,000	38,556,000	311,760,000
1	10	3,600,032,718	38,556,000	340,101,818
2	20	3,704,888,040	38,556,000	368,443,636
3	30	3,812,797,401	38,556,000	396,785,455

RM Remanufacturing Center, *R* Recycling Center

Table 17 Sensitivity analysis results for operating costs at recycling center

Scenario	Operating cost increase R (%)	Network cost ($)	Operating cost R ($)	Operating cost RM ($)
0 (base case)		3,498,145,000	38,556,000	311,760,000
1	10	3,583,465,610	38,784,000	311,850,000
2	20	3,670,867,212	39,056,000	312,070,000
3	30	3,760,400,555	39,225,000	312,260,000

RM Remanufacturing Center, *R* Recycling Center

center is higher than the transportation cost of sending the WT components to a remanufacturing center.

During the model run, for the first scenario, the supply increased throughout the time horizon. For the third scenario, in order to see the effect of different supplies, the supply was randomly increased and decreased to force the model to run under extreme conditions. In addition, with sensitivity analysis, the processing and transportation costs were increased to see the individual effect of each to total network cost.

Comparing the results of these different scenarios show that this current reverse logistics network fits all scenarios quite well, with the potential to be adjusted to fit the strategic change of recycling and remanufacturing options. The key issue is the availability of data related to physical conditions of WTs. If decision maker does not have the data to decide which WTs need to be sent to recycling or remanufacturing center, results from the modeling provides guidance in decision making by quantifying the difference, in terms of transportation and operating cost of reverse logistics of WTs.

4.5 Wind Turbine Reverse Logistics Network Problem with Total Profit Objective (WTRLN-TP)

To be able to better analyze the results and determine if recycling or remanufacturing of existing WTs are profitable, the objective function was modified by adding the total profit from selling the remanufactured components and also by recycling the three main components of WTs. For the sake of simplicity in our analysis, we consider three main components of WTs, blades/tower, generator and gearbox. The cost data for each component are gathered from several different works of literature and are summarized in Table 7. The objective function includes total profit, which comes from remanufacturing, recycling minus the transportation, and operation and installation cost of each center. All other constraints remain the same as in the original problem.

$$\begin{aligned}\text{Maximize} & \sum_t \sum_c \sum_i \sum_m PURM_{mct} * X2_{imct} * (1+dr)^{-t} + \sum_t \sum_c \sum_i \sum_r PURR_{rct} * X3_{irct} * (1+dr) - t\_ \\ & \{\sum_i \sum_t FCI_{it} * (Y_{it} - Y_{i,t-1}) * (1+dr)^{-t} + \sum_m \sum_t FCM_{mt} * (Z_{mt} - Z_{m,t-1}) * (1+dr)^{-t} \\ & + \sum_r \sum_t FCR_{rt} * (U_{rt} - U_{r,t-1}) * (1+dr)^{-t} + \sum_t \sum_c \sum_w \sum_i T_{wict} * \theta * X1_{wict} (1+dr)^{-t} \\ & + \sum_t \sum_c \sum_i \sum_r T_{irct} * \theta * X3_{irct} * (1+dr)^{-t} + \sum_t \sum_c \sum_i \sum_m T_{imct} * \theta * X2_{imct} * (1+dr)^{-t} \\ & + \sum_t \sum_c \sum_m \sum_s T_{msct} * \theta * X4_{msct} * (1+dr)^{-t} + \sum_t \sum_c \sum_i \sum_m OPM_{cmt} * X2_{imct} * (1+dr)^{-t} \\ & + \sum_t \sum_c \sum_i \sum_r OPR_{crt} * X3_{irct} * (1+dr)^{-t} + \sum_t \sum_c \sum_w \sum_i OPI_{cit} * X1_{wict} * (1+dr)^{-t}\}\end{aligned} \tag{22}$$

subject to the following constraints: (2)–(16).

where $PURM_{mct}$ is the price of component c at remanufacturing center m at time period t ($), and $PURR_{rct}$ is the price of component c at remanufacturing center r at time period t ($).

Table 18 Net profit

Scenario	RM demand ratio	Total profit ($)	Operating cost R ($)	Operating cost RM ($)	Transportation cost to R ($)	Transportation cost to RM ($)
0 (base case)		251,078,100	6,437,500	692,050,000	7,835,069	207,188,400
1	10	251,578,100	6,437,500	692,350,000	7,822,210	207,201,200
3	20	253,250,800	5,500,000	693,925,000	6,744,017	208,279,400
4	50	255,133,800	4,687,500	695,550,000	5,727,190	209,296,200

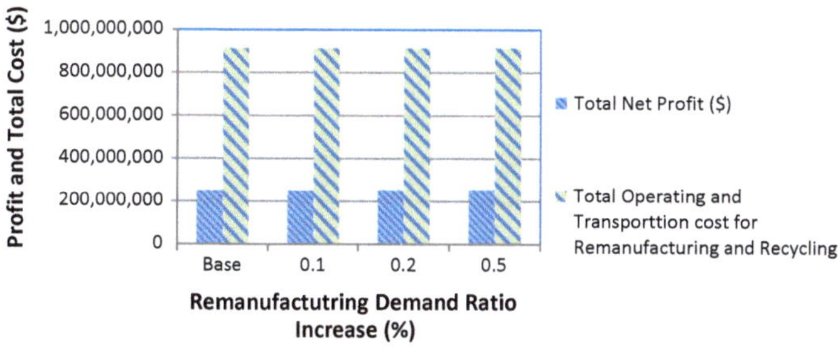

Fig. 8 Total network versus total operation and transportation cost

For the scenario analysis, the remanufacturing market demand is changed while keeping recycling demand constant. It can be seen that by increasing the remanufacturing market demand, total profit increases. Despite the increase in operating cost and transportation cost of remanufacturing activities, with increasing remanufacturing demand, there is an approximately 1% to 2% net profit increase. The results of these scenarios are presented in Table 18 and Fig. 8.

5 Conclusions

This chapter has presented a cost minimization model for minimizing the total transportation cost and operating cost of multi-type components for a multi-period reverse logistics network design for EOL WTs by using a scenario study approach. In comparison to previous literature on addressing EOL WTs, none of those studies addressed the RLN design for WTs. The proposed model will help the decision maker to choose the most suitable disposal method with the remanufacturing and recycling alternatives. Together with a baseline run of the current situation, various scenarios are modeled. The results of this study show that due to the high operating cost at remanufacturing center, sending most WTs to remanufacturing centers is costlier than sending them to recycling centers. In addition, it was shown that transportation cost depends on the amount of flow that has been sent to the recycling or remanufacturing center.

In addition, to help the decision maker, the ratio factor was added during the initial inspection/sorting phase. Even though, this would help the decision maker see what would be the reverse logistics network cost of different recycling and remanufacturing ratios, it would be essential to use reliable data to determine which WT components should be remanufactured or recycled. Reliability data can help decision makers decide on which option to use for EOL WTs; for instance, if the reliability of some of the EOL WT component is higher than the required threshold (i.e., 96%), then remanufacturing would be the best option, because the component has the ability to be brought back to "as good as new" condition. In addition, assuming that the reliability of certain components is lower than the required reliability, this may provide an idea of how much investment is required for the remanufacturing operation. By comparing the cost for each option, the decision maker could decide whether to remanufacture or recycle the WT components. In the future, it would be interesting to use reliability data for expanding the reverse logistics network for WTs.

The real-world reverse logistics network for WT EOL can be more complicated than the one considered in this paper. As such, some additions to the model are proposed, in order to extend the current MILP formulation to more realistic real-world RLN structures for a WT EOL network, including the following:

- Consider multiple types of WTs to evaluate dynamic situations.
- Incorporate landfilling (waste disposal) and inventory holding costs within the model.
- Include the randomly selected location of potential inspection, recycling, remanufacturing, and secondary market to make the model more widely applicable.
- Utilize complex stochastic programming techniques for developing a reverse logistics network to better account for the stochastic nature of the problem.

The main objective of many models developed and analyzed in the area of RLN optimization, logistics management, and transportation systems analysis is to minimize costs. Most recently, there is interest to incorporate environmental and social effects into the objective function. Opening remanufacturing or recycling centers definitely creates more job opportunities for local communities as well as reduces the negative effect of manufacturing new WTs. Adding an environmental constraint to the model by estimating carbon dioxide emissions due to transportation of the EOL WT or determining the correlation between remanufacturing WTs versus new WT manufacturing would be another contribution. Combining economic and environmental constraints would help to determine how to control CO_2 emissions by selecting the shortest distance between inspection and recycling/remanufacturing centers. Considering the positive environmental effects of remanufacturing, one should include the environmental constraints, modify the model objective, and run the model as a multi-objective problem.

References

Ancona, D., & McVeigh, J. (2001, August 29). Wind turbine—Materials and manufacturing fact sheet. Prepared for the Office of Industrial Technologies, U.S. Department of Energy by Princeton Energy Resources International, LLC. Retrieved July 12, 2014 from http://www.perihq.com/documents/WindTurbine-MaterialsandManufacturing_FactSheet.pdf.

Arvesen, A., & Hertwich, E. G. (2012). Assessing the life cycle environmental impacts of wind power: A review of present knowledge and research needs. *Renewable and Sustainable Energy Review, 16,* 5994–6006.

Atlantic County Utilities Authority. (2016). Title of page here. http://www.acua.com/disposal-recycling/.

Bauer, D., Diamond, D., Li, J., Sandalow, D., Telleen, P., & Wanner, B. (2010). U.S. Department of Energy critical materials strategy. Technical Report for U.S. Department of Energy. Retrieved 10 June, 2014 from http://www.osti.gov/scitech/servlets/purl/1000846.

Centers of Excellence. (2009). Environmental scan: Wind turbine technicians in California. Retrieved May 5, 2014 from http://www.coeccc.net/environmental_scans/wind_scan_sw_09.pdf.

Cho, R. (2012). Rare earth metals: Will we have enough? Blog from the Earth Institute, Columbia, University. Retrieved 7 July, 2014 from http://blogs.ei.columbia.edu/2012/09/19/rare-earth-metals-will-we-have-enough.

Geyer, R., & Jackson, T. (2004). Supply loops and their constraints: The industrial ecology of recycling and reuse. *California Management Review, 46*(2), 55–73.

Ghenai C. (2012). *Life cycle analysis of wind turbine* (Ocean and Mechanical Engineering Department, Florida Atlantic University). Retrieved 13 March, 2014 from http://cdn.intechopen.com/pdfs/29930.pdf.

Haapala, K. R., & Prempreeda, P. (2014). Comparative life cycle assessment of 2.0 MW wind turbines. *International Journal of Sustainable Manufacturing, 3*(2), 170–185.

Hahn, M., & Gilman, P. (2014). Offshore wind market and economic analysis. Retrieved 11 March. 2015 from http://energy.gov/sites/prod/files/2015/09/f26/2014-Navigant-Offshore-Wind-Market-Economic-Analysis.pdf.

Hulshorst, W. (2008). Repowering and used wind turbines. *Leonardo Energy*. Retrieved 23 June, 2013 from http://www.leonardo-energy.org/sites/leonardo-energy/files/root/pdf/2008/repowering-wind.pdf.

Krikke, H. R., van Harten, A., & Schuur, P. C. (1999). Business case Océ: Reverse logistics network re-design for copiers. *OR Spektrum, 21*(3), 381–409.

Knight, S. (2011). The gearbox repair market continues to grow. Retrieved 23 June, 2014 from http://www.windpowermonthly.com/article/1086978/gearbox-repair-market-continues-grow.

Lehner, F., & Roastogi, A. (2012). Securing the supply chain for wind and solar energy (RE-SUPPY). Final report. Report by E4tech and Avalon Consulting for the International Energy Agency. Retrieved 23 March, 2013 from http://iea-retd.org/wp-content/uploads/2012/11/RE-SUPPLY-final-report.pdf.

Life cycle assessment of electricity produced from onshore sited wind power plants based on Vestas V82-1.65 MW turbines. (2006). Retrieved 22 June, 2015 from https://www.vestas.com/~/media/vestas/about/sustainability/pdfs/lca%20v82165%20mw%20onshore2007.pdf.

Longitude Store.com. (2014). The haversine formula. Retrieved 10 July, 2014 from http://www.longitudestore.com/haversine-formula.html.

McDermott, M. (2009). Remanufactured wind turbines halve the price of new: Community wind power projects take note. Retrieved 13 June, 2014 from http://www.treehugger.com/renewable-energy/remanufactured-wind-turbines-halve-the-price-of-new-community-wind-power-projects-take-note.html[cited.

Milanese, A. (2009). Recyclability of wind turbines, current and future: Technical, economic and environmental. Dissertation. Master of Science in Renewable Energy and Resource

Management, The University of Glamorgan. Retrieved 14 June, 2015 from https://www.renooble.com/blog/wp-content/uploads/2012/04/Master_Thesis_on_recyclability_of_wind_turbines.pdf.
Post, W. (2013, January 10). Energy from wind turbines actually less than estimated? *The Energy Collective*. Retrieved 14 July, 2014 from http://www.theenergycollective.com/willem-post/169521/wind-turbine-energy-capacity-less-estimated.
Renew Energy Maintenance. (2012). Remanufacturing. Retrieved 24 May, 2015 from http://renewenergy.com/services/remanufacturing/.
Rydh, C. J., Jonsson, M., & Lindahl, P. K. (2004). Replacement of old wind turbines assessed from energy, environmental and economic perspectives. Retrieved 24 July, 2015 from http://citeseerx.ist.psu.edu/viewdoc/download?doi=10.1.1.524.8291&rep=rep1&type=pdf[cited.
Repowering Solutions. (2011). Remanufactured wind turbines—New opportunities for the wind industry. Retrieved 13 March, 2014 from http://www.repoweringsolutions.com/english/sales_brochure/Brochure_refurbished_wind_turbines.pdf.
Sandia National Laboratories. (2003). Cost study for large wind turbine blades: WindPACT blade system design studies. Sand Report. Retrieved 17 March, 2014 from http://windpower.sandia.gov/other/031428.pdf.
Sosa Skrainka, M. (2012). Analysis of the environmental impact on remanufacturing wind turbines. Master's Thesis, Rochester Institute of Technology.
Tegen, S., Hand, M., Maples, B., Lantz, E., Schwabe, P., & Smith, A. (2012). 2010 cost of wind energy review. National Renewable Energy Laboratory. Technical Report NREL/TP-5000-52920. Retrieved 11 April, 2015 from http://www.nrel.gov/docs/fy12osti/52920.pdf.
U.S. Department of Labor. (2010). Careers in wind energy. Retrieved 12 May, 2014 from http://www.bls.gov/green/wind_energy.
U.S. Inflation Calculator. (2008–2015). Current U.S. inflation rates: 2006–2016. Retrieved 20 March, 2016 from http://www.usinflationcalculator.com/inflation/current-inflation-rates/.
Walton, S., & Parker, D. (2008). The potential for remanufacturing of wind turbines. Product Group Study, Centre for Remanufacturing and Reuse. Retrieved April, 2015 from http://www.remanufacturing.org.uk/pdf/story/1p259.pdf?.

Maintenance Outsourcing Contracts Based on Bargaining Theory

Maryam Hamidi and Haitao Liao

Abstract We address a maintenance outsourcing problem where the owner of a piece of critical equipment plans on outsourcing preventive and failure replacement services to a service agent. The owner (i.e., customer) and the agent negotiate on the maintenance policy and spare part ordering strategy in the service contract. We first provide the classical Nash bargaining solution to the problem and analytically determine the optimal threat values the decision makers can use in negotiation. We then extend the model and show how the decision makers can increase their profits through a price discount scheme, which requires the total profit to be achieved at the maximum level. The total maximum profit is analytically determined, and the effects of the price discount scheme and threats on the individual and total profits are illustrated through a numerical study.

Keywords Maintenance outsourcing contract · Nash bargaining solution · Threat point · Price discount scheme

1 Introduction

Maintenance costs can account for 15–70% of the expenditures of companies and can even exceed companies' annual net profits (Ding and Kamaruddin 2014). To ensure the operational availability of critical equipment subject to failure and to facilitate maintenance and replacement activities, optimal preventive maintenance and spare parts ordering policies have been extensively studied (Nakagawa 2008; Jardine and

M. Hamidi (✉)
Department of Industrial Engineering, Lamar University, Beaumont, TX 77710, USA
e-mail: mhamidi@lamar.edu

H. Liao
Department of Industrial Engineering, University of Arkansas, Fayetteville, AR 72701, USA
e-mail: liao@uark.edu

A. Matsumoto (ed.), *Optimization and Dynamics with Their Applications*,
DOI 10.1007/978-981-10-4214-0_12

Tsang 2013; Nakagawa 2014). However, most of these studies assume that maintenance is performed in-house by the owner of the equipment.

Nowadays, maintenance outsourcing is a major trend in many business environments (Martin 1997). In particular, it is a common practice for hospital equipment, aircraft engine and brakes, mining machinery, and manufacturing processes (Tarakci et al. 2006). One of the main advantages of outsourcing is the cost reduction in operation, labor, and spare parts inventory. Besides, it let companies focus more on their core businesses (Wang 2010). According to Campbell (1995), 35% of north American businesses have considered outsourcing as some of their maintenance needs. For example, Federal Aviation Administration announced in 2007 that major air carriers outsourced an average of 64% of their maintenance expenses as opposed to 37% in 1996 (McFadden and Worrells 2012).

A challenging problem in maintenance outsourcing is how to design a contract agreeable to both parties (Martin 1997; Hartman and Laksana 2009). To deal with such processes with multiple decision makers, one of the most popular approaches is the use of game theory. In the literature, game theory has been used in many different areas (Jackson and Pascual 2008; Hamidi et al. 2014). Ashgarizadeh and Murthy (2000) employed the non-cooperative Stackelberg leader-follower concept to model maintenance outsourcing contracts and determined the agent's optimal pricing strategy and number of customers to service as well as the customers' optimal contract option. Hamidi et al. (2016) studied non-cooperative and cooperative game theoretic models for leasing contracts.

The related literature has overwhelmingly showed cases where decision makers negotiate over different contract terms (Nagarajan and Bassok 2008). Gurnani and Shi (2006) and Nagarajan and Sošić (2008) provided reviews of such contracts. Bajari et al. (2009) analyzed a comprehensive data set of building construction contracts and observed that almost half of the contracts were developed through negotiation. The study suggested that more complicated projects were more likely to be negotiated. Nash bargaining solution (Nash 1950) is the most frequently used solution concept in conflict resolution. There are several reasons for its popularity. First, if the conflict is considered as individual decision problem for the players who want to maximize their expected profits, then under certain conditions the Nash bargaining solution provides common optima (Matsumoto and Szidarovszky 2016). This solution is the only outcome of a bargaining problem satisfying certain fairness axioms (Cross 1965), and if one considers the dynamic bargaining process with offer dependent break-down probabilities (Szidarovszky 1999), then the process converges to the non-symmetric Nash bargaining solution and in special cases to the classical Nash solution (Szidarovszky 1999). So this solution models a fictitious bargaining process between the players. It is not an agreement between the players, it is the expected outcome if they follow a certain bargaining process. The above mentioned facts are reasons that we choose the Nash bargaining model.

We use bargaining game-theoretic approach to design contracts for the case where the owner of a piece of equipment (i.e., the customer of the service agent) plans on outsourcing preventive and failure replacement services to a service agent (who can be the original equipment manufacturer or a third party service provider). The customer decides on the preventive replacement age of the equipment and the agent decides on the ordering time for the required spare part. We will first study the classical Nash bargaining solution. Like many other interactive settings, such as litigation, international and political relations, each party can start bargaining by threatening the other player in order to improve his own position and decrease the other player's position (Anbarci et al. 2002). This will make the other player more reluctant to risk a conflict in negotiation (Harsanyi 1986; Myerson 1991). In our model, the customer makes threat against the agent using the replacement policy, and the agent threats the customer by spare part availability. We will analytically characterize the customer's and agent's threat values by maximin (max-min) values (Roth 1982; Myerson 1991; Thomas 2003).

However, this contract is not efficient since it does not maximize the total profit of the players (Cachon 2003). A possible solution is for the decision makers to cooperate when determining the terms of contract (Leng and Parlar 2005; Karsten et al. 2012; Schaarsberg et al. 2013; Matsumoto and Szidarovszky 2016), so that an outcome better than the classical Nash bargaining solution be achieved for them. We will extend the original bargaining process by including the preventive and failure replacement prices as decision variables and requiring that the total profit is on its maximum level. The solution of this extended bargaining process divides the excess profit equally among the players and also provides those values of the preventive and failure replacement prices, which will automatically lead to these payoff values. We will analytically determine the maximum total profit of the customer and agent, and through a four-step procedure, we show how the total maximum profit can be obtained if the agent adjusts the service charges. A numerical study shows how the policies and profit allocation alter through the use of threats and price discount.

The remainder of this paper is organized as follows. Section 2 provides a description of the problem and derives the payoff functions of the customer and the agent. Sections 3 and 4 describe how to model and solve negotiation through the classical Nash bargaining process and also its modified version through price discount, where Sects. 3.1 and 3.2 determine the threat points of the agent and customer, respectively. Section 4.1 calculates the maximum total profit, and Sect. 4.2 determines the price discount contract. In Sect. 5, we numerically examine the effect of negotiation with and without price discount on the outcome of the contract. Finally, Sect. 6 concludes the paper and outlines the directions for future research.

2 Problem Description and Model Formulation

The owner of a piece of equipment makes revenue R per unit time when the equipment is in operation and makes no revenue when it fails. The time to failure of the equipment, denoted by X, has a known probability density function (pdf) $f(x)$, cumulative distribution function (cdf) $F(x)$, and reliability function $\bar{F}(x) = 1 - F(x)$. The equipment's failure rate $\lambda(x) = f(x)/\bar{F}(x)$ is an increasing function of time (i.e., increasing failure rate).

The owner outsources preventive and failure replacement services to a service agent and thus becomes a customer of the service agent. If the agent and the customer come to an agreement, the service agent is responsible for doing preventive replacement at equipment age T_R and failure replacement whenever the equipment fails based on the contract. For both cases, spare parts are required to fulfill the service, and the equipment is as good as new after replacement. The service agent orders a spare part after time T_O followed by each service, and the lead time L is fixed. The service agent can hold at most one spare part with inventory holding cost of C_i per unit of time. We assume that $T_O + L \leq T_R$ (Armstrong and Atkins 1996) and $T_O \geq 0$ to ensure that the inventory is empty upon the arrival of a new spare part. When a replacement service is requested, there are two possibilities. If a spare part is on-hand, the agent does an immediate replacement; otherwise replacement is delayed until the ordered spare part arrives and the agent has to pay shortage cost S per unit time to the customer to compensate the downtime loss. The agent charges the customer P_p and P_f for each preventive and failure replacement, respectively, where $P_p \leq P_f$. In this paper, we first assume that the charges are exogenously determined by the market, and later relax this assumption by making them negotiable between the customer and the agent. For the agent, the cost C_f for performing each failure replacement is higher than the one C_p for preventive replacement. It is worth pointing out that Murthy and Yeung (1995) assumed a zero lead time and uniformly distributed repair time, but we consider a fixed lead time and instantaneous repairs.

In establishing the maintenance service contract, decision variables (i.e., terms to be specified in the contract) are the preventive replacement age T_R for the customer and the spare part reordering time T_O for the agent. According to game theoretic terminology, the two decision variables are called the strategies of the two players. Particularly, the set of simultaneous strategies is given by:

$$S = \{(T_R, T_O) \mid T_O \geq 0\,,\ T_O + L \leq T_R\}. \tag{1}$$

The expected profits per unit time can be naturally considered as the payoff functions of the players, which will be derived next.

Nomenclature

T_R	Preventive replacement age (decision variable of the customer)	Π_c^B	Customer's Nash bargaining profit
T_O	Spare part order time (decision variable of the agent)	Π_a^B	Agent's Nash bargaining profit
Π_c	Payoff function of the customer	$\bar{\Pi}_c$	Customer's extended Nash bargaining profit
Π_a	Payoff function of the agent	$\bar{\Pi}_a$	Agent's extended Nash bargaining profit
Π_c°	Threat value of the customer	Π	Total profit
Π_a°	Threat value of the agent	Π^*	Maximum total profit

2.1 Customer's Payoff Function

The customer's long-run profit per unit of time can be determined based on the renewal reward theorem (Ross 2013; Murthy and Yeung 1995), which can be expressed as the expected cycle profit divided by the expected cycle length. We define a service cycle length as the time interval between the installation of a new part and its replacement. Under the assumption that $T_O + L \leq T_R$, three scenarios may occur in a cycle. The first one is that the equipment fails before the agent receives a spare part, i.e., $X < T_O + L$. In this case, replacement is delayed until the agent receives the part at $T_O + L$, so the corresponding cycle length is $T_O + L$. The customer's profit in such a cycle is $RX + S(T_O + L - X) - P_f$, as the agent must compensate the shortage cost to the customer. The second scenario is that a failure occurs after the arrival of the ordered part while before the preventive replacement, i.e., $T_O + L < X < T_R$. In this case, the agent replaces the failed part immediately, so the cycle length is X and the customer's profit is $RX - P_f$. Lastly, when the equipment does not fail before the scheduled preventive replacement time, i.e., $X > T_R$, the agent performs preventive replacement at T_R, so the cycle length is T_R and the customer's profit is $RT_R - P_p$.

Considering these scenarios, the expected cycle profit for the customer can be expressed as:

$$EPC = \int_0^{T_O+L} (Rx + S(T_O + L - x) - P_f)f(x)dx + \int_{T_O+L}^{T_R} (Rx - P_f)f(x)dx + \int_{T_R}^{\infty} (RT_R - P_p)f(x)dx.$$

After simplification, we have:

$$EPC = R\int_0^{T_R} \bar{F}(x)dx + S\int_0^{T_O+L} F(x)dx - P_f F(T_R) - P_p \bar{F}(T_R).$$

On the other hand, the expected cycle length can be expressed as:

$$ECL = \int_0^{T_O+L} (T_O + L)f(x)dx + \int_{T_O+L}^{T_R} xf(x)dx + \int_{T_R}^{\infty} T_R f(x)dx = T_R - \int_{T_O+L}^{T_R} F(x)dx.$$

As a result, the long-run profit per unit time for the customer is given by:

$$\Pi_c = \frac{EPC}{ECL} = \frac{R\int_0^{T_R} \bar{F}(x)dx + S\int_0^{T_O+L} F(x)dx - P_f F(T_R) - P_p \bar{F}(T_R)}{T_R - \int_{T_O+L}^{T_R} F(x)dx}. \quad (2)$$

2.2 Agent's Payoff Function

The expected cycle profit for the agent can also be determined based on the three scenarios: if $X < T_O + L$, the agent's profit is $P_f - C_f - S(T_O + L - X)$; if $T_O + L < X < T_R$, the profit is $P_f - C_f - (X - T_O - L)C_i$, and it should be noted that the agent pays the holding cost in such a cycle; if $X > T_R$, the profit is $P_p - C_p - (T_R - T_O - L)C_i$. Therefore, the expected cycle profit for the agent can be expressed as:

$$\begin{aligned} EPA = &\int_0^{T_O+L} (P_f - C_f - S(T_O + L - x))f(x)dx \\ &+ \int_{T_O+L}^{T_R} (P_f - C_f - (x - T_O - L)C_i)f(x)dx \\ &+ \int_{T_R}^{\infty} (P_p - C_p - (T_R - T_O - L)C_i)f(x)dx. \end{aligned}$$

After simplification, we have:

$$EPA = (P_f - C_f)F(T_R) + (P_p - C_p)\bar{F}(T_R) - S\int_0^{T_O+L} F(x)dx - C_i\int_{T_O+L}^{T_R} \bar{F}(x)dx,$$

and the agent's long-run profit per unit time can be expressed as:

$$\begin{aligned} \Pi_a &= \frac{EPA}{ECL} \\ &= \frac{(P_f - C_f)F(T_R) + (P_p - C_p)\bar{F}(T_R) - S\int_0^{T_O+L} F(x)dx - C_i\int_{T_O+L}^{T_R} \bar{F}(x)dx}{T_R - \int_{T_O+L}^{T_R} F(x)dx}. \end{aligned} \tag{3}$$

3 Nash Bargaining Solution

The decision-making process involves bargaining, where the customer and agent determine their strategies T_R and T_O via negotiation. We model the bargaining process by the Nash bargaining model (Nash 1953). Nash presented six axioms that all bargaining solutions should satisfy and proved that there is a unique solution that meets these axioms (Nash 1950). In particular, the six axioms are: (1) symmetry meaning that if the players are identical, they receive identical payoffs; (2) feasibility requiring that the players can distribute only existing amount of profit; (3) Pareto optimality showing that if both players can increase their payoffs, this solution has to be included in the agreement; (4) independence from monotone increasing linear transformations stating that changing the unit in which the payoffs are computed cannot change the solution; (5) rationality, so players do not agree with payoff values which are below payoffs they can receive without bargaining; (6) independence from unfavorable strategies meaning that if additional constraints restrict the feasible set and the solution still remains feasible, then the solution must remain the same. A thorough explanation of the axioms can be seen in Roth (1979). The Nash bargaining solution can be obtained using different concepts as well. If the bargaining process is considered as individual decision problem for each player under uncertain choice of the other player, then the Nash bargaining solution is the common decision maximizing both expected profits (Matsumoto and Szidarovszky 2016), and if a modified version of the alternating offer bargaining process is considered with offer dependent break-down probabilities, then the Nash solution is the limit, the final outcome of the process. So a fictitious bargaining process is considered instead of cooperative agreement between the players.

In our case, the Nash bargaining solution (Π_c^B, Π_a^B) is defined as the profits of the customer and agent, which maximize the product of the differences between the payoff functions and fixed disagreement payoffs given by the following optimization problem:

$$\max \quad \left(\Pi_c - \Pi_c^\circ\right)\left(\Pi_a - \Pi_a^\circ\right) \tag{4}$$

$$\text{subject to} \quad \Pi_c \geq \Pi_c^\circ, \tag{5}$$

$$\Pi_a \geq \Pi_a^\circ, \tag{6}$$

$$(\Pi_c, \Pi_a) \in H, \tag{7}$$

where Π_c and Π_a are customer's and agent's payoffs given by (2) and (3), respectively, Π_c° and Π_a° are the disagreement points, and H is the payoff set $H = \{(\Pi_c, \Pi_a)|(T_R, T_O) \in S\}$ where S is the strategy set defined in (1).

It is important to mention that the disagreement payoff vector $(\Pi_c^\circ, \Pi_a^\circ)$ is defined as the guaranteed payoff obtained by the players in case they disagree to negotiate or negotiation breaks down. To determine the disagreement point, different alternatives are available. One is to let $(\Pi_c^\circ, \Pi_a^\circ) = (0, 0)$, another possibility is a non-cooperative equilibrium, and a third one is to select it as a threat point (Myerson 1991). Here we assume that the customer needs to outsource the maintenance services, and the agent is the only service provider, and on the other hand, the equipment owner is the major customer of the service agent. That is, some sort of business has to take place between the customer and the agent, so $(0, 0)$ is not the best choice. Here, we consider the threat point as the disagreement point for the players (Anbarci et al. 2002). The main purpose of making threat is to increase the cost of possible conflict to the other player, in order to make him negotiate and agree on mutually satisfactory strategies (Harsanyi 1986). However, if the customer and agent have the possibility to avoid each other and to have business with others, the disagreement payoffs for both players can be considered as zero. Constraints (5)–(6) assert that neither player should get less than $(\Pi_c^\circ, \Pi_a^\circ)$ in the bargaining, since this is the profit they could get without negotiation. So the Nash bargaining payoff is at or above this security level for each player. Figure 1 shows how posing threat shrinks the feasible set H.

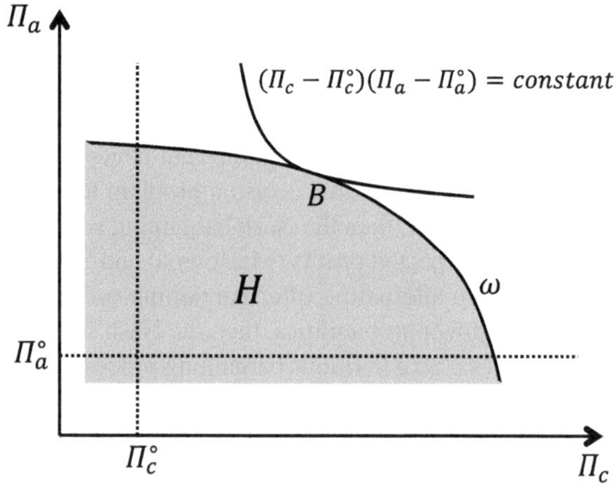

Fig. 1 Nash bargaining solution

The optimal solution of problem (4)–(7) is Pareto optimal. That is, there is no other feasible solution which is better than the Nash bargaining solution for one player and not worse for the other player. In other words, it ensures that all other solutions, which make one player better off, make the other player worse off. In this case, the Pareto frontier, ω, can be determined by $\underset{T_R,T_O}{max}(1-\theta)\,\Pi_c(T_R,T_O)+\theta\,\Pi_a(T_R,T_O)$ where $\theta \in [0,1]$. At $\theta = 0$, the objective function considers only the payoff function of the customer, at $\theta = 1$ it considers only the agent, and for values between 0 and 1 it considers some trade-off between the profits. For each constant $\theta \in [0,1]$, the optimum strategy, (T_R^θ, T_O^θ), and the corresponding profits $\Pi_c(T_R^\theta, T_O^\theta)$ and $\Pi_a(T_R^\theta, T_O^\theta)$ can be calculated, and the entire Pareto frontier can be determined. The Nash Bargaining solution (T_R^B, T_O^B) selects the unique point from the Pareto frontier, which maximizes the objective function in (4) while satisfying constraints (5) and (6). The parameters of the contract are (T_R^B, T_O^B), which determine the profits of the customer and agent as $\Pi_c^B = \Pi_c(T_R^B, T_O^B)$ and $\Pi_a^B = \Pi_a(T_R^B, T_O^B)$. Figure 1 also illustrates a Pareto frontier, ω, and the corresponding Nash profits of the customer and agent $B = (\Pi_c^B, \Pi_a^B)$.

We next derive $(\Pi_c^\circ, \Pi_a^\circ)$, the threat point of the customer and agent, in Sects. 3.1 and 3.2, respectively, by the maximin (max-min) value. However, because of the very different non-algebraic properties of the Nash bargaining solution (Anbarci et al. 2002), it is difficult to derive the solution of (4)–(7) analytically. Instead, we will look closely into this solution using simulation in a numerical study.

3.1 Threat Point for the Agent

To make the agent reluctant to cause conflict in negotiation, the customer can adopt the threat strategy $T_{Ra}^\circ = \underset{T_R}{arg\,min}\,\Pi_a$ in case of disagreement, which causes the greatest damage to the agent. Given the threat of the customer, the agent improves his bargaining position by choosing strategy $T^{\circ}_{Oa} = \underset{T_O}{arg\,max}\, \Pi_a(T^{\circ}_{Ra})$, which maximizes his payoff (maximin strategy). Therefore, the threat profit Π°_a (maximin profit) is the guaranteed payoff for the agent in the worst case, and the agent won't agree on any less profit when he negotiates. Technically, the agent's threat payoff, Π°_a, and the corresponding optimal strategies (Harsanyi 1956) can be determined by solving the following problem:

$$\Pi^{\circ}_a = \max_{T_O} \min_{T_R} \frac{(P_f - C_f)F(T_R) + (P_p - C_p)\bar{F}(T_R) - S\int_0^{T_O+L} F(x)dx - C_i\int_{T_O+L}^{T_R} \bar{F}(x)dx}{T_R - \int_{T_O+L}^{T_R} F(x)dx}$$
$$\text{subject to} \quad 0 \leq T_O \leq T_R - L. \tag{8}$$

where the objective function is the payoff function of the agent, Π_a. To solve this problem, we consider a two-step optimization process. In the first step, we find the customer's threat strategy against the agent T°_{Ra}; in the second step, we determine the reordering time T°_{Oa} that maximizes the agent's payoff function with the value of T°_{Ra} determined in the first step.

In particular, the first step solves:

$$\min_{T_R} \frac{(P_f - C_f)F(T_R) + (P_p - C_p)\bar{F}(T_R) - S\int_0^{T_O+L} F(x)\,dx - C_i\int_{T_O+L}^{T_R} \bar{F}(x)\,dx}{T_R - \int_{T_O+L}^{T_R} F(x)\,dx}$$
$$\text{subject to} \quad T_O + L \leq T_R.$$

We assume that the profits the agent obtains from failure replacement and preventive replacement are the same (i.e., $P_f - C_f = P_p - C_p$). As a result, the numerator of the derivative of the objective function with respect to T_R divided by $\bar{F}(T_R)$ is:

$$D(T_O) = -(P_p - C_p) + S\int_0^{T_O+L} F(x)\,dx - C_i(T_O + L),$$

which is independent of T_R. The following proposition provides the optimum T_R and the corresponding conditions (see Appendix for the proof).

Proposition 1 *For any fixed T_O, the value of T_R that minimizes the agent's payoff function is: $T^{\circ}_{Ra} = \infty$ if $D(T_O) < 0$, $T^{\circ}_{Ra} = T_O + L$ if $D(T_O) > 0$, and all $T_R \geq T_O + L$ if $D(T_O) = 0$.*

The next step is to determine the maximum value of Π_a with respect to T_O. Due to the tedious computation required to obtain analytical results, we will only consider one parameter set corresponding to the case of $D(T_O) < 0$, and the other cases of $D(T_O) \geq 0$ can be considered similarly. Substituting $T^{\circ}_{Ra} = \infty$ into Π_a, the problem to be solved is:

$$\max_{T_O} \quad \frac{P_f - C_f - S\int_0^{T_O+L} F(x)\,dx + C_i(T_O + L)}{\mu + \int_0^{T_O+L} F(x)\,dx} - C_i \tag{9}$$

subject to $\quad T_O \geq 0,$

where $\mu = \int_0^{\infty} xf(x)\,dx$. The numerator of the derivative of the objective function with respect to T_O is:

$$G(T_O) = C_i\left(\mu + \int_0^{T_O+L} F(x)\,dx\right) - F(T_O + L)\left(P_f - C_f + C_i(T_O + L) + S\mu\right). \tag{10}$$

The following theorem provides the optimum T_O and the corresponding conditions (see Appendix for the proof).

Theorem 1 *The objective function in (9) is unimodal and pseudo-concave in T_O. For $T^{\circ}_{Ra} = \infty$, the value of T_O that maximizes the agent's payoff function satisfies $G(T^{\circ}_{Oa}) = 0$.*

In this case, the following proposition provides the threat point (maximin payoff) for the agent, which can be obtained by substituting (10) into the objective function in (9).

Proposition 2 *The threat value for the agent is $\Pi^{\circ}_a = -S + C_i\frac{\bar{F}(T^{\circ}_{Oa}+L)}{F(T^{\circ}_{Oa}+L)}$ if $D(T^{\circ}_{Oa}) < 0$, and the corresponding threat strategies are $T^{\circ}_{Ra} = \infty$ and $G(T^{\circ}_{Oa}) = 0$.*

Proposition 2 shows that the customer threatens the agent by the threat strategy $T^{\circ}_{Ra} = \infty$ (do not do preventive replacement), which hurts the agent as much as possible by minimizing agent's payoff. Given the threat of the customer, the agent improves his bargaining position by choosing strategy T°_{Oa} which maximizes his payoff, and the corresponding threat payoff $\Pi^{\circ}_a = \Pi_a(T^{\circ}_{Ra}, T^{\circ}_{Oa})$ is the agent's security payoff in the bargaining process.

3.2 Threat Point for the Customer

The agent threatens the customer that if he causes disagreement in negotiation, the agent will implement the threat strategy, $T^{\circ}_{Oc} = \arg\min_{T_O} \Pi_c$, which hurts the customer as much as possible by minimizing customer's payoff. The purpose of the agent by making threat against the customer is to make him more reluctant to risk a conflict in negotiation. Given the threat of the agent, the customer improves his bargaining position by choosing strategy $T^{\circ}_{Rc} = \arg\max_{T_R} \Pi_c(T^{\circ}_{Oc})$ which maximizes his payoff. The customer's threat profit, Π°_c, can be determined by solving the following problem:

$$\Pi_c^\circ = \max_{T_R} \min_{T_O} \quad \frac{R\int_0^{T_R} \bar{F}(x)dx + S\int_0^{T_O+L} F(x)dx - P_f F(T_R) - P_p \bar{F}(T_R)}{T_R - \int_{T_O+L}^{T_R} F(x)dx} \tag{11}$$
$$\text{subject to} \quad 0 \le T_O \le T_R - L,$$

where the objective function is the customer's payoff function (2). In order to solve the problem, we consider a two-step optimization approach: in the first step, we find T_{Oc}° as the ordering time that minimizes the customer's payoff function for a given value of T_R; and next, we determine T_{Rc}°, as the optimum preventive replacement time that maximizes the customer's payoff function with the value of T_{Oc}° determined in the first step (Danskin 1966). In the first step, the following problem is solved:

$$\min_{T_O} \quad \frac{R\int_0^{T_R} \bar{F}(x)\,dx + S\int_0^{T_O+L} F(x)\,dx - P_f F(T_R) - P_p \bar{F}(T_R)}{T_R - \int_{T_O+L}^{T_R} F(x)\,dx}$$
$$\text{subject to} \quad 0 \le T_O \le T_R - L.$$

Clearly, the numerator of the derivative of the objective function with respect to T_O divided by $F(T_O + L)$ is independent of T_O:

$$W(T_R) = (S - R)\int_0^{T_R} \bar{F}(x)\,dx + P_f F(T_R) + P_p \bar{F}(T_R). \tag{12}$$

The following proposition provides the optimum T_O and the corresponding condition (see Appendix for the proof).

Proposition 3 *For any fixed T_R, the value of T_O that minimizes the customer's payoff function is $T_{Oc}^\circ = 0$ when $W(T_R) > 0$, and $T_{Oc}^\circ = T_R - L$ when $W(T_R) < 0$.*

Proposition 3 indicates that the agent can pose threat against the customer by adjusting spare part availability. That is, the agent can delay a failure replacement when the customer prefers an instant replacement, or he can do an instant replacement when the customer prefers a delayed one. In other words, the spare part availability can affect the number of failure replacements and the customer's profit.

The next step for the customer is to choose the action T_{Rc}° that maximizes the worst-case payoff. We will only consider the case for $W(T_R) > 0$, and the case for $W(T_R) < 0$ can be considered similarly. By substituting $T_{Oc}^\circ = 0$ into Π_c, the problem for the customer to solve is:

$$\max_{T_R} \quad \frac{R\int_0^{T_R} \bar{F}(x)\,dx + S\int_0^{L} F(x)\,dx - P_f F(T_R) - P_p \bar{F}(T_R)}{T_R - \int_L^{T_R} F(x)\,dx}$$
$$\text{subject to} \quad T_R - L \ge 0. \tag{13}$$

The numerator of the derivative of the objective function with respect to T_R divided by $\bar{F}(T_R)$ is:

$$B(T_R) = (P_p - P_f)\lambda(T_R)(T_R - \int_L^{T_R} F(x)\,dx) + (R - S)\int_0^L F(x)\,dx + P_f F(T_R) + P_p \bar{F}(T_R). \tag{14}$$

The following theorem gives the optimality conditions (see Appendix for the proof).

Theorem 2 *The objective function in (13) is unimodal and pseudo-concave in T_R. For $T^{\circ}_{Oc} = 0$, the value of T_R that maximizes the customer's payoff function satisfies $B(T^{\circ}_{Rc}) = 0$.*

The following proposition provides the threat point for the customer, which can be proved by substituting $B(T_R) = 0$ into the objective function in (13).

Proposition 4 *The threat point for the customer is $\Pi^{\circ}_c = (P_p - P_f)\lambda(T^{\circ}_{Rc}) + R$ if $W(T^{\circ}_{Rc}) > 0$, and the corresponding threat strategies are $T^{\circ}_{Oc} = 0$ and $B(T^{\circ}_{Rc}) = 0$.*

Proposition 4 shows that the agent can threaten the customer with the threat strategy $T^{\circ}_{Oc} = 0$ (doing failure replacement when the customer prefers a delayed replacement), which minimizes the customer's payoff. Given the threat of the agent, the customer improves his bargaining position by choosing strategy T°_{Rc} that maximizes his payoff in this worst case scenario, and the corresponding threat payoff of the customer Π°_c is his security payoff in bargaining. Based on the obtained threat points the Nash bargaining solution can be obtained by solving the optimum problem (4)–(7) which provides the corresponding strategies and payoff values of the players as the solution of the bargaining process.

In many bargaining situations the players realize that by considering the interest of each other simultaneously their payoffs can be increased in comparison to the Nash overall profit, which will be divided among the players in a fair, mutually acceptable way. Another way of establishing the highest possible overall profit is that one player selects a strategy such that with the corresponding best response of the other player they get the maximum possible overall profit.

4 Extended Nash Bargaining Solution

In the previous section, we focused on the results for classical Nash bargaining model, but this solution generally does not lead to the maximum total profit for the customer and agent. An alternative solution is for the decision makers to act cooperatively in order to increase their total profit (Nagarajan and Sošić 2008). Next, we will study a second bargaining problem in which the total profit of the players equals its maximum level, and both players get higher payoff than in the case of Nash bargaining solution. In this modified bargaining process the following optimization problem has to be solved:

$$\begin{aligned} \max \quad & \left(\Pi_c - \Pi_c^B\right)\left(\Pi_a - \Pi_a^B\right) \\ \text{subject to} \quad & \Pi_c \geq \Pi_c^B, \\ & \Pi_a \geq \Pi_a^B, \\ & \Pi_c + \Pi_a = \Pi^*, \end{aligned} \tag{15}$$

where Π_c^B and Π_a^B are the payoffs at the Nash bargaining solution, and Π^* is the maximum total profit of the players. The last constraint requires that the total profit of players is at its maximal level.

In the previous discussion the decision variables were preventive replacement age, T_R, and spare part order time, T_O. So the prices of replacement services were considered given and were not part of negotiation process. In this alternative model, we require that the overall profit of the players is at maximum level, and the subject of the bargaining is a set of four decision variables: T_R, T_O and the failure and preventive replacement prices, P_f and P_p. That is, Π_c and Π_a now depend on the four decision variables. Next, we will first determine the maximum total profit Π^* and the corresponding strategies (T_R^*, T_O^*) in Sect. 4.1, which satisfy the last constraint of problem (15), and second we will solve the problem in Sect. 4.2 by determining the profits $(\bar{\Pi}_c, \bar{\Pi}_a)$ and the corresponding discounted prices $(\bar{P}_p, \bar{P}_f)$.

4.1 Maximum Total Profit

In a cooperative regime, the players choose the set of strategies (T_R^*, T_O^*), that solves $\Pi^* = max\ \{\Pi_c + \Pi_a\}$, where Π_c and Π_a are the payoffs of the customer and the agent given in (2) and (3), respectively. We determine the optimal values by solving:

$$\begin{aligned} \Pi^* = & \max_{T_O, T_R} \quad \{\Pi_c + \Pi_a\} = \frac{R\int_0^{T_R} \bar{F}(x)\,dx - C_i\int_{T_O+L}^{T_R} \bar{F}(x)\,dx - C_f F(T_R) - C_p \bar{F}(T_R)}{T_R - \int_{T_O+L}^{T_R} F(x)\,dx} \\ & \text{subject to} \quad 0 \leq T_O \leq T_R - L. \end{aligned} \tag{16}$$

In this formulation, the customer and agent seek to jointly maximize their total profit. We consider a two-step optimization process to solve the problem: in the first step, we find T_R^* as the optimum preventive replacement age that maximizes the total payoff function for a given value of T_O; in the second step, we determine T_O^* as the optimum ordering time that maximizes the total payoff function for the value of T_R^* determined in the first step. One can see that the numerator of the derivative of the objective function with respect to T_R is:

$$K(T_R) = -(C_f - C_p)\lambda(T_R)(T_R - \int_{T_O+L}^{T_R} F(x)\,dx) + C_f F(T_R)$$
$$+ C_p \bar{F}(T_R) + R\int_0^{T_O+L} F(x)\,dx - C_i(T_O + L).$$

The following proposition gives the optimum T_R value and the corresponding condition (see Appendix for the proof).

Proposition 5 *For any fixed T_O, the value of T_R that maximizes the total payoff function is $T_R^* = T_O + L$, when $K(T_R = T_O + L) < 0$.*

The next step is to determine the maximum of $\Pi_c + \Pi_a$ with respect to T_O. After substituting $T_R^* = T_O + L$ into $\Pi_c + \Pi_a$, problem (16) becomes:

$$\max_{T_O} \quad \frac{R\int_0^{T_O+L} \bar{F}(x)\,dx - C_f F(T_O + L) - C_p \bar{F}(T_O + L)}{T_O + L} \tag{17}$$
$$\text{subject to} \quad T_O \geq 0.$$

The numerator of the derivative of the objective function with respect to T_O is:

$$H(T_R = T_O + L) = [R\bar{F}(T_R) - (C_f - C_p)f(T_R)]T_R - R\int_0^{T_R} \bar{F}(x)\,dx + C_f F(T_R) + C_p \bar{F}(T_R). \tag{18}$$

Here we assume $Rf(T_R) + f'(T_R)(C_f - C_P) > 0$ to ensure that the objective function in (17) is a unimodal and pseudo-concave function in T_O. The following proposition addresses the optimum solution (see Appendix for the proof).

Proposition 6 *The total maximum profit is $\Pi^* = R\bar{F}(T_R^*) - (C_f - C_p)f(T_R^*)$, and the corresponding strategies are $(T_R^*, T_O^*) = (T_O^* + L, H(T_O^*) = 0)$ when $K(T_R^*, T_O^*) < 0$.*

Choosing strategies (T_R^*, T_O^*) by the customer and agent guarantees that the total profit is at its maximum level Π^*, satisfying the last constraint of problem (15).

4.2 Price Discount Contract

Now, we define $\Delta\Pi$ as:

$$\Delta\Pi = \Pi^* - \Pi^B > 0 \tag{19}$$

where $\Pi^B = \Pi_c^B + \Pi_a^B$ is the sum of the Nash bargaining profits. Notice that $\Delta\Pi$ is the difference between the total maximum profit and the total profit of the players at the Nash bargaining solution. We call $\Delta\Pi$ the joint profit gain, which is the extra profit the players jointly achieve by cooperation. The outcome of problem (15) can be described as follows.

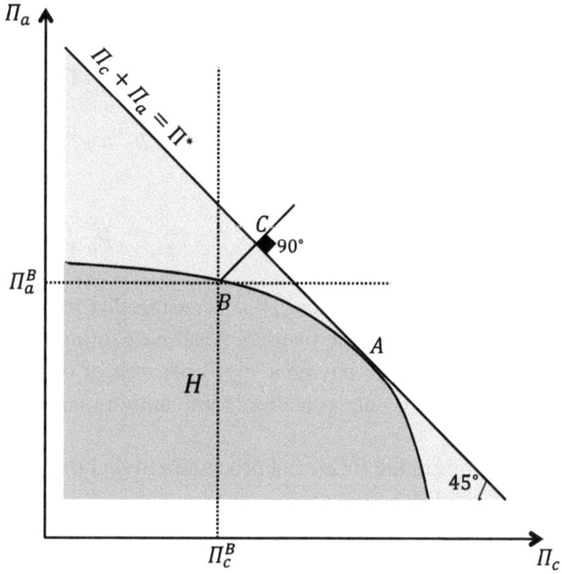

Fig. 2 Extended Nash bargaining solution

Proposition 7 *The solution of the modified Nash bargaining process shows that the profits of the customer and agent respectively are:*

$$\bar{\Pi}_c = \Pi_c^B + \frac{\Delta\Pi}{2} = \frac{\Pi^* + \Pi_c^B - \Pi_a^B}{2},$$
$$\bar{\Pi}_a = \Pi_a^B + \frac{\Delta\Pi}{2} = \frac{\Pi^* + \Pi_a^B - \Pi_c^B}{2}, \tag{20}$$

The proof can be found in Muthoo (1999). Proposition 7 states that when the players cooperate then they can equally increase their payoffs in comparison to the Nash bargaining solution by equally sharing the joint profit gain, $\Delta\Pi$. It can be proven that this solution also coincides with the Shapley values that are also based on certain fairness axioms (Shapley 1952). This solution is illustrated in Fig. 2. The shaded region H is the feasible set of the original Nash bargaining problem, point A gives Π^*, the maximum overall profit on this region. In the modified bargaining process we introduce the $-45°$ line passing through point A, and the feasible set is expanded to the entire triangle under this line, $\Pi_c + \Pi_a = \Pi^*$. The original Nash bargaining solution is point $B = (\Pi_c^B, \Pi_a^B)$, where no player accepts any profit less than this, and the solution of the modified bargaining process is point $C = (\bar{\Pi}_c, \bar{\Pi}_a)$. Since the players have equal gains, as stated by Proposition 7, the BC segment has unit slope.

In addition to the optimal profits, the solution of problem (15) presents the preventive and failure replacement prices as follows.

Proposition 8 *The negotiated preventive and failure replacement prices are given by:*

$$\bar{P}_p = P_p - \tau T_R^*, \tag{21}$$

$$\bar{P}_f = P_f - \tau T_R^*, \tag{22}$$

where

$$\tau = \Pi_a(T_R^*, T_O^*, P_p, P_f) - \bar{\Pi}_a. \tag{23}$$

The discounted prices $(\bar{P}_p, \bar{P}_f)$ ensure that the customer's and agent's profits are $\Delta\Pi/2$ higher than the original Nash bargaining profits. The discounted prices can be interpreted as a payment $\tau \geq 0$ per unit of time from the agent to the customer. Because of this interpretation some authors call this kind of price setting as side payment.

We present the following procedure to find the proper bargaining profits and decision variables.

1. Use Propositions 2 and 4 to compute threat point $(\Pi_c^\circ, \Pi_a^\circ)$.
2. Compute the Nash bargaining solution of problem (4)–(7) with profits (Π_c^B, Π_a^B) and the corresponding strategies (T_R^B, T_O^B), also derive $\Pi^B = \Pi_c^B + \Pi_a^B$.
3. If the players wish to cooperate, use Proposition 6 to compute the maximum total profit Π^* and the corresponding strategies (T_R^*, T_O^*), also derive the joint profit gain $\Delta\Pi = \Pi^* - \Pi^B$.
4. Use Propositions 7 and 8 to calculate the extended profits $(\bar{\Pi}_c, \bar{\Pi}_a)$ and the negotiated preventive and failure maintenance prices, $\bar{P}_p$ and $\bar{P}_f$.

Therefore, if the agent chooses discounted prices $\bar{P}_p$ and $\bar{P}_f$ and the players select the preventive replacement time and the spare part order time (T_R^*, T_O^*), then both players enjoy higher profits than in the original Nash bargaining solution, and their total profit is at its maximum level.

5 Numerical Examples

In this section, numerical examples are presented to illustrate the application of the two bargaining models for the customer and agent. We begin with a base case where the equipment's time to failure distribution is assumed to be Weibull with pdf $f(x) = \frac{\beta}{\alpha}(\frac{x}{\alpha})^{\beta-1}e^{-(x/\alpha)^\beta}$, with scale parameter $\alpha = 40$, and shape parameter $\beta = 3$. Other parameters assumed for the problem setting are order lead time $L = 10$ days, inventory cost $C_i = \$10$/day, equipment's generated revenue $R = \$30$/day, failure replacement cost $C_f = \$300$, preventive replacement cost $C_p = \$100$, price of each failure replacement $P_f = \$600$, price of each preventive replacement $P_p = \$400$, and shortage cost $S = \$30$/day.

In Tables 1 and 2, we study the sensitivity of the Nash bargaining solution and the extended Nash bargaining solution to the scale parameter α and revenue R. For the Nash bargaining model, we numerically obtain the threat profit of the customer Π_c° and the threat profit of the agent Π_a°. Next, the bargaining solutions to the problem in (4)–(7) are calculated, where T_R^B is the preventive replacement age of the equipment, T_O^B is the time the agent has to order the spare part after each replacement, $\Pi_c^B = \Pi_c(T_R^B, T_O^B)$ is the profit of the customer, $\Pi_a^B = \Pi_a(T_R^B, T_O^B)$ is the profit of the agent, and $\Pi^B = \Pi_c^B + \Pi_a^B$ is the total profit of the players. For the extended Nash bargaining solution, we numerically solve the problem in (16) and derive the maximum total profit Π^* and the optimal preventive replacement age and ordering time, $(T_R^*, T_O^*) = \underset{T_R, T_O}{arg\,max}\,(\Pi_c + \Pi_a)$. The negotiated preventive and failure maintenance prices $(\bar{P}_p, \bar{P}_f)$ and the expected payoffs $\bar{\Pi}_c = \Pi_c(T_R^*, T_O^*, \bar{P}_p, \bar{P}_f)$ and $\bar{\Pi}_a = \Pi_a(T_R^*, T_O^*, \bar{P}_p, \bar{P}_f)$ are also presented.

One can see that the threat profits and the bargaining outcomes are highly sensitive to the scale parameter of the equipment α and also generated revenue R. The customer has a higher bargaining position when he owns a more reliable equipment. Also the higher the revenue generated by the equipment, the higher the bargaining position of the customer. For example, when the scale parameter increases from $\alpha = 30$ to 60 (mean time to failure increases from 26.78 days to 53.57), the threat profit of the customer increases from $\Pi_c^\circ = \$8.36$/day to 19.14, and his bargaining profit increases from $\Pi_c^B = \$9.99$/day to 19.97.

In Tables 3 and 4, we evaluate the results of the extended Nash bargaining model. We define the relative increase in the agent's profit in the price discount contract over the general Nash bargaining contract as:

$$\Delta\Pi_a = \frac{\bar{\Pi}_a - \Pi_a^B}{\Pi_a^B} 100\%.$$

The relative increase in profit of the customer and total profit are also determined in the same manner. Furthermore, we define ΔT_R and ΔP_p, the relative changes in preventive replacement age and price as:

$$\Delta T_R = \frac{T_R^* - T_R^B}{T_R^B} 100\%,$$

$$\Delta P_p = \frac{\bar{P}_p - P_p}{P_p} 100\%,$$

Tables 3 and 4 show the advantages of price-discount contract over the general Nash bargaining model. As can be seen, price discount contract increases the maintenance quality of equipment, since the preventive maintenance age in price-discount contract is shorter than the general bargaining contract, $\Delta T_R < 0$. Besides, this scheme is economically beneficial to both parties. At $\alpha = 50$, although the agent

Table 1 Nash bargaining and extended Nash bargaining strategies and payoffs for different scale parameter values

Nash bargaining							
Scale parameter α	Customer's threat Π_c° (\$/day)	Agent's threat Π_a° (\$/day)	Maintenance policy T_R^B (day)	Order time T_O^B (day)	Customer's profit Π_c^B (\$/day)	Agent's profit Π_a^B (\$/day)	Total profit Π^B (\$/day)
30	8.36	6.34	24	14	9.99	9.16	19.15
40	13.73	3.65	31.9	21.9	14.96	6.09	21.05
50	16.97	2.05	39.9	29.9	17.97	4.20	22.17
60	19.14	0.98	47.8	37.8	19.97	2.97	22.94
Extended nash bargaining							
Scale parameter α	Preventive price $\bar{P}_p$ (\$)	Failure price $\bar{P}_f$ (\$)	Maintenance policy T_R^* (day)	Order time T_O^* (day)	Customer's profit $\bar{\Pi}_c$ (\$/day)	Agent's profit $\bar{\Pi}_a$ (\$/day)	Total profit Π^* (\$/day)
30	277.88	477.88	15.99	5.99	10.86	10.03	20.9
40	262.08	462.08	20.32	10.32	15.89	7.02	22.92
50	247.35	447.35	24.41	14.41	18.96	5.18	24.14
60	235.06	435.06	28.32	18.32	20.99	3.99	24.98

Table 2 Nash bargaining and extended Nash bargaining strategies and payoffs for different revenue values

Nash bargaining

Revenue R (\$/day)	Customer's threat Π_c° (\$/day)	Agent's threat Π_a° (\$/day)	Maintenance policy T_R^B (day)	Order time T_O^B (day)	Customer's profit Π_c^B (\$/day)	Agent's profit Π_a^B (\$/day)	Total profit Π^B (\$/day)
20	3.73	3.65	31.1	21.1	5.75	6.54	12.30
30	13.73	3.65	31.9	21.9	14.96	6.09	21.05
40	23.73	3.65	35	22.1	24.16	5.21	29.38
50	33.02	3.65	37	13	33.48	5.53	39.01

Extended nash bargaining

Revenue R (\$/day)	Preventive price $\bar{P}_p$ (\$)	Failure price $\bar{P}_f$ (\$)	Maintenance policy T_R^* (day)	Order time T_O^* (day)	Customer's profit $\bar{\Pi}_c$ (\$/day)	Agent's profit $\bar{\Pi}_a$ (\$/day)	Total profit Π^* (\$/day)
20	269.37	469.37	21.73	11.73	6.23	7.01	13.25
30	262.08	462.08	20.32	10.32	15.89	7.02	22.92
40	253.02	453.02	19.3	9.3	25.78	6.83	32.62
50	255.80	455.80	18.51	8.51	35.16	7.21	42.37

Table 3 Comparison of the extended Nash bargaining model for different scale parameters

α	ΔT_R [%]	ΔP_p [%]	ΔP_f [%]	$\Delta\Pi_c$ [%]	$\Delta\Pi_a$ [%]	$\Delta\Pi$ [%]
30	−33.37	−30.52	−20.35	8.70	9.49	9.13
40	−36.30	−34.47	−22.98	6.21	15.27	8.88
50	−38.82	−38.16	−25.44	5.50	23.49	8.93
60	−40.75	−41.23	−27.48	5.10	34.34	8.93

Table 4 Comparison of the extended Nash bargaining model for different revenues

R \$/day	ΔT_R [%]	$\Delta \bar{P}_p$ [%]	$\Delta \bar{P}_f$ [%]	$\Delta\Pi_c$ [%]	$\Delta\Pi_a$ [%]	$\Delta\Pi$ [%]
20	−30.12	−32.65	−21.77	8.34	7.18	7.81
30	−36.30	−34.47	−22.98	6.21	15.27	8.88
40	−44.85	−36.74	−24.49	6.70	31.09	11.06
50	−49.97	−36.04	−24.03	5.01	30.37	8.61

decreases preventive maintenance price by $\Delta P_p = 38.16\%$ and failure maintenance price by $\Delta P_f = 25.44\%$, the customer decreases the preventive maintenance interval by $\Delta T_R = 38.82\%$, and this increases the profit of customer by $\Delta\Pi_c = 5.50\%$ and the profit of the agent by $\Delta\Pi_a = 23.49\%$.

6 Conclusions and Future Research Directions

In this paper, we studied a problem when an equipment owner outsources preventive and failure replacement services to a service agent, where the players bargain to determine the terms of contract. We considered the Nash bargaining solution to compute the bargaining profit of players and determined the optimal threat strategies a player can pose against the other player in order to increase his bargaining position. We next extended the Nash bargaining solution, where the players achieved their maximum total profit. Our numerical examples illustrated the feasibility and advantage of using such extended contract in maintenance service outsourcing. Our result showed that posing threat can dramatically increase the profit of the player with a higher bargaining position (the customer in our example), moreover both the customer and agent can benefit in the extended bargaining model.

The analysis in this paper is focused on a two-person game in the context of contract negotiation. An interesting direction for future research is to consider cases where there are several players with different payoff parameters leading to multi-player games. The development of such models will provide insights into the effect of increased competition on maintenance outsourcing contracts.

Appendix

Proof of Proposition 1. If $D(T_O) < 0$ then Π_a is a decreasing function, and its minimum is at $T^{\circ}_{Ra} = \infty$. If $D(T_O) > 0$, then Π_a is an increasing function of T_R and achieves the minimum at $T^{\circ}_{Ra} = T_0 + L$. This completes the proof. □

Proof of Theorem 1. One can see that:

$$\frac{\partial G(T_O)}{\partial T_O} = -f(T_O + L)\Big((P_f - C_f) + C_i(T_O + L) + S\mu\Big) < 0$$

as $P_f > C_f$ and $f(T_O + L) > 0$. Therefore, $G(T_O)$ has at most one zero point, and if it exists it must be a maximum, otherwise, the function is monotonic. In other words, the objective function is unimodal and pseudo-concave in T_O. If there is a strategy T°_{Oa} at which $G(T^{\circ}_{Oa}) = 0$, this strategy must be the unique maximum of (9). This completes the proof. □

Proof of Proposition 3. Notice that $S \leq R$, so depending on the sign of W, there are two possibilities for the optimum strategy of T°_{Oc} for a given value of T_R. For any fixed T_R, if $W(T_R) > 0$, Π_c is increasing in T_O. According to the constraint $0 \leq T_O \leq T_R - L$, minimum of Π_c is at $T^{\circ}_{Oc} = 0$. If $W \leq 0$, Π_c is decreasing in T_O, so one minimum is at $T^{\circ}_{Oc} = T_R - L$. This completes the proof. □

Proof of Theorem 2. The derivative of $B(T_R)$ with respect to T_R is:

$$\frac{\partial B(T_R)}{\partial T_R} = \lambda'(T_R)(P_p - P_f)\left[T_R - \int_L^{T_R} F(x)\,dx\right] < 0$$

because $P_p < P_f$, $\lambda(T_R)$ is an increasing function of T_R (i.e., $\lambda'(T_R) > 0$), and $T_R = \int_0^{T_R} 1\,dx > \int_0^{T_R} F(x)\,dx > \int_L^{T_R} F(x)\,dx$. Therefore, $B(T_R)$ is a decreasing function of T_R. Because there is a unique T°_{Rc} strategy such that $B(T^{\circ}_{Rc}) = 0$, where the derivative of (13) is zero as well, this strategy gives the unique maximum of (13). □

Proof of Proposition 5. One can see that:

$$\frac{\partial K(T_R)}{\partial T_R} = -\lambda'(T_R)(C_f - C_p)(T_R - \int_{T_O+L}^{T_R} F(x)dx) < 0, \tag{24}$$

since $C_f > C_p$ and the equipment has an increasing failure rate function ($\lambda'(T_R) > 0$). This means that $K(T_R)$ is a decreasing function in T_R where $T_R \geq T_O + L$. If $K(T_R = T_O + L) < 0$ then $K(T_R)$ is always negative, and so the objective function (16) is a decreasing function of T_R and achieves the maximum at $T^*_R = T_O + L$. □

Proof of Proposition 6. Assuming $Rf(T_R) + f'(T_R)(C_f - C_P) > 0$, there is a unique T^*_O strategy such that $H(T^*_O) = 0$, where the derivative of (17) is zero as well, and this strategy gives the unique maximum of (17). Substituting $H(T_R) = 0$ in the objective function in (17) yields the value of Π^* stated in Proposition 6. □

Proof of Proposition 8. The agent transfers $\tau = \Pi_a(T_R^*, T_O^*, P_p, P_f) - \bar{\Pi}_a$ to the customer by adjusting the service prices, where $\bar{\Pi}_a = \Pi_a(T_R^*, T_O^*, \bar{P}_p, \bar{P}_f)$. Substituting $T_R^* = T_O^* + L$ into agent's payoff function (3) and assuming $P_p - C_p = P_f - C_f$, we have:

$$\Pi_a(T_R^*, T_O^*, P_p, P_f) = \frac{(P_p - C_p) - S\int_0^{T_R^*} F(x)dx}{T_R^*}$$

and

$$\bar{\Pi}_a = \frac{(\bar{P}_p - C_p) - S\int_0^{T_R^*} F(x)dx}{T_R^*}$$

Equating $\tau = \Pi_a(T_R^*, T_O^*, P_p, P_f) - \bar{\Pi}_a$ and solving for $\bar{P}_p$ lead to the adjusted prices. □

References

Anbarci, N., Skaperdas, S., & Syropoulos, C. (2002). Comparing bargaining solutions in the shadow of conict: How norms against threats can have real effects. *Journal of Economic Theory*, *106*(1), 1–16.

Armstrong, M. J., & Atkins, D. R. (1996). Joint optimization of maintenance and inven-tory policies for a simple system. *IIE Transactions*, *28*(5), 415–424.

Ashgarizadeh, E., & Murthy, D. (2000). Service contracts: A stochastic model. *Mathe-matical and Computer Modelling*, *31*(10), 11–20.

Bajari, P., McMillan, R., & Tadelis, S. (2009). Auctions versus negotiations in procure-ment: an empirical analysis. *Journal of Law, Economics, and Organization*, *25*(2), 372–399.

Cachon, G. P. (2003). Supply chain coordination with contracts. *Handbooks in Operations Research and Management Science*, *11*, 227–339.

Campbell, J. D. (1995). Outsourcing in maintenance management: A valid alternative to self-provision. *Journal of Quality in Maintenance Engineering*, *1*(3), 18–24.

Cross, J. G. (1965). A theory of the bargaining process. *The American Economic Review*, *55*(1/2), 67–94.

Danskin, J. M. (1966). The theory of max-min, with applications. *SIAM Journal on Applied Mathematics*, *14*(4), 641–664.

Ding, S.-H., & Kamaruddin, S. (2014). Maintenance policy optimization literature review and directions. *The International Journal of Advanced Manufacturing Technology*, *76*(5–8), 1263–1283.

Gurnani, H., & Shi, M. (2006). A bargaining model for a first-time interaction under asymmetric beliefs of supply reliability. *Management Science*, *52*(6), 865–880.

Hamidi, M., Liao, H., & Szidarovszky, F. (2014). A game-theoretic model for outsourcing maintenance services. Proceedings of 2014 annual reliability and maintainability symposium (RAMS) (pp. 1–6).

Hamidi, M., Liao, H., & Szidarovszky, F. (2016). Non-cooperative and cooperative game-theoretic models for usage-based lease contracts. *European Journal of Operational Research*, *255*(1), 163–174.

Harsanyi, J. C. (1956). Approaches to the bargaining problem before and after the theory of games: A critical discussion of Zeuthen's, Hicks', and Nash's theories. *Econometrica*, *24*(2), 144–157.

Harsanyi, J. C. (1986). *Rational behavior and bargaining equilibrium in games and social situations*. Cambridge, U.K.: Cambridge University Press.
Hartman, J. C., & Laksana, K. (2009). Designing and pricing menus of extended warranty contracts. *Naval Research Logistics (NRL)*, *56*(3), 199–214.
Jackson, C., & Pascual, R. (2008). Optimal maintenance service contract negotiation with aging equipment. *European Journal of Operational Research*, *189*(2), 387–398.
Jardine, A. K., & Tsang, A. H. (2013). *Maintenance, replacement, and reliability: Theory and applications*. Boca Raton, FL: CRC Press.
Karsten, F., Slikker, M., & van Houtum, G.-J. (2012). Inventory pooling games for expensive, low-demand spare parts. *Naval Research Logistics (NRL)*, *59*(5), 311–324.
Leng, M., & Parlar, M. (2005). Game theoretic applications in supply chain management: A review1, 2. *INFOR Information Systems and Operations Research*, *43*(3), 187–220.
Martin, H. (1997). Contracting out maintenance and a plan for future research. *Journal of Quality in Maintenance Engineering*, *3*(2), 81–90.
Matsumoto, A., & Szidarovszky, F. (2016). *Game theory and its applications*. Tokyo: Springer.
McFadden, M., & Worrells, D. S. (2012). Global outsourcing of aircraft maintenance. *Journal of Aviation Technology and Engineering*, *1*(2), 4.
Murthy, D., & Yeung, V. (1995). Modelling and analysis of maintenance service contracts. *Mathematical and Computer Modelling*, *22*(10), 219–225.
Muthoo, A. (1999). *Bargaining theory with applications*. Cambridge, U.K.: Cambridge University Press.
Myerson, R. B. (1991). *Game theory: Analysis of conict*. Cambridge, MA: Harvard University Press.
Nagarajan, M., & Bassok, Y. (2008). A bargaining framework in supply chains: The assembly problem. *Management Science*, *54*(8), 1482–1496.
Nagarajan, M., & Sošić, G. (2008). Game-theoretic analysis of cooperation among supply chain agents: Review and extensions. *European Journal of Operational Research*, *187*(3), 719–745.
Nakagawa, T. (2008). *Advanced reliability models and maintenance policies*. London: Springer.
Nakagawa, T. (2014). *Random maintenance policies*. London: Springer.
Nash, J. (1953). Two-person cooperative games. *Econometrica*, *21*(1), 128–140.
Nash, J. F. (1950). The bargaining problem. *Econometrica*, *18*(2), 155–162.
Ross, S. M. (2013). *Applied probability models with optimization applications*. Mineola, NY: Dover Publications.
Roth, A. E. (1979). *Axiomatic models of bargaining*. Berlin: Springer.
Roth, A. E. (1982). A note on the maximin value of two-person, zero-sum games. *Naval Research Logistics Quarterly*, *29*(3), 521–527.
Schaarsberg, M. G., Borm, P., Hamers, H., & Reijnierse, H. (2013). Game theoretic analysis of maximum cooperative purchasing situations. *Naval Research Logistics (NRL)*, *60*(8), 607–624.
Shapley, L. S. (1952). *A value for N-person games*. Santa Monica, CA: RAND Corporation.
Szidarovszky, F. (1999a). A new Characterization of the non-symmetric Nash solution. *Applied Mathematics and Computation*, *106*(1), 63–68.
Szidarovszky, F. (1999b). A stochastic bargaining process and corresponding one-shot solution concept. *International Game Theory Review*, *1*(02), 159–168.
Tarakci, H., Tang, K., Moskowitz, H., & Plante, R. (2006). Incentive maintenance out-sourcing contracts for channel coordination and improvement. *IIE Transactions*, *38*(8), 671–684.
Thomas, L. C. (2003). *Games*. Theory and Applications: Courier Corporation, North Chelmsford, MA.
Wang, W. (2010). A model for maintenance service contract design, negotiation and optimization. *European Journal of Operational Research*, *201*(1), 239–246.

Agricultural Production Planning in a Fuzzy Environment

M.R. Salazar, R.E. Fitz and S.F. Pérez

Abstract A model for the planification of agricultural production is proposed in the Alto Rio Lerma Irrigation District (ARLID) located in the state of Guanajuato in Mexico. The ARLID have a limited water supply from ground and surface sources, as well as area restrictions. In addition, producer faces the problem of high price uncertainty, which affect seriously the amount of expected profit in each season. Therefore, farmers need to distribute their available land between crops in a fuzzy environment. A multiobjective linear programming model in a fuzzy environment (MLFM) is proposed to approach the problem described above. Ten price scenarios are considered according to the records of the last 10 years, these prices were given in the same scale base 2009. In the results all available area, 112,000 ha, was used. Each price scenario generate one objective to be maximize, some price scenarios generate high profits, while others low profits. The results obtained in the MLFM produce the best expected benefit, 4,820 million of pesos, when prices behave as random variables. For Winter season land is distributed mainly between red tomato and wheat, while in the Spring season between corn and wheat. Sorghum was the only second crop to be sown. The model applied in this particular problem of agricultural planification, show the best land use distribution when market fluctuations are expected.

Keywords Multiobjective decision making · Irrigation district · Price uncertainty · Fuzzy optimization

M.R. Salazar (✉) · R.E. Fitz · S.F. Pérez
Universidad Autónoma Chapingo, Km 38.5 Carr., Chapingo Edo., 56230 México-Texcoco, Mexico
e-mail: raquels60@hotmail.com

A. Matsumoto (ed.), *Optimization and Dynamics with Their Applications*,
DOI 10.1007/978-981-10-4214-0_13

1 Introduction

More than 50% of Mexico is considered as a semiarid region where water availability is limited by the low annual precipitations below 500 mm. In addition 74% of the rainfall is concentrated in 4 months from June to September. This fact has forced the construction of large infrastructure for water uptake. The surface with irrigation infrastructure in México is 6.5 million hectares distributed among 85 irrigation districts and 39,492 irrigation units (Martinez 2013). The National Water Commission (CONAGUA) recognizes the irrigation districts as geographic areas where irrigation service is provided by hydro-agricultural infrastructure such as storage vessels, direct referrals, pumping stations, wells, canals and roads.

The Alto Rio Lerma Irrigation District (ARLID) is the biggest in the Lerma Chapala Basin as shown in Fig. 1. It is located in the southern part of the state of Guanajuato. INEGI (2010) reported that 10 states of Mexico generated 65% of total gross domestic product and Guanajuato is one of them. This Irrigation District provides water to 11 crop production modules which are legal civil associations with concessions granted by the Government that allow them to use the irrigation infrastructure and water (Kloezen et al. 1997), for the purpose of growing crops.

The ARLID has a total area of 112,772 ha under irrigation from which 77,697 ha have surface water, 7,421 ha are irrigated from official wells and 27,654 ha from particular wells. The main Autumn-Winter crops are wheat, barley, beans, broccoli which need 4–5 irrigations. Spring and Summer crops are sorghum, corn, beans and broccoli, which use one or two irrigations and wait for the rain. The available surface water amount was 872 MCM (Millions of cubic meters) and groundwater availability 330 MCM in 1999, with a water availability reduction of 10% every year (Pérez et al. 2011). However, the National Water Commission (NWC) for the period 2013–2014, announce a water availability of 800 MCM in the Irrigation District 011 for 46, 000 ha in Winter, and 77,000 ha in Spring and 2° crops with a total area availability of 112,000 ha (Dominguez 2013).

Fig. 1 The Alto Rio Lerma irrigation district in guanajuato

There are differences in irrigation patterns of different crops; the main and most obvious difference is based on the season, given that during the Spring-Summer cycle water comes from rainfall, whereas for the Autumn-Winter cycle it almost entirely comes from irrigation sources.

Due to the necessity to optimize the water management in agriculture, many studies have been done in Irrigation Districts in Mexico. Ortega et al. (2009) proposed a model for resource optimization for the Irrigation District 005 located in Chihuahua which was a linear programming model to maximize benefit subject to water and land constraints, with four scenarios of water availability. Four crop patterns were found that maximize the benefits.

Zetina et al. (2013) pointed out the use of linear programming models by many authors (Liu et al. 2007; Jabeen et al. 2006; Godínez et al. 2007; Florencio et al. 2002; Garrido et al. 2004) to obtain an economic valuation of water. Particularly, Florencio et al. (2002) modeled scenarios with linear programming in the Alto Río Lerma Irrigation District to estimate the economic value of water, the obtained economic price of water varied between 0.54 and 2.28 pesos per m^3 for surface water, and between 0.66 and 1.25 for underground water, values that are higher than the current prices paid.

Another approach used in Irrigation Districts in México is Multiobjective Optimization to help in the decision making process. Sanchez et al. (2006) applied multiobjective decision making in the Irrigation District 017 using the following criteria: productivity of water for irrigation, increase of the conduction efficiency, increase in the global efficiency of the irrigation district. They concluded that this tool is very useful when we have conflicting objectives related with water management. Salazar et al. (2005) applied multicriteria decision making to an irrigation district in mexico considering economic and environmental objectives and they found a compromise solution to minimize the environmental damage of the region. Also Salazar et al. (2010) approach a problem for water distribution between agriculture, industry and domestic usage in the Mexican Valley using a multiobjective linear model and they propose three policy scenarios and two priority orders of importance but the final decision depends on the decision maker.

A specific problem arises in agricultural production by the uncertainty in prices which affect dramatically the farmer´s income. Figure 2 shows the price variation in the main agricultural products in ARLID, all prices were converted to prices base 2009 using the inflation rate and agricultural price index.

Specially for horticultural products price can vary even in a daily basis, so in addition with many production problems, farmers have to deal with high uncertainty in agricultural prices. Therefore, the purpose of this study is to concentrate our attention in how price variation can affect the farmer's net income and how to deal with this problem, using a modelling technique to maximize the net income in a fuzzy environment.

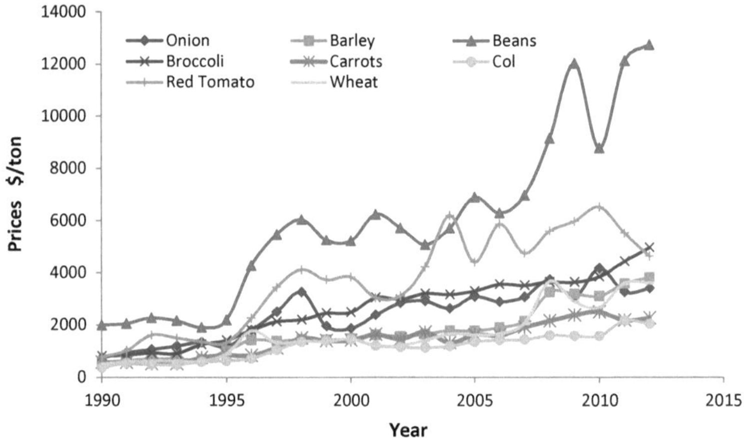

Fig. 2 Price tendency in the main crops in ARLID

2 Mathematical Model

In agricultural production profit coefficients depend on the climatological variables so farmers have to deal with uncertain factors during the growing season, stochastic analysis could be the most natural model to deal with prices as random variables. However to define the probability density functions for the random variables could be difficult and not realistic for the producers. An alternative approach is proposed by Itoh et al. (2003) for minimum value maximization of the total gains subject to the probability distributions of the n-dimensional profit coefficients which were treated as a discrete random vector. Garg and Raj (2010) improved the Itoh model by applying fuzzy multiobjective linear programming to the same problem and they concluded that the results obtained in their algorithm are clearly superior to the approach of Itoh because it is free of an arbitrary parameter value d as taken by Itoh.

The Garg and Raj model can be used when the profit coefficients are crisp discrete random variables. The fuzzy multiobjective linear programming approach is summarized below.

The first step is to solve the problem under different probabilistic scenarios one by one using linear programming techniques taking one objective function with constraints at a time while ignoring the other probabilistic cases. Then the lower and upper bounds ($z_k^{'}$ and z_k^{*}) for each objective function z_k(x) are obtained, where x is the decision vector.

Next a linear and nondecreasing membership function is applied based on the concept of preference or satisfaction:

$$\mu_k(x) = \begin{cases} 1 & \text{if} \quad z_k(x) > z_k^* \\ \frac{z_k(x) - z_k'}{z_k^* - z_k'} & \text{if} \quad z_k' \le z_k(x) \le z_k^* \\ 0 & \text{if} \quad z_k(x) < z_k' \end{cases} \tag{1}$$

Then multiobjective fuzzy linear programming is transformed into a linear programming problem introducing the maximum and minimum objective values into the constraints.

Using the above approach, in this paper we will find the most profitable scenario for the Alto Rio Lerma Irrigation District (ARLID) in Mexico, given uncertain crop prices.

The fuzzy multiobjective linear programming model for the ARLID is described below. Assume we have K price predictions, which are uncertain. First we ignore the uncertainty and construct deterministic model. For each price prediction we have the objective function

$$z_k = \sum_{j=1}^{n} \left[(Y_j P_j^{(k)} - C_j) \right] * X_j \tag{2}$$

where

z_k net benefit in pesos ($)
n number of crops
Y_j yield of crop j
$P_j^{(k)}$ predicted Price of crop j in scenario k (peso/ton)
C_j production cost of crop j (peso/ha)
X_j decision variable, area of crop j (ha)

These objectives define a multiobjective optimization problem. The constraints are as follows:

(a) Winter crops require more irrigation water, so the irrigated area in Winter is limited

$$\sum_{j=1}^{W} X_j \le A_W \tag{3}$$

where W is the number of Winter crops

(b) Area limitation for Spring and Summer crops

$$\sum_{j=W+1}^{n} X_j \le A_S \qquad (A_S > A_W) \tag{4}$$

(c) Annual water usage is bounded by its availability

$$\sum_{j=1}^{W} CWR_j X_j \leq WA \tag{5}$$

where
CWR_j = water requirement of crop j per year

(d) Some crops need to be irrigated by groundwater and some by surface water only. Let G and S define the sets of such crops, respectively. Then we have to add two constraints to the model:

$$\sum_{j \varepsilon G} CWR_j X_j \leq WA_G \tag{6}$$

$$\sum_{j \varepsilon S} CWR_j X_j \leq WA_s \tag{7}$$

(e) Each crop needs a minimum area by the minimum possible demands:

$$X_j \geq X_j^{\min} \tag{8}$$

In applying the model introduced by Garg and Raj we have to identify lower and upper bounds for each objective. The upper bound is selected as the maximum value z_k^* of the objective. It can be obtained by using the simplex method, since all objectives and constraints are linear. We can also define lower bounds which are the minimal acceptance levels of the objective functions. If $z_k^{\min}$ is the minimal value of objective z_k then the lower bound z_{k*} can be selected as a value in interval $[z_k^{\min}, z_k^*]$. The following fuzzy membership function is chosen:

$$\mu(z_k) = \begin{cases} \frac{z_k - z_{k*}}{z_k^* - z_{k*}} & if \quad z_{k*} \leq z_k \leq z_k^* \\ 0 & if \quad z_k < z_{k*} \end{cases} \tag{9}$$

So the multiobjective model can be replaced by minimizing the minimal membership $\alpha = \min_k \mu\ (z_k(x))$ value of the objectives.

Then we have the following new objective and constraints:

$$\max \quad \alpha \tag{10}$$

Subject to

$$\frac{\sum_{j=1}^{n} a_j^{(k)} X_j - z_{k*}}{z_k^* - z_{k*}} \geq \alpha \qquad (k = 1, 2, \ldots, M) \tag{11}$$

where

$$a_j^{(k)} = Y_j P_j^{(k)} - C_j \tag{12}$$

Notice that constraints (10) are also linear

$$\sum_{j=1}^{n} a_j^{(k)} X_j - \alpha(z_k^* - z_{k*}) \geq z_{k*} \qquad (k = 1, 2, \ldots, M) \tag{13}$$

Finally, the fuzzy model is summarized below:

$$\begin{aligned}
&Max \quad \alpha \\
&Subject \quad to \\
&\quad \sum_{j=1}^{W} X_j \leq A_W \\
&\quad \sum_{j=W+1}^{n} X_j \leq A_S \qquad\qquad (A_S > A_W) \\
&\quad \sum_{j=1}^{W} CWR_j X_j \leq WA \\
&\quad \sum_{j \varepsilon G} CWR_j X_j \leq WA_G \\
&\quad \sum_{j \varepsilon S} CWR_j X_j \leq WA_S \\
&\quad X_j \geq X_j^{\min} \\
&\quad \sum_{j=1}^{n} a_j^{(k)} X_j - \alpha(z_k^* - z_{k*}) \geq z_{k*} \qquad (k = 1, \ldots, M)
\end{aligned}$$

The decision variables are $X_j (1 \leq j \leq n)$ and α.
Notice this is a linear programming problem.

3 Model Parameters and Data

According to past price records ten probabilistic price scenarios (in base 2009) are considered in Table 1 for different crops. The yield and production costs are also presented.

From 10 years records in the ARLID, each crop has a minimum area sown displayed in Table 2. According to NWC the water availability for the year 2013–2014 is 800 Mm^3. The most important season for irrigation is the Winter season

Table 1 Crop prices scenarios, yield and production costs in ARLID year 2009

	Crop	Crop prices scenarios $/ton										Yield ton/ha	Costs $/ha
		1	2	3	4	5	6	7	8	9	10		
Winter	Onion	909	1246	1548	1966	2957	2719	2966	3805	4222	3416	32.3	2736
	Barley	686	818	786	1560	1634	1847	1959	3320	3146	3835	4.24	1694
	Beans	2426	2667	2202	5490	5958	5909	6456	9327	8873	12790	1.1	1985
	Broccoli	987	1083	1445	2616	3104	3271	3645	3717	3889	4984	13.1	26338
	Carrots	698	591	862	1488	1524	1357	1646	2199	2530	2276	25.5	15000
	Col	440	572	675	1530	1195	1230	1459	1625	1603	2046	33.6	15000
	Red tomato	953	1898	1572	4032	3248	6407	6027	5712	6599	4645	32.7	13030
	Wheat	619	724	708	1542	1263	1712	1718	3746	2721	3619	5.3	1771
Spring	Onion	909	1246	1548	1966	2957	2719	2966	3805	4222	3416	32.3	2736
	Barley	686	818	786	1560	1634	1847	1959	3320	3146	3835	4.24	1694
	Beans	2426	2667	2202	5490	5958	5909	6456	9327	8873	12790	1.1	1985
	Broccoli	987	1083	1445	2616	3104	3271	3645	3717	3889	4984	13.1	26338
	Carrots	698	591	862	1488	1524	1357	1646	2199	2530	2276	25.5	15000
	Corn	744	896	760	1585	1560	1732	2060	2868	2844	4021	7.38	8174
	Red tomato	953	1898	1572	4032	3248	6407	6027	5712	6599	4645	32.7	13030
	Sorghum	416	517	471	1105	1242	1371	1604	2352	2292	3422	3	603
	Wheat	619	724	708	1542	1263	1712	1718	3746	2721	3619	5.3	1771
2nd crops	Beans	2426	2667	2202	5490	5958	5909	6456	9327	8873	12790	1.1	1985
	Corn	744	896	760	1585	1560	1732	2060	2868	2844	4021	7.38	8174
	Sorghum	416	517	471	1105	1242	1371	1604	2352	2292	3422	3	603

The production costs data are taken from Ortega et al. (2009)

Table 2 Crop water requirements for different water sources

	Crop	MA[1] (Has)	CWR[2] (m^3)	Crops S and G[3]
Winter	Onion	256	7860	G
	Barley	2082	7870	S
	Beans	29	6950	S
	Broccoli	1057	9400	G
	Carrots	79	7000	G
	Cauliflower	273	9400	G
	Red tomato	93	3700	G
	Wheat	8902	5800	S
Spring	Onion	9	5590	G
	Barley	0	6500	S
	Beans	199	5150	S
	Broccoli	104	6870	G
	Carrots	22	7150	G
	Corn	161	6410	S
	Red tomato	54	6200	G
	Sorghum	313	8200	S
	Wheat	288	5800	S
2nd crops	Beans	0	16670	S
	Corn	0	6410	S
	Sorghum	415	2200	S

[1]Minimum area
[2]Annual crop water requirement
[3]Crops irrigated by Surface (S) and Groundwater (G)

from November to May when the irrigation requirement is high. The total surface water availability is 584 Mm^3; and groundwater availability is 216 Mm^3.

4 Results

Each price scenario in Table 1 generates a different objective to be maximized. Using a linear programming solver we obtained the maximum values for each objective. The corresponding values of the other objectives are shown in Table 3, the values in the diagonal represent the maximum values for objectives 1–10. In column 2 we notice that when objective 1 is maximized objective three also reached the same value, and similar cases happened in the other objectives except for objectives 3 and 5. A more detailed description is provided in Table 4, where each

Table 3 Optimal profit values for ten prices scenarios in millions of pesos

	Z1	Z2	Z3	Z4	Z5	Z6	Z7	Z8	Z9	Z10
Z1	1054	1493	1842	2691	3738	3651	3949	5705	5808	5232
Z2	915	2368	1885	5874	4626	9495	8941	9156	10153	7549
Z3	1054	2274	2055	5231	4739	8227	7935	8633	9464	7285
Z4	915	2368	1885	5874	4626	9495	8941	9156	10153	7549
Z5	1039	2259	2038	5156	4755	8168	7898	8344	9413	7141
Z6	915	2368	1885	5874	4626	9495	8941	9156	10153	7549
Z7	915	2368	1885	5874	4626	9495	8941	9156	10153	7549
Z8	915	2368	1885	5874	4626	9495	8941	9156	10153	7549
Z9	696	2179	1648	5723	4543	9354	8925	8891	10162	7763
Z10	696	2179	1648	5723	4543	9354	8925	8891	10162	7763
Zmax-Zmin	358	875	407	3182	1016	5843	4992	3451	4354	2530

column represents the optimal cropping pattern for each maximized objective. In the Winter season red tomato and wheat changes the sown area according to the price scenarios, the rest of the crops in this season remains with the same area. In the spring season four crops: onion, barley, corn, red tomato, and wheat have area variation for different price scenarios. The most drastic situation occurs with barley, for this crop only scenario 5 could be suitable, in the other cases the recommendation is zero area for this crop. Finally, the second crops, corn is the one that has drastic variations, only price scenarios 9 and 10 are favorable for sowing this crop.

The last column in Table 4 represent the optimal cropping pattern scenario under price uncertainty, solved by using Multiobjective Fuzzy Linear Programming (MFLP). This result represents the best outcome the farmers can expect given the price uncertainty. In other words, the result is obtained by minimizing the maximum loss in this situation. These results are very similar to scenarios 6–10 until the row 14 (carrots in Spring). The MFLP results have a substantial increase in the area dedicated to corn and wheat in the Spring season.

The last row in Table 4 show the maximum net income for each scenario including the Fuzzy Multiobjective scenario, we can observe that the maximum net income is obtained for scenario 9 because red tomato is one of the most profitable crops reaching its maximum price in this scenario. Another important observation is that the maximum net income for farmers can vary from 1,054 to 10,162 millions of pesos, which represents a large range in net returns of 9,108 million of pesos. For the Fuzzy Multiobjective Model (last row last column) the maximum net income is 4,820 million pesos, which is not as high as in scenario 9 but it is the benefit the farmers can expect to distribute the available land between crops.

Table 5 reports the best use of the available land under price uncertainty, for Winter season the most important crops are red tomato and wheat and for the Spring season corn and wheat, for second crops only sorghum.

Table 4 Optimal cropping pattern scenarios for profit maximization

	Crop	1	2	3	4	5	6	7	8	9	10	Fuzzy
Winter	Onion	256	256	256	256	256	256	256	256	256	256	260
	Barley	2082	2082	2082	2082	2082	2082	2082	2082	2082	2082	2080
	Beans	29	29	29	29	29	29	29	29	29	29	30
	Broccoli	1057	1057	1057	1057	1057	1057	1057	1057	1057	1057	1060
	Carrots	79	79	79	79	79	79	79	79	79	79	80
	Cauliflower	273	273	273	273	273	273	273	273	273	273	270
	Red tomato	93	33322	33322	33322	33322	33322	33322	33322	33322	33322	33320
	Wheat	35967	8902	8902	8902	8902	8902	8902	8902	8902	8902	8900
Spring	Onion	35668	9	13673	9	13673	9	9	9	9	9	10
	Barley	0	0	0	0	50771	0	0	0	0	0	0
	Beans	199	199	199	199	199	199	199	199	199	199	200
	Broccoli	104	104	104	104	104	104	104	104	104	104	100
	Carrots	22	22	22	22	22	22	22	22	22	22	20
	Corn	161	161	161	161	161	161	161	161	26218	26218	34080
	Red tomato	54	12374	54	12374	54	12374	12374	12374	12374	12374	50
	Sorghum	313	313	313	313	313	313	313	313	313	313	310
	Wheat	35228	52403	51059	52403	288	52403	52403	52403	288	288	30810
2nd crops	Beans	0	0	0	0	0	0	0	0	0	0	0
	Corn	0	0	0	0	0	0	0	0	26057	26057	0
	Sorghum	415	415	415	415	415	415	415	415	415	415	420
	Max Z*	**1054**	**2368**	**2055**	**5874**	**4755**	**9495**	**8941**	**9156**	**10162**	**7763**	**4820**

*Maximum net benefit in millions of mexican pesos

Table 5 Results for the crop planning model with optimal profit under price uncertainty

Winter									
Crops	Onion	Barley	Beans	Broccoli	Carrots	Cauliflower	Red tomato	Wheat	
Has	260	2080	30	1060	80	270	33320	8900	
Spring									
Crops	Onion	Barley	Beans	Broccoli	Carrots	Corn	Red tomato	Sorghum	Wheat
Has	10	0	200	100	20	34080	50	310	30810
2nd crops									
Crops	Beans	Corn	Sorghum						
Has	0	0	420						

5 Conclusions

This work applied multiobjective linear programming in a fuzzy environment to solve the problem of maximizing profits in the Alto Rio Lerma Irrigation District when prices are uncertain. Given historical data of price variations the different crops, ten price scenarios were proposed. The most profitable scenarios were 6 and 9 however there is a low chances that the farmer can get these prices. Planning the agricultural production is a difficult task because the farmers have to face many uncontrollable factors. Therefore, we need to distribute the available land in a less risky environment. The approach we apply here provides a compromise solution or the best solution the farmers can take to maximize the minimum value of the total gains given the random prices.

The total area occupied by the crops for all seasons was 112,000 ha, with a weighted profit of 4820 millions of pesos. For winter season the most profitable crops were red tomato and wheat, while in the Spring season corn and wheat and for the second crops sorghum was more profitable.

References

Domínguez, C. (2013). Asignan 800 Mm^3 al Estado. Periodico Correo. Guadalajara Jal, Mexico/Noviembre 29, 2013. https://www.google.com.mx/#q=Cuca+Dom%C3%ADnguez+%2FNoviembre+29%2C+2013.

Florencio, C. V., Valdivia, A. R., & Scott, C. A. (2002). Producti-vidad del agua en el Distrito de Riego 011 Alto Río Lerma. *Agrociencia, 36–004,* 483–493.

Garg, A., & Raj, S. S. (2010). *Optimization under uncertainty in agricultural production planning.* India: Department of Mathematics, Banaras Hindu University.

Garrido, C. A., Palacios, E. V., Calatrava, J. L., Chávez, J. M., & Exebio, A. G. (2004). La importancia del valor, costo y precios de los recursos hídricos en su ges:ón. Proyecto Regional de Co-operación Técnica para la formación de Economía y Polí cas Agrarias y de Desarrollo

Rural en América La na. FODEPAL. Colaboración de Universidad Politécnica de Madrid - Co-legio de Postgraduados en Ciencias Agrícolas - Universidad Politécnica de Cartagena. 49 p.

Godínez, M. L., García, J. A. S., Fortis, M. H., Mora, J. S. F., Martínez, M. A. D., Valdivia, R. A. et al. (2007). Valor económico del agua en el sector agrícola de la Comar-ca Lagunera. *Terra Latinoamericana, 25*(1), 51–59.

INEGI. (2010). Sistema de Cuentas Nacionales: Producto Interno Bruto por Entidad Federativa, 2003–2008, base 2003. Versión 2. Instituto Nacional de Estadística y Geografía, Aguascalientes, Ags. México. 283.

Itoh, T., Ishiib, H., & Nanseki, T. (2003). A model of crop planning under uncertainty in agricultural management. *International Journal of Production Economics, 81–82,* 555–558.

Jabeen, S., Ashfaq, M., & Ahmad I, B. (2006). Linear program modeling for determining the value of irrigation water. *Journal of Agriculture and Social Sciences, 2*(2), 101–105.

Kloezen, W., Garcés, R. C., & Johnson, S. H. (1997). Impact assessment of irrigation management transfer in the Alto Rio Lerma Irrigation District, Mexico. International Irrigation Management Institute. Research Report 15, Colombo, Sri Lanka.

Liu, X., Chen, X., & Wang, S. (2007). Evaluating and predicting shadow prices of water resources in China and its nine major river basins. *Water Resource Management, 23,* 1467–1478.

Martínez, P. R. (2013). *Evaluación del Distrito de Riego 011, Alto Rio Lerma a 20 años de su transferencia.* Colegio de Postgraduados, Montecillo, Texcoco, Edo de México: Tesis de Maestría.

Ortega, G. D., Mejía, S. E., Palacios, V. E., Pimentel, L. R., & García, E. A. (2009). Model for the Optimization of Resources for an Irrigation District. *Terra Latinoamericana, 27,* 219–226.

Pérez, E. R., Jara D. K. A., & Santos B. A. (2011). Agricultural pollution and costs in the irrigation district 011, Guanajuato, Revista Mexicana de Ciencias Agrícolas Pub. Esp. *1*, 69–84.

Salazar, M. R., Stone, J., Yakowitz, D., & Slack, D. (2005). Multicriteria analysis in an irrigation district in Mexico. *Journal of Irrigation and Drainage Engineering ASCE Nov/Dec, 2005,* 514–524.

Salazar, M. R., Szidarovszky, F., López, C. I., & Rojano, A. A. (2010). Multiobjective linear model optimize water distribution in Mexican Valley. *Journal of Optimization Theory and Applications, 144,* 557–573. doi:10.1007/s10957-009-9608-2.

Sánchez, C. I., Macías, R. H., Heilman, P., González, C. G., Mendoza, M. S., Inzunza, M. A., et al. (2006). Planeación multiobjetivo en los distritos de riego en México. Aplicación de un sistema de auxilio para la toma de decisiones. *Ingeniería Hidráulica en México XXI, 3,* 101–111.

Zetina, E. A., Mora, F. J., Martínez, D. M., Cruz, J. J, & Téllez, D. R. (2013). Economic value of water in Irrigation District 044, Jilotepec, Estado de México. *ASyD, 10*, 139–156.

Optimal Replacement Decisions with Mound-Shaped Failure Rates

Qiuze Yu, Huairui Guo and Miklos Szidarovszky

Abstract Time-to-failure distributions with mound-shaped failure rates are examined, and the existence of optimal preventive replacement policies is studied. Sufficient and necessary conditions are derived for the existence and the uniqueness of the optimal solutions. The cases of lognormal, log-logistic, log-gamma and log-Weibull variables are discussed in detail. Examples of lognormal cases are provided for illustration.

1 Introduction

Determining optimal preventive replacement policies is one of the most important tasks of reliability engineering. Many different models have been proposed (Jardine 2006; Jardine and Tsang 2006; Elsayed 1996; Wang 2002; Wang and Pham 2006; Misra 2008; Nakagawa and Yasui 1987). In industrial applications, the reward renewal model is probably the most popular approach (Jardine 2006; Jardine and Tsang 2006). The reward renewal model offers a mathematical method to determine optimal preventive replacement policies by minimizing the expected cost per unit time. In the case of distributions with increasing failure rates, such as normal, Gumbel and Weibull distribution (when the shape parameter is larger than 1), there is always a unique optimum, and a preventive replacement should be conducted at this optimum. However, if the failure rate does not increase monotonically, is there still an optimum preventive replacement time? Failures caused by fatigue are often modeled by lognormal distributions; however, the failure rate of a lognormal dis-

Q. Yu
Wuhan University, Wuhan, China
e-mail: yuhenry007@whu.edu.cn

H. Guo (✉) · M. Szidarovszky
ReliaSoft Corporation, Tucson, AZ, USA
e-mail: guohuairui@hotmail.com

M. Szidarovszky
e-mail: Miklos.Szidarovszky@hbmprenscia.com

A. Matsumoto (ed.), *Optimization and Dynamics with Their Applications*,
DOI 10.1007/978-981-10-4214-0_14

tribution always increases first and then starts to decrease. In this paper, we will consider distributions with mound-shaped failure rates including lognormal, log-logistic, log-gamma and log-Weibull variables.

After a brief review of the reward renewal optimization model, general formulas will be introduced for the exponential transformation of random variables, which will be then used to obtain the cumulative distribution function (*cdf*), probability density function (*pdf*) and failure rate of each of the distribution types under consideration in this paper. These results will be the basis for examining the existence of finite optima and their computations in the considered cases.

2 Model for Optimum Preventive Replacement

Let T denote the time to failure of an object with *pdf* $f(t)$, *cdf* $F(t)$, reliability function $R(t)$, and failure rate $\rho(t)$. If the preventive replacement is scheduled at time period $\bar{t}$, then the object is replaced unless it has failed before, in which case, it is replaced immediately. Assume that the average replacement cost C_f at a failure is higher than the average cost C_p of a scheduled preventive replacement. The expected cost of replacement is:

$$C_pR(\bar{t}) + C_f(1 - R(\bar{t})) \tag{1}$$

and the expected time until replacement is:

$$\int_0^{\bar{t}} tf(t)dt + \bar{t}R(\bar{t})dt = \int_0^{\bar{t}} R(t)dt \tag{2}$$

Therefore, the expected cost per unit time is given as (Jardine 2006; Jardine and Tsang 2006):

$$G(t) = \frac{C_pR(t) + C_f(1 - R(t))}{\int_0^t R(\tau)d\tau} \tag{3}$$

which has to be minimized by finding the optimum replacement time t. Notice that $\lim_{t\to 0} G(t) = \infty$ and $\lim_{t\to\infty} G(t) = C_f/E(T)$. In order to find the optimum, the derivative of $G(t)$ has to be examined, which has the same sign as:

$$(C_f - C_p)R(t)\left[\frac{f(t)}{R(t)}\int_0^t R(\tau)d\tau + R(t) - \frac{C_f}{C_f - C_p}\right] \tag{4}$$

Since $C_f > C_p$ and $R(t) > 0$, the sign of $G'(t)$ is the same as the sign of function:

$$k(t) = \rho(t)\int_0^t R(\tau)d\tau + R(t) - \frac{C_f}{C_f - C_p} \tag{5}$$

The shape of this function can be examined if we can analyze the sign of its derivative:

$$k'(t) = \rho'(t)\int_0^t R(\tau)d\tau \tag{6}$$

Therefore, the monotonic properties of $k(t)$ are the same as those of $\rho(t)$. If $\rho(t)$ is mound-shaped, then the same holds for $k(t)$, and both $\rho(t)$ and $k(t)$ have maximum at the same value of t.

If $E(T) = \infty$, then $G(t) \to 0$ as $t \to \infty$, and since $G(t) > 0$ for all $t > 0$, there is no finite optimum and $G(t)$ has its infimum at $t = \infty$. So we will assume that $E(T)$ exists and is finite.

Notice that mound-shaped failure rates occur with lognormal, log-logistic ($\sigma < 1$), log-Weibull ($\beta > 1$) and log-gamma ($\alpha > 1$) distributions, where:

$$\rho(0) = \lim_{t \to \infty} \rho(t) = 0$$

Furthermore, $\rho(t)$ increases for $t < t^*$ and decreases as $t > t^*$, where $t^* > 0$ is a distribution dependent value. Clearly:

$$k(0) = 0 + R(0) - \frac{C_f}{C_f - C_p} < 0$$

and

$$\lim_{t \to \infty} k(t) = 0 + 0 - \frac{C_f}{C_f - C_p} < 0$$

Based on the value of $k(t^*)$, there are two cases:

- If $k(t^*) \leq 0$, then $k(t) < 0$ for all $t \neq t^*$, so $G(t)$ strictly decreases and no finite optimum exists. $G(t)$ has its infimum at infinity.
- If $k(t^*) > 0$, then there are values $t_1 \in (0, t^*)$ and $t_2 \in (t^*, \infty)$ such that $k(t_1) = k(t_2) = 0$, $k(t) < 0$ if $t < t_1$ or $t > t_2$, and $k(t) > 0$ as $t_1 < t < t_2$.

Therefore, $\bar{t} = t_1$ is the local minimum and t_2 is the local maximum. If $G(\bar{t}) \leq C_f/E(T)$, then $\bar{t}$ has global minimum; otherwise, no finite optimum exists.

Since the distribution types to be examined in this paper are obtained by exponential transformation from well-known distributions, we will briefly review how the *pdf*, *cdf*, reliability function and failure rate can be determined.

3 Exponential Transformation of Random Variables

Let Y denote a random variable with *pdf* $f_Y(y)$, *cdf* $F_Y(y)$, reliability function $R_Y(y)$ and failure rate $\rho_Y(y)$. Assume that $f_Y(y) > 0$ is defined in interval $I_Y = (\alpha, \beta)$ and $T(y)$ is a strictly increasing function in I_Y. Introduce the transformed variable $X = T(Y)$, then its *cdf* can be obtained as $F_X(x) = 0$ if $x < T(\alpha)$, and $F_X(x) = 1$ if $X > T(\beta)$. If $x \in I_X = (T(\alpha), T(\beta))$, then:

$$\begin{aligned} F_X(x) &= P(X < x) = P(T(y) < x) = P(y < T^{-1}(x)) \\ &= F_y(T^{-1}(x)) = F_y(\bar{T}(x)) = F_Y(y) \end{aligned} \tag{7}$$

where $\bar{T}(x)$ denotes the inverse function of $T(y)$. The *pdf* of X is therefore:

$$f_X(x) = F_Y'(\bar{T}(x))\bar{T}'(x) = f_Y(y)\frac{1}{T'(y)} \tag{8}$$

The reliability function is:

$$R_X(x) = 1 - F_X(x) = 1 - F_Y(y) = R_Y(y) \tag{9}$$

and the failure rate function becomes:

$$\rho_X(x) = \frac{f_X(x)}{R_X(x)} = \frac{f_Y(y)}{T'(y)R_Y(y)} = \frac{\rho_Y(y)}{T'(y)} \tag{10}$$

Exponential transformation of random variables guarantees positivity of the resulting variable. If the domain of Y is the interval $(-\infty, \infty)$, then the transformed variable is $X = \exp(Y)$, and if the domain of Y is only interval $(0, \infty)$, then the values of $X = \exp(Y) - 1$ run through the entire set of the positive real numbers. So in such cases, $T(y)$ is either $\exp(y)$ or $\exp(y) - 1$ with common derivative $T'(y) = \exp(y)$, and the inverse function $\bar{T}(y)$ is either $\ln x$ or $\ln(x+1)$. Thus, in both cases:

$$F_X(x) = F_Y(y),\ R_X(x) = R_Y(y),\ f_X(x) = \frac{f_Y(y)}{\exp(y)},\ \text{and}\ \rho_X(x) = \frac{\rho_Y(y)}{\exp(y)} \tag{11}$$

Since transformation $T(y)$ is strictly increasing, the monotonic properties of these functions in x are the same as their monotonic properties in y.

4 Particular Distributions

In this section, we will discuss the existence of the optimal replacement time for lognormal, log-logistic, log-gamma and log-Weibull distributions.

4.1 Lognormal Distribution

If Y is a normal distribution with parameters μ and σ, then the failure time $T = \exp(Y)$ is lognormal. From Eq. (11), we can get:

$$F(t) = \Phi\left(\frac{\ln t - \mu}{\sigma}\right), \quad R(t) = 1 - \Phi\left(\frac{\ln t - \mu}{\sigma}\right)$$
$$f(t) = \frac{1}{t\sigma}\phi\left(\frac{\ln t - \mu}{\sigma}\right)$$

where Φ and ϕ are the standard normal *cdf* and *pdf*. So the lognormal failure rate is as follows:

$$\rho(t) = \frac{\frac{1}{\sigma\sqrt{2\pi}}\exp\left(-\frac{x^2}{2} - \sigma x - \mu\right)}{1 - \Phi(x)} \tag{12}$$

with $x = \frac{\ln t - \mu}{\sigma}$. The failure rate functions for lognormal distributions with $\mu = 0$ and different σ values are shown in Fig. 1.

The derivative of $\rho(t)$ has the same sign as:

$$\exp\left(-\frac{x^2}{2} - \sigma x - \mu\right)(1 - \Phi(x))(\rho_N(x) - (x + \sigma))$$

where ρ_N is the standard normal failure rate. It is known that $\rho_N' < 1$, where ρ_N strictly increases and approaches the horizontal axis as $x \to -\infty$ and the 45° line as $x \to +\infty$. Therefore, there is a unique value x^* such that $\rho_N(x^*) = x^* + \sigma$, and $\rho_N(x) > x + \sigma$ as $x < x^*$ and $\rho_N(x) < x + \sigma$ as $x > x^*$. Hence, the maximum of $\rho(t)$ and $k(t)$ occurs at $t^* = \exp(\sigma x^* + \mu)$. The value of $k(t^*)$ is used to determine the existence of the optimal replacement time as given in Sect. 2. If $k(t^*) \leq 0$, then there is no optimum, and if $k(t^*) > 0$, then we have to solve the monotonic equation $k(t) = 0$ in interval $(0, t^*)$. Let $\bar{t}$ denote the solution. If $G(\bar{t}) \leq C_f/E(T)$, then $\bar{t}$ is the global minimum; otherwise, there is no optimum.

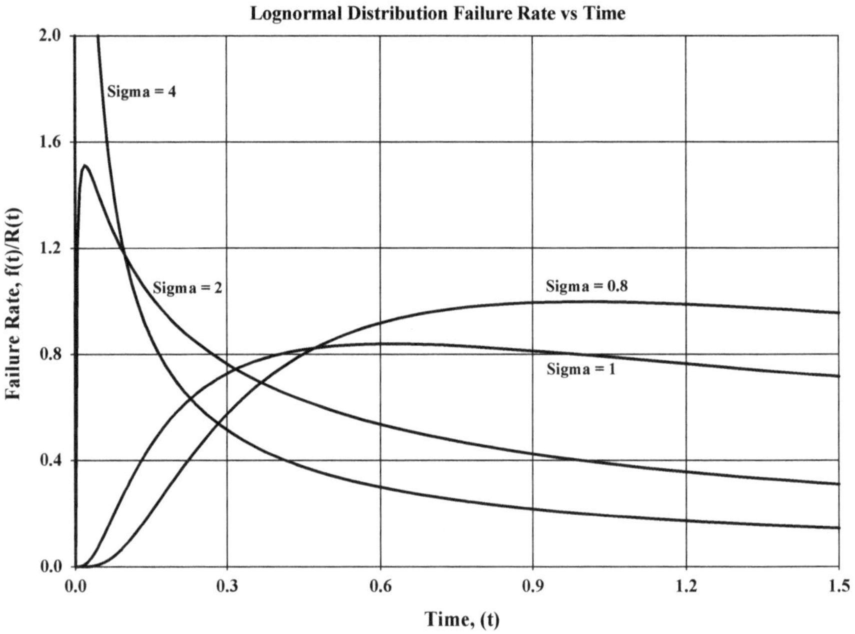

Fig. 1 Mound-shaped failure rate for lognormal distribution

4.2 *Log-Logistic Distribution*

For a log-logistic distribution, the *cdf* and *pdf* are:

$$F(t) = \frac{\exp(x)}{\exp(x)+1}, \quad f(t) = \frac{\exp(x)}{\sigma t(\exp(x)+1)^2}$$

with $x = \frac{\ln t - \mu}{\sigma}$. Its failure rate is

$$\rho(t) = \frac{\exp(x)}{\sigma t(\exp(x)+1)} \tag{13}$$

If $\sigma \geq 1$, then $\rho(t)$ decreases; otherwise, it is mound-shaped. It is easy to see that $\rho'(t)$ has the same sign as:

$$\exp(x)(1-\sigma-\sigma\exp(x))$$

This is negative if $\exp(x) > \frac{1-\sigma}{\sigma}$, positive as $\exp(x) < \frac{1-\sigma}{\sigma}$, and $\varrho(t)$ has maximum at $x^* = \ln\frac{1-\sigma}{\sigma}$ with the corresponding value of $t^* = \exp(\sigma x^* + \mu)$. The optimum may exist if $k(t^*) > 0$, and the local minimizer $\bar{t}$ is obtained by solving the monotonic equation $k(t)=0$ in interval $(0,t^*)$, and $\bar{t}$ gives global minimum if $G(\bar{t})\leq C_f/E(T)$.

4.3 Log-Gamma Distribution

In the case of the log-gamma variable, $T=e^Y-1$, where Y is a gamma variable, the *cdf* and *pdf* of T are as follows:

$$F(t)=F_Y(\ln(t+1)),\quad f(t)=\frac{\lambda\,(\lambda\,\ln(t+1))^{\alpha-1}e^{-\lambda\,\ln(t+1)}}{\Gamma(\alpha)(t+1)}$$

where F_Y is the gamma *cdf* with parameters α and λ. The failure rate is given as

$$\rho(t)=\frac{z^{\alpha-1}\exp(-(\lambda+1)z)}{\int_z^\infty \mu^{\alpha-1}\exp(-\lambda\mu)d\mu} \tag{14}$$

with $z=\ln(t+1)$. If $\alpha\leq 1$, then $\rho(t)$ strictly decreases, and if $\alpha>1$, then $\rho(t)$ is mound-shaped. $\rho^{'}(t)$ has the same sign as:

$$z^{\alpha-2}\exp(-(\lambda+1)z)(\alpha-1-(\lambda+1)z)[\int_z^\infty \mu^{\alpha-1}\exp(-\lambda\,\mu)d\mu+\frac{z^\alpha\exp(-\lambda\,z)}{\alpha-1-(\lambda+1)z}]$$

Let $g(z)$ denote the bracketed term in the above equation. Clearly:

- $g(0)>0,$
- $\lim_{z\to\frac{\alpha-1}{\lambda+1}-0}g(z)=+\infty,\ \lim_{z\to\frac{\alpha-1}{\lambda+1}+0}g(z)=-\infty$
- $\lim_{z\to\infty}g(z)=0$

and

$$g^{'}(z)=\frac{z^{\alpha-1}\exp(-\lambda\,z)}{(\alpha-(\lambda+1)z)^2}(-(\lambda+1)z^2+(\alpha-1)z+(\alpha-1))$$

which has the same sign as the second factor, the quadratic polynomial. Its roots are:

$$z_{1,2}=\frac{\alpha-1\pm\sqrt{(\alpha-1)^2+4(\alpha-1)(\lambda+1)}}{2(\lambda+1)}$$

where $z_1<0$ and $z_2>\frac{\alpha-1}{\lambda+1}$. Therefore, $g^{'}(z)>0$ if $z<z_2$, and $g^{'}(z)<0$ as $z>z_2$. Therefore, $g(z)$ is strictly increasing as $z<z_2$, with a pole at $z=\frac{\alpha-1}{\lambda+1}$, and strictly decreasing for $z>z_2$, so $g(z_2)$ has to be positive. The shape of $g(z)$ for $\alpha=2$ and

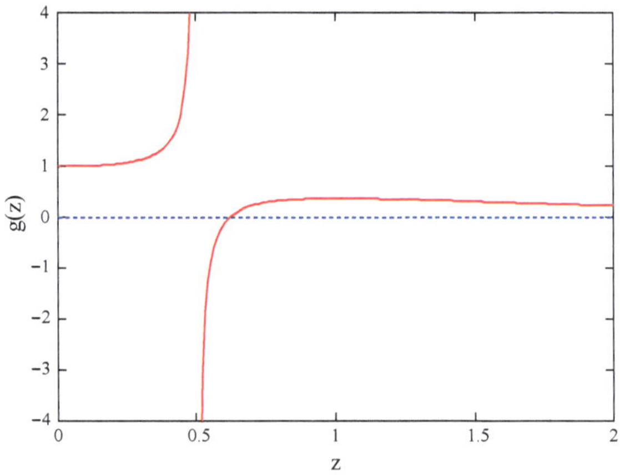

Fig. 2 Plot of $g(z)$ for $\alpha = 2$ and $\lambda = 1$

$\lambda = 1$ is shown in Fig. 2, where $z_2 = 1$ and $\frac{\alpha-1}{\lambda+1} = 0.5$. Since $g(z_2) > 0$, there is a unique value between $\frac{\alpha-1}{\lambda+1}$ and z_2 such that $g(z^*) = 0$. Notice that $\rho^{'}(t) > 0$ as $z < z^*$, and $\rho^{'}(t) < 0$ as $z > z^*$. Therefore, z^* and the corresponding value $t^* = \exp(z^*) - 1$ gives the maximum of $\rho(t)$ as well as the maximum of $k(t)$. If $k(t^*) \leq 0$, then there is no finite optimum; otherwise, there is a unique value $\bar{t}$ in interval $(0, t^*)$ such that $k(\bar{t}) = 0$, which gives global minimum if $G(\bar{t}) \leq C_f / E(T)$. For computing the value of $z^* \in \left(\frac{\alpha-1}{\lambda+1}, z_2\right)$, the monotonic equation:

$$\int_z^{\infty} \mu^{\alpha-1} \exp(-\lambda \mu) d\mu + \frac{z^{\alpha} \exp(-\lambda z)}{\alpha - 1 - (\lambda+1)z} = 0$$

needs to be solved, and then $\bar{t} \in (0, t^*)$ is obtained by solving $k(t) = 0$.

4.4 Log-Weibull Distribution

In the case of log-Weibull variable, $T = e^Y - 1$, where Y is a Weibull variable, the *cdf* and *pdf* of T are as follows:

$$F(t) = 1 - \exp\left(-\left(\frac{\ln(t+1)}{\eta}\right)^{\beta}\right),$$

$$f(t) = \frac{\beta}{\eta(t+1)}\left(\frac{\ln(t+1)}{\eta}\right)^{\beta-1} \exp\left(-\left(\frac{\ln(t+1)}{\eta}\right)^{\beta}\right)$$

By introducing the new variable $z = \ln(t+1)$ as before, the failure rate becomes:

$$\rho(z) = \beta\left(\frac{z}{\eta}\right)^{\beta-1}\frac{1}{\eta e^{z}} = \frac{\beta}{\eta^{\beta}}z^{\beta-1}e^{-z} \tag{15}$$

If $\beta \leq 1$, then this function is decreasing, and if $\beta > 1$, then it is mound-shaped. Since

$$\rho'(z) = \frac{\beta}{\eta^{\beta}}z^{\beta-2}e^{-z}(\beta-1-z)$$

the maximum of $\rho(z)$ as well as the maximum of $k(t)$ occurs at $z^{*} = \beta - 1$ or at $t^{*} = e^{\beta-1} - 1$. So if $k(t^{*}) \leq 0$, then no optimum exists; otherwise, the global minimum is at the unique solution $\bar{t}$ of $k(t) = 0$ in interval $(0, t^{*})$ if $G(\bar{t}) \leq C_f/E(T)$.

5 Examples

Since lognormal is probably the most popularly used distribution in life data analysis, several examples of lognormal distributions are given in this section to illustrate the theory discussed in the previous sections.

Assume that a component's failure time distribution is a lognormal distribution with $\mu = 1$ and the value of σ is given in each of the following examples. The average preventive maintenance cost is $C_p = 400$, and the average corrective maintenance cost is $C_f =$ 5,000.

Example 1: Assume that $\sigma = 2$. We will determine the optimum replacement time for this component.

Following the method given in Sect. 4.1, we have $z^{*} = -1.937$ and $k(t^{*}) = -0.082$. Remember that $k(t)$ has the same sign as the derivative of the cost function $G(t)$. Since $k(t^{*}) < 0$, the cost function $G(t)$ is always decreasing with t; therefore, it does not have an optimum for this case. Notice that:

$$\lim_{t\to\infty} G(t) = \frac{C_f}{E(T)} = \frac{C_f}{\exp(\mu + \sigma^2/2)} = 676.676$$

The graph of $k(t)$ is shown in Fig. 3.

The graph of the cost function $G(t)$ is illustrated in Fig. 4.

Cleary there is no optimal replacement time interval.

Example 2: Assume next that $\sigma = 0.8$. We will find the optimum replacement time for this component.

Following the equations given in Sect. 4.1, we have $z^{*} = -6.589 \times 10^{-3}$ and $k(t^{*}) = 0.203$. $k(t)$ has the same sign as the derivative of the cost function $G(t)$. The

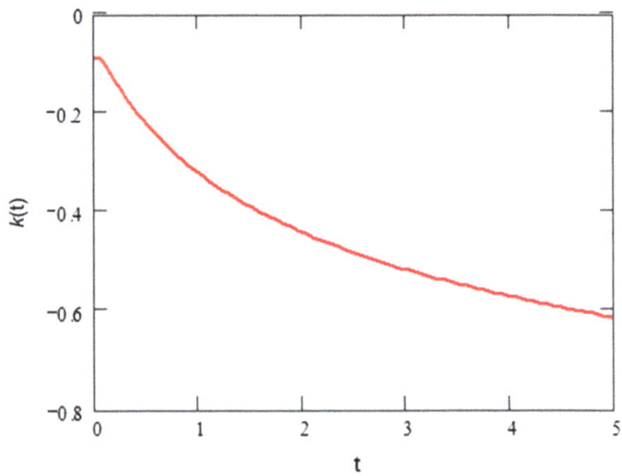

Fig. 3 Plot of $k(t)$ for $\sigma = 2$

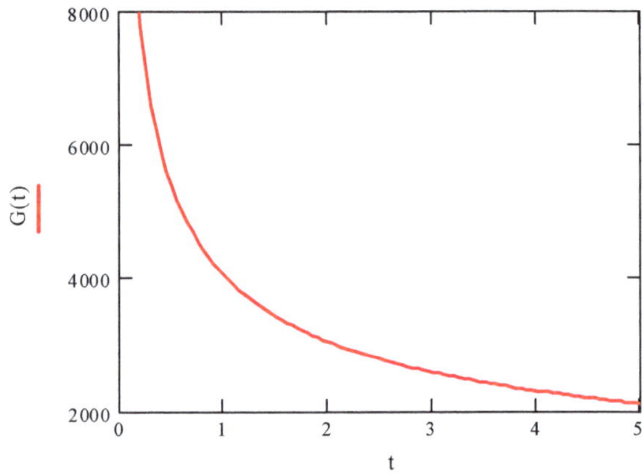

Fig. 4 Plot of $G(t)$ for $\sigma = 2$

two roots for $k(t) = 0$ in Eq. (5) are $t_1 = 0.273$ and $t_2 = 2.611$. $G(t_1) = 2.379 \times 10^3$ gives the local minimum and $G(t_2) = 3.716 \times 10^3$ is the local maximum. Since $G(t_1)$ is less than $\lim_{t \to \infty} G(t) = \frac{C_f}{\exp(\mu + \sigma^2/2)} = 3.631 \times 10^3$, the unique solution for the optimal replacement is therefore t_1. The graph of $k(t)$ is shown in Fig. 5, and that of $G(t)$ is illustrated in Fig. 6.

It is clear that there is an optimal replacement time interval for this example.

Example 3: Assume now that $\sigma = 1.05$.

Following the equations of Sect. 4.1, we have $z^* = -0.582$ and $k(t^*) = 0.027$. $k(t)$ has the same sign as the derivative of the cost function $G(t)$. The two roots for $k(t) = 0$ in Eq. (5) are $t_1 = 0.29$ and $t_2 = 0.898$. $G(t_1) = 3.414 \times 10^3$ gives the local minimum, while $G(t_2) = 3.58 \times 10^3$ is the local maximum. Since $G(t_1)$ is larger than $\lim_{t \to \infty} G(t) = 2.881 \times 10^3$, there is no optimal replacement time. The graphs of $k(t)$ and $G(t)$ are shown in Figs. 7 and 8.

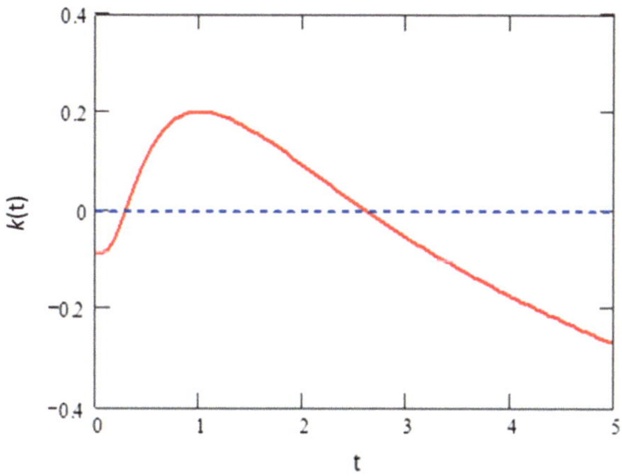

Fig. 5 Plot of $k(t)$ for $\sigma = 0.8$

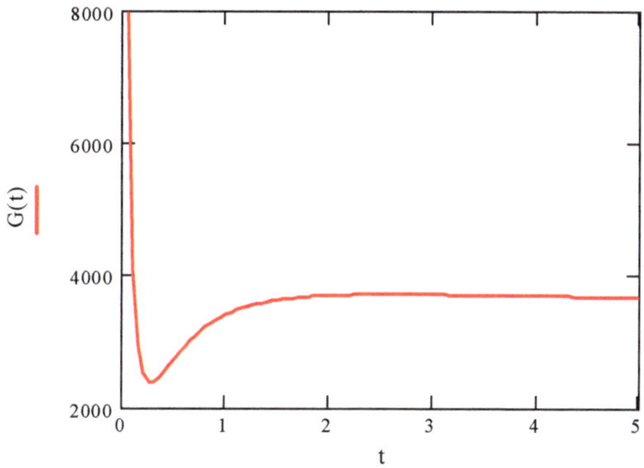

Fig. 6 Plot of $G(t)$ for $\sigma = 0.8$

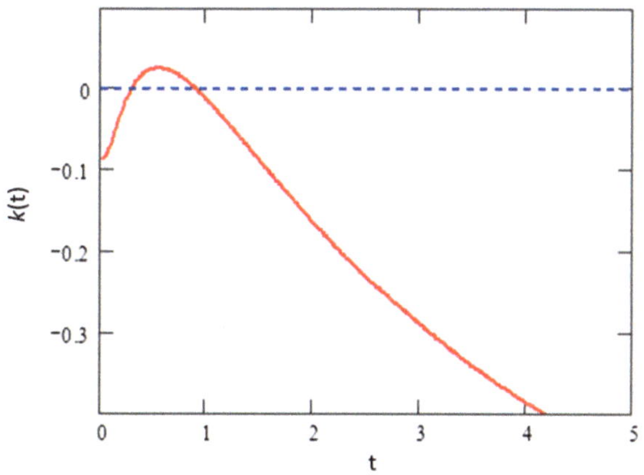

Fig. 7 Plot of $k(t)$ for $\sigma = 1.05$

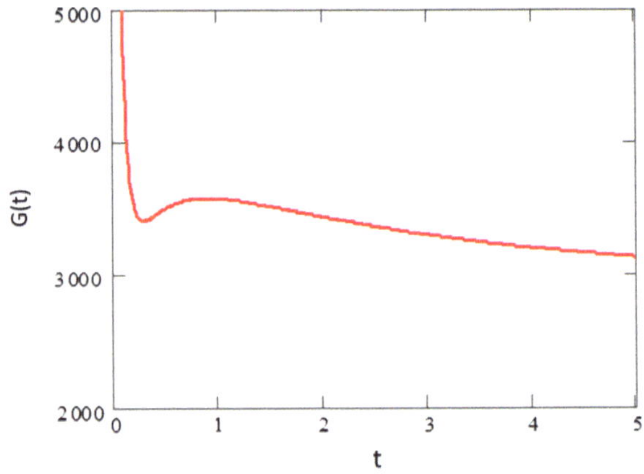

Fig. 8 Plot of $G(t)$ for $\sigma = 1.05$

Obviously, $G(t)$ does not have finite optimum.

The examples above illustrate the three possible cases for the lognormal distribution. From the theory discussed in this paper, it can be seen that the existence of the optimal solution depends on the distribution parameters μ, σ, the average preventive maintenance cost C_p and the average corrective maintenance cost C_f. For the other distribution types with mound-shaped failure rates, as discussed in Sect. 4 and as illustrated in the above three examples, the optimal replacement time interval may or may not exist.

6 Conclusions

It is well known that an optimal replacement interval usually exists for components with increasing failure rates. However, for some of the widely used failure time distributions, such as lognormal and log-logistic, their failure rates are mound-shaped. It is unclear if there is an optimal replacement interval for components with such distributions. In this article, the solution of the optimal replacement model, which minimizes the cost per unit time, was investigated in the cases of mound-shaped failure rates, including the cases of lognormal, log-logistic, log-gamma and log-Weibull distributions. A sufficient and necessary condition was derived for the existence of the optimal solution, and a simple algorithm was introduced to find the optimal solution, which is based on solving single-dimensional monotonic equations for which standard methods are available (Yakowitz and Szidarovszky 1989).

References

Elsayed, E. (1996). *Reliability engineering*. New York: Addison Wesley Longman.
Jardine, A. (2006). Optimizing maintenance and replacement decisions. In *Proceeding of the Annual Reliability and Maintainability Symposium, Newport Beach, CA, 23–26 Jan 2006.*
Jardine, A., & Tsang, A. (2006). *Maintenance, replacement, and reliability: Theory and applications*. Boca Raton, FL: Taylor and Francis.
Misra, K. (2008). *Handbook of performability engineering*. London: Springer.
Nakagawa, T., & Yasui, K. (1987). Optimal policies for a system with imperfect maintenance. *IEEE Transactions on Reliability*, *R-36*, 631–633.
Wang, H. (2002). A survey of maintenance policies of deteriorating systems. *European Journal of Operational Research, 139,* 469–489.
Wang, H., & Pham, H. (2006). *Reliability and optimal maintenance*. London: Springer.
Yakowitz, S., & Szidarovszky, F. (1989). *Introduction to numerical computations*. New York: Macmillan.

A System Dynamics Approach to Simulate the Restoration Plans for Urmia Lake, Iran

Mahdi Zarghami and Mohammad AmirRahmani

Abstract Increasing water demand is threatening many hydro-environmental systems, particularly lakes in arid regions. The goal of this research is to assess restoration plans for a drying saline lake, Urmia Lake, in Iran. In order to include interactions of different lake sub-systems, effectiveness of the plans, as a challenging question for decision makers, is studied by a system dynamics model. The plans that are studied and modeled are increasing irrigation efficiency, decreasing irrigated land area, cloud seeding, inter-basin water transfer projects, and using refined wastewater. Here, it is found that increasing irrigation efficiency by 4% annually and controlling irrigated lands would have around 60% effect in revitalizing the lake to its ecological level, among those considered restoration plans. By linking potential policies to their outcomes, this modeling effort is a step toward supporting the consensus to restore the lake.

1 Introduction

About 1.2 billion people are estimated to work in water-dependent sectors globally; on the other hand, about 1.6 billion people are facing serious economic water shortages (WWAP 2016). Several water management practices, like water transfer, efficient usage, and demand reduction are proposed by managers to address economic water shortages. However, due to resources limitations and interconnectivity of the actions, there is usually not a straightforward and consensus based solution. This inconsistency on the decisions by stakeholders, hence, has resulted in large hydro-political fragmentations and conflicts among water stakeholders (Cooper et al. 2015). Robust negotiations could play key role in solving this dilemma (Islam and Susskind 2015), nevertheless, as long as the ambiguity of these problems and

M. Zarghami (✉) · M. AmirRahmani
Faculty of Civil Engineering, University of Tabriz, 51666 Tabriz, Iran
e-mail: mzarghami@tabrizu.ac.ir; zarghaami@gmail.com

M. AmirRahmani
e-mail: m.amirrahmani@gmail.com

A. Matsumoto (ed.), *Optimization and Dynamics with Their Applications*,
DOI 10.1007/978-981-10-4214-0_15

potential solutions and impacts persist, engagement of main patrons in an appropriate action is unexpected (e.g. climate talks, genetically modified foods, and polar code among others). The resulting delay may end in loss of humanitarian, financial, and natural resources.

The degradation of inland lakes typically has many known and unknown causes: usually with analogy among them (AghaKouchak et al. 2015). Diverting water for industrial, agricultural and domestic uses and climate change are among the reasons found for the degradation of inland lakes. The Amu-Darya and Syr-Darya rivers have almost ceased to flow to the Aral Sea due to irrigation withdrawals in Central-Asia. This resulted in a 23 m drop in the level of the lake between 1940 and 2000. Two shallow separated remained lakes accelerated the desertification in the region and consequently caused a deep social, ecological and economic crisis (Nihoul 2004). Although some restoration plans have been implemented, the Aral Sea has not been revitalized due to lack of co-operation between regional governments (Barghouti 2006). Lake Chad is another striking example of a drying lake. Decreasing precipitation and the growth of agriculture are almost equally blamed for dramatic decline in the level of the lake (Coe and Foley 2001). The neighboring countries of Lake Chad have agreed to divert water from the Oubangui River to the lake basin to save Lake Chad, but consequences are not clear.

Decision makers can see the outcomes of different policies in hydro-environmental management by employing simulation models. Identifying interventions and feedbacks among systems elements is an essential part of sustainable development solutions (Simonovic 2009). The System dynamics (SD) approach, which is designed for complex dynamic problems, is an effective tool for policy analysis studies of social, economic and hydro-environmental systems (Forrester 1973; Simonovic 2009; Arshadi and Bagheri 2014). SD allows modelers to conduct multi-scenario, multi-attribute analyses that result in relative comparisons of many alternative management strategies over time (Sahlke and Jacobson 2005). It provides a process to facilitate negotiations, joint understanding of water problems and cause-and-effect relationships between stakeholders' actions. A few hydro-environmental research studies have been conducted on lakes using the SD approach. For example, Guo et al. (2001) used this method to evaluate the environmental, social, and economic impacts of Chinese government policy on the quality of water in Erhai Lake. Liu et al. (2008) used SD to understand the effect of urban population increase and economic development on the eutrophication of a lake. Hassanzadeh et al. (2012) simulated Urmia Lake basin by a SD model to determine how much each known cause had been co-respondent in the level fall of the lake. Mirchi (2013) developed an eco-environmental dynamic model to study the recovery of Lake Allegan under several policies.

The recent decline in water levels in Urmia Lake, located in Northwestern Iran, is an environmental disaster in progress. The main reasons of this decline in water levels are still being challenged by academics and researchers. Rapid population growth, improper crop patterns, insufficient irrigation systems and water mismanagement are also suggested as causes of water overuse (Madani 2014; Merufinia et al. 2014). Analysis of facts indicates during 1984–2005, water consumption in irrigated lands has been increased by almost 150% in the basin whereas population is estimated to have risen by almost 50% (WRI 2013a). Delju et al. (2013) analyzed climate variability by statistical methods and found that mean precipitation has decreased by 9.2% and average maximum temperature has increased by 0.8 °C. Also, they used an index to study drought in the region and concluded that since 1996, the watershed has faced moderate to extreme droughts. Fathian et al. (2014) estimated trends of precipitation, temperature and measured flows in Urmia Lake basin using non-parametric statistical tests. High correlation between decreasing trend of streamflows in the basin and increasing of temperature is an evidence for contribution of climate change.

The goal of this research is to propose a framework to test the effectiveness of recovery solutions for dead and drying lakes to facilitate the understanding of effects of plans. To achieve this goal, a simulation model is first developed to model different drivers and processes for the Urmia Lake case study in order to assess plans to achieve its sustainability for current and future generations. The main advantage of this approach is that by simulating consequences of different policies on the lake management, "fixes that fail" might be reduced. For example, the Urmia Lake Restoration Program (ULRP) has suggested nineteen solutions; these plans include hard and soft solutions, such as increasing irrigation efficiency, decreasing irrigated lands area, cloud seeding, inter-basin water transfer projects, and using refined wastewater. Estimating environmental and economic consequences of each plan is highly enlightening. This paper, based on available data and modeling limitations, simulates six restoration plans for this case study and compares their separate and combined effectiveness. Moreover, a simulation model can help stakeholders reach an agreement and build trust. Lack of co-operation and trust are the most important crisis in achieving robust restoration strategy and implementation of it (Zarghami and Alemohammad 2015). Hence, achieving a joint fact-finding of the complex system of lakes is essentials to have effective water management. Hence this research —by providing a tool to visualize the effects of the restoration plans—could help to build trust among conflicting users to address the current fragmentation.

In the rest of chapter, the Urmia Lake case is presented in Sect. 2. In Sect. 3, the methodology of SD is used to model the Urmia Lake problem. In Sect. 4 the results of applying six restoration plans are provided and finally, Sect. 5 concludes the research.

2 Problem Description: Urmia Lake Case Study in Iran

Urmia Lake is the largest inland lake of the country and one of the largest saline lakes in the world (Fig. 1). Within the lake basin is a highly influential and valuable aquatic ecosystem. Approximately 800 species of birds, mammals and inland plants, including the unique Artemia sp., inhabit the lake and its wetlands (adopted from Asem et al. 2014). Because of its unique natural and ecological features, the lake and its surrounding wetlands have been designated as a National Park, Ramsar

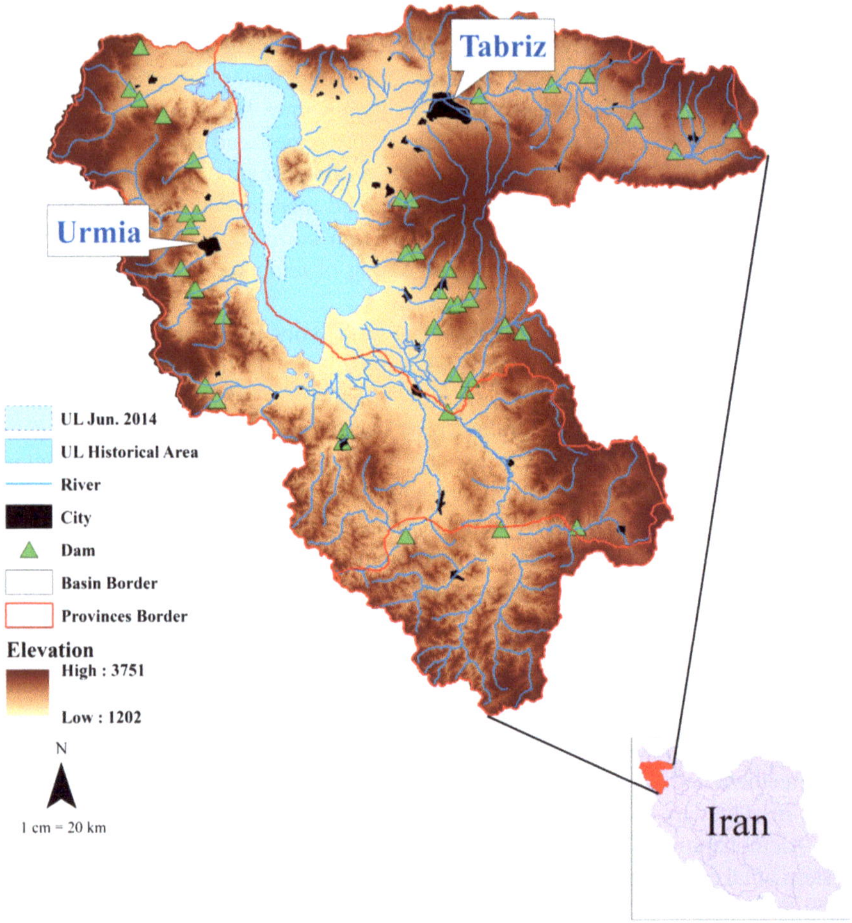

Fig. 1 Map of Urmia Lake basin

Site, and a UNESCO Biosphere Reserve (CIWP 2008). The lake basin, as a socio-ecological region, has experienced extreme water shortages in recent years (Alipour 2006; Zarghami 2011; Khatami 2013; AghaKouchak et al. 2015). The groundwater level in some parts of the basin has dropped by up to 16 m. As presented in Fig. 2 the water level of the lake and its surrounding area are now below the critical level (1274.1 m above sea level based on Abbaspour and Nazaridoust 2007). Wind-blown salty dusts from dry areas of the lakebed can become a serious threat to the health of the people residing in the area if the water inflow falls short of its minimum level of 3 billion cubic meters annually to sustain the lake.

Given the severity of the situation, saving drying Urmia Lake is currently one of the top priorities of national and several international organizations (including United Nations Development Programme, United Nations Environment

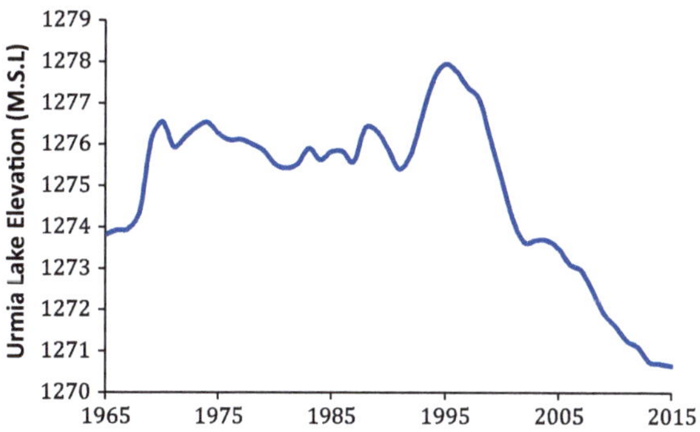

Fig. 2 Mean annual water level in Urmia Lake (*Data source* Iran Water Resources Management Company)

Programme, and the Japan International Cooperation Agency). However, there is a lack of comprehensive research to evaluate the effects of alternative restoration plans on the lake. This research, therefore, provides a SD framework to determine the feasibility and compare the effectiveness of plans in order to help better to understand the effects of them. This research deals with two important questions. The first set of questions concerns the condition of the lake in the near future without action, while the second evaluates effective plans to address the current conditions. Theoretically, the first question leads to an argument about "the tragedy of the commons," and the second heads an argument to prevent "fixes that fail." The question of this paper is focused on the second one that which plans would be most effective and efficient in sustaining the basin. These options will be studied in two groups:

- *Soft solutions*: Increasing water efficiency, water allocation adjustments like reuse of refined wastewater for environmental needs.
- *Hard solutions*: Reducing agricultural area, cloud seeding, inter-basin water transfer.

These proposed solutions need to be compared and contrasted objectively and SD tools can be used to simulate their effect on the lake.

3 SD Model Framework

Studying dynamic behavior of a complex system with interacting subsystems is simplified by SD. Analyzing a hydro-environmental system without the systems thinking can lead to an unsustainable management (Mirchi 2013). SD, with its capacity to model casual relationships and system thinking, is well suited for simulating and understanding water resources and environmental systems. Hence, the Urmia Lake System Dynamics (ULSD) model is developed as a prototype for comprehensive analysis of the Urmia Lake's behavior and possible solutions to restore it.

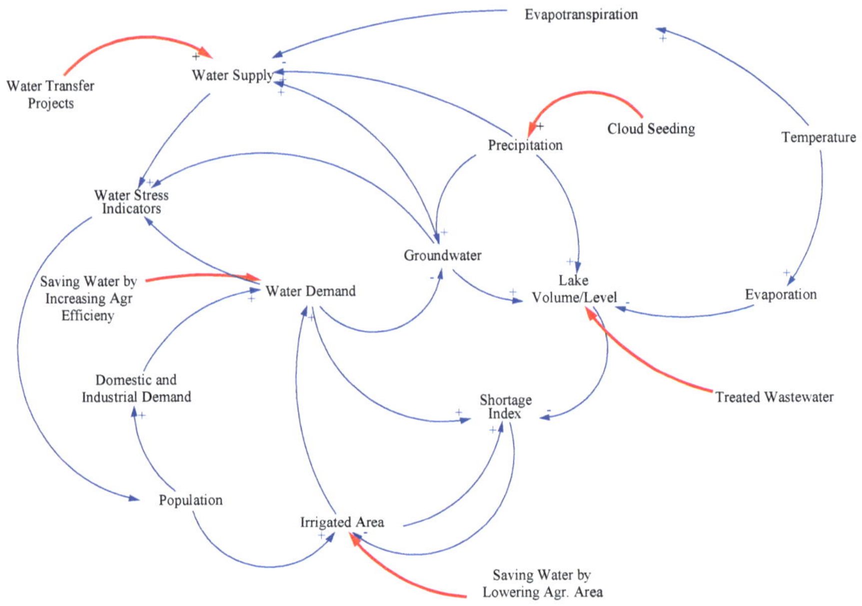

Fig. 3 Casual loop diagram of ULSD model

Defining key variables based on approaches and goals is a crucial step in developing the model. Usually, lake volume and level are variables which could indicate the environmental condition. For instance, salt concentration of water or aquatic population could be estimated by volume of the lake. Hence, lake volume and level are key variables, and they are employed to indicate the lake condition.

Determining system goals and variables, and visualizing them using a casual loop diagram, is essential to understand nature of the model and interactions of the variables. Figure 3 demonstrates interactions of variables in the SD model of the lake. Modeling circular feedbacks of variables is the core of system thinking. This loop diagram could be drawn based on former studies and based on interviews with the experts working on different aspects of the system. In other words, in this step, drivers of complex system of a lake are identified and linked by arrows.

The data are obtained and collected from the regional water authorities as well as the agricultural departments and ULRP. Table 1 demonstrates main variables and their data sources.

Variables which affect the key variable (Lake Level/Volume) are easy to identify. Inlet to the lake includes direct precipitation on the lake. A simple water balance equation also could be used to estimate direct inflow to the lake. Direct evaporation from the lake is the most crucial outlet of the terminal lakes. In this case study, water in the lake is salty; hence, there is no water extraction or diversion from the lake.

Table 1 Main variables of the model and sources for their historical data

Variable	Source
Irrigated area	WRI (2013a)
Lake level	Urmia Lake Restoration Program
Population	WRI (2013b)
Precipitation	Rahmani and Zarghami (2013), updated from IWRMa for 2011–2015.
Temperature	Rahmani and Zarghami (2013), updated from IRIMOb for 2011–2015.
Evaporation	Calculated based on modified pan evaporation data from WRI
Evapotranspiration	Calculated based on Belaney and Criddle (1950)

The social system of the lake is influenced by the basin population. Usually the more people that live in the basin, the more water is consumed. In addition, when more water is available and the lake is not in disastrous condition, population growth rate and immigration to the basin tend to be positive. The gross domestic product of the basin partially depends on agriculture; this economic interaction links the hydro-environmental and social systems. Economic growth could affect the population and the tendency of farmers to increase their cultivated land.

News about the lake in the media is a proxy which can demonstrate the public sensitivity to critical condition of the lake and the disaster happening; The water news in the media is not a precise indicator of water situations (as shown in Hurlimann and Dolnicar 2012 and Garcia et al. 2016) however it is really an effective measure of population variables, especially the interest of people in migrating in or out of the basin. In this research, the number of alarming news on the lake is measured by web search and the result in Fig. 4 shows a strong inverse correlation with the lake level shortfall.

To couple social awareness and hydrological conditions of the basin, shortage index, *SI*, is introduced. Then the following equation is developed to define the shortage index for the lake:

$$SI = \begin{cases} 0 & H \geq \overline{H} \\ \frac{H}{\overline{H}} & H_E < H < \overline{H} \\ 2\frac{H}{H_E} & H < H_E \end{cases} \tag{1}$$

H is water depth in the lake, $\overline{H}$ is long-term average depth and H_E is ecologic water depth in the lake. The aim of proposing restoration plans in this research is to increase the lake level and then reduce the shortage index. In reverse, increasing the shortage index will motivate the managers to pursue restoration plans. In Fig. 3, dashed arrows demonstrate the variables which are added to the casual model in case of using restoration plans. More detailed information about restoration plans are explained in Sect. 3.2.

The last step in developing the model is articulating it with mathematical equations in a software environment and calibrating it. VENSIM software, which is

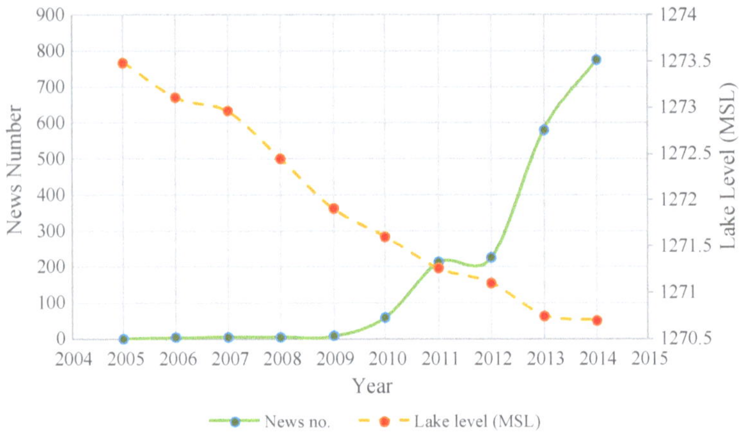

Fig. 4 The inverse correlation among the Urmia Lake level (meters above mean sea level, MSL) and public awareness (number of news based on web search)

free for educational purposes, is selected for this research. In this study, a dynamic water balance approach is employed to develop the model because of its simplicity and operationality. The water balance equation which is based on the principle of conservation of water mass in a terminal lake is demonstrated as:

$$S(t) = \int_{t_0}^{t_n} [GW(t) + R(t) + SW(t) + I(t) - E(t)]dt + V(t_0) \tag{2}$$

where $S(t)$ is lake storage, $GW(t)$ is groundwater volume inlet to the lake, $R(t)$ is rainfall volume inlet to the lake including the cloud seeing effect, $SW(t)$ is surface water volume inlet to the lake including inter basin water transfer and refined wastewater, $I(t)$ *is water inflow to the lake,* $E(t)$ is evaporation volume from lake and $V(t_0)$ is initial lake volume; all at monthly periods of t.

The evaporation volume of the lake is calculated as:

$$E(t) = C_{pan} C_{salt} EF(t) A(t) \tag{3}$$

$EF(t)$ is the rate of evaporation of fresh water from the pan in that area, and $A(t)$ is the lake wet area at time t. Since evaporation from natural water body is less than the pan, then C_{pan} or pan coefficient is used to amend it. A coefficient C_{salt} is also employed to adjust difference between evaporation from fresh and salty water.

To estimate domestic water demand, population is multiplied by average water consumption per capita in the basin. Industrial demand is roughly assumed to be a percentage of domestic demand. Irrigated area multiplied by average agricultural water consumption per area in the basin which then equals to water demand in agriculture sector. Evapotranspiration in the basin is calculated based on Belaney and Criddle (1950) formula.

3.1 Calibration and Validation

To test the model, its results are compared with observations. In addition, the behavior of a model under different circumstances can help determine if the model is behaving properly. Therefore, two tests, behavior reproduction and behavior anomaly (Sterman 2000), are employed to test the model.

For behavior reproduction, lake volume changes are best suited for testing. The calculated volume of the lake from its volume-elevation table for the period of 1991–2011 is used to calibrate the model. Observed lake level and modeled values are plotted in Fig. 5. The correlation measure (R^2) is 0.95 and normalized root-mean-square error (RMSE) for the lake level change is 0.10, which show acceptable fitness of modeled values to the observed data.

For behavior anomaly, two key variables (irrigation area and precipitation) are assumed zero to evaluate the behavior of the system. These two variables are important because they are representing the main drivers of water supply and demand. Figure 6 represents the change in the volume of the lake by assuming zero irrigation and zero precipitation. In the first case the lake will be full without any decline and in the second case the test represents that lake will be completely dried. Then both cases are reasonable and it shows that model has acceptable behavior.

Variables of population, news, farmed area and groundwater are also validated by two mentioned tests (behavior reproduction and behavior anomaly tests), which their report is omitted to save the chapter space. Although the modeling approach of

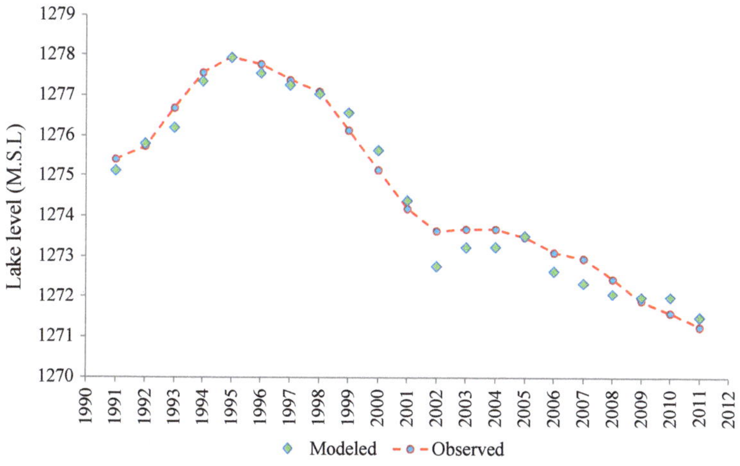

Fig. 5 Observed and modeled lake volume (1990–2011)

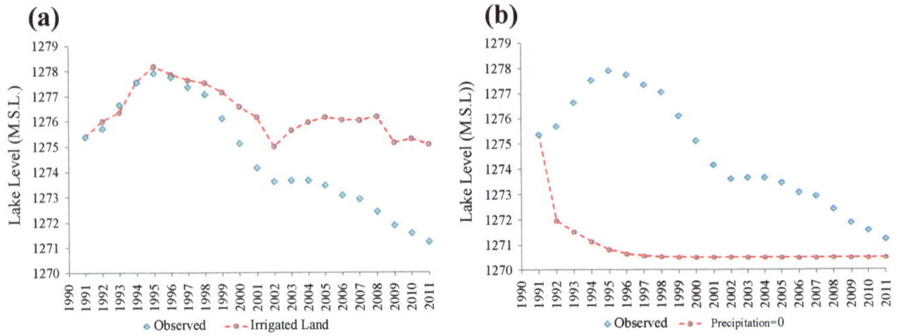

Fig. 6 Behavioral anomaly tests on lake volume for the period 1990–2011 in two conditions: **a** irrigated area = 0, **b** precipitation = 0

this paper differs from that of the Regional Council on Lake Urmia Basin Management (2012), the results of this modeling approach confirm their findings.

3.2 Restoration Plans

The scientific approach to recovering dead and drying lakes is to investigate the causes of the disaster and to stop them, if they are anthropogenic. Sometimes, eliminating source of the problem is not beneficial in bringing back the lake then additional means are needed. In addition, political, social and economic factors could prevent the removal of the drivers. Hence, six lake restoring plans which are among suggestions of ULRP for saving the lake are considered. There are also other plans suggested by the scholars (such as not cultivating water intensive crops) however due to the modeling limitations they are not considered here but will be the topic of future work. The modeled plans are explained as below and also summarized in Table 2.

Plan 1—Increasing irrigation efficiency: Most academics and engineers agree that low efficiency of agricultural water use is problematic and one of the reasons for water overuse (Ardakanian 2005; Madani 2014 among others). Hence, increasing water efficiency is often considered as a solution for water scarcity problems. For the case of Urmia Lake, it is assumed in a 10 year plan that water efficiency could be raised from 30 to 70% in the basin (i.e., 4% annual increase of this parameter).

Plan 2—Reducing irrigated land: Rapid growth in water consumption for agricultural section is a major factor in some cases of drying lakes (Aral Sea, Urmia Lake, etc.). Therefore, plan 2 is to stop the growth of irrigated area and to decrease it then it could be a solution which is followed in this plan (Zarghami and Alemohammad 2015). The land reduction variable linearly decreases based on the

Table 2 Name of restoration plans and their characteristics

Plans	Aim	Horizon of implementation	Annual effect	Tolerance
P.1	Increasing irrigation efficiency	2015–2025	4%[a]	2–6%
P.2	Reducing irrigated land	Based on drought intensity	0–5%[a]	0–10%
P.3	Cloud seeding	2015–2030	7%	5–9%
P.4	Transfer from Zaab	2020–2030	600 MCM[b]	500–700 MCM
P.5	Transfer from Aras	2020–2030	140 MCM[b]	120–160 MCM
P.6	Wastewater	2018–2030	250 MCM	200–300 MCM

[a] not compounding
[b] transfer to the basin and not directly to the lake

shortage index, however it is assumed to have maximum reduction of 5% (constant and not compounded) because of the social resistance regarding this plan.

Plan 3—Cloud seeding: Cloud seeding (to increase precipitation over an area) could make an effective contribution toward reducing water stress. Consequently, adding some water to the basin by this plan could help to restore the Urmia Lake. Although cloud seeding impact is uncertain based on the seeding method, climate of the region, and type of clouds, it is assumed that cloud seeding could enhance the annual precipitation by up to 7%. This value is the rough average of numbers found in the literature (e.g. Curic et al. 2007; Silverman 2010; Acharya et al. 2011, DeFelice et al. 2014).

Plans 4 and 5—Inter basin water transfers: The other way of adding water to a basin is via inter-basin water transfer projects. Two main water transfer projects, considered by ULRP, are from the Zaab and Aras basins. The Zaab plan, in south of Urmia Lake, could transfer 600 million cubic per year (MCM/Y) to the basin to supply increasing water demands and not directly to the lake. Also there is a plan to transfer 140 MCM/Y from the Aras basin, north of Urmia Lake (Zarghami 2011). However, these projects are under construction/study and supposed to be in action at least from 2020. Another issue is that these transfers will be partially allocated to the current water demands other than the lake need. Then they are just to alleviate the local and growing demands within the basin however their water return could be a source.

Plan 6—Wastewater: Reuse of refined domestic and industrial wastewater is a common solution to the water shortage in many regions. Hence, refining this water and draining it directly to the lakes is among restoration plans. This project is supposed to contribute 250 MCM/Y to Urmia Lake volume.

In addition to the level of the lake that serves best for measuring the restoration success, two other indicators were selected to check if the plans lead the basin towards sustainability. Relative water stress indicator (RWSI) computes the demands on available water resources in a basin as

$$RWSI = \frac{DIA}{Q} \quad (4)$$

where *DIA* is total demand (MCM) in a basin including domestic, industrial and agricultural water needs and *Q* is total available surface and ground water (MCM). $RWSI > 0.4$ for a basin indicates a highly stressed and critical condition (Vörösmarty et al. 2005). Second, groundwater dependency (*GD*) expresses the relative contribution of groundwater to basin water supply (Vrba et al. 2007). This indicator can be computed as follow:

$$GD = \frac{\text{total groundwater abstraction}}{\text{total surface and ground water supplies}} \quad (5)$$

A favorable value for *GD* is assumed less than 0.25. Basins with high dependency to groundwater have a *GD* value over 0.50.

4 Results

Here the results of simulating ULSD model from 2015 to 2030 are presented. The results will be shown in two parts. First, the effect of each plan will be presented and discussed individually. Second, the combined effect of the plans will be shown in the first year of possible restoration of the lake.

4.1 Individual Effect of Restoration Plans

Figure 7 illustrates the lake level changes under each plan in comparison to the no action condition. Plan 1 follows the policy of increasing water efficiency by 4% annually. Hence, water efficiency increases to 70% by 2025 in the basin. It is seen that improving this parameter could not save the lake on its own, because unintended effect of this policy is to increase the irrigated area in addition to the crop density by farmers to compensate the renovation costs of their irrigation systems. It is estimated that *RWSI* and *GD* are almost 0.8 and 0.4 in the basin under this plan. Hence, implementing only this plan neither restores the lake nor leads the basin to sustainability because of the socio-economic responses of the stakeholders.

Reducing irrigation area (Plan 2) is shown to have no considerable difference with the no action plan. Although decreasing area, results in 33% less agricultural area after 15 years, however farmers' reaction to this plan, might be to increase crop density, therefore gradually makes this plan ineffective. While this plan cannot bring back the lake, *RWSI* and *GD* are estimated to be almost 0.9 and 0.3 in the

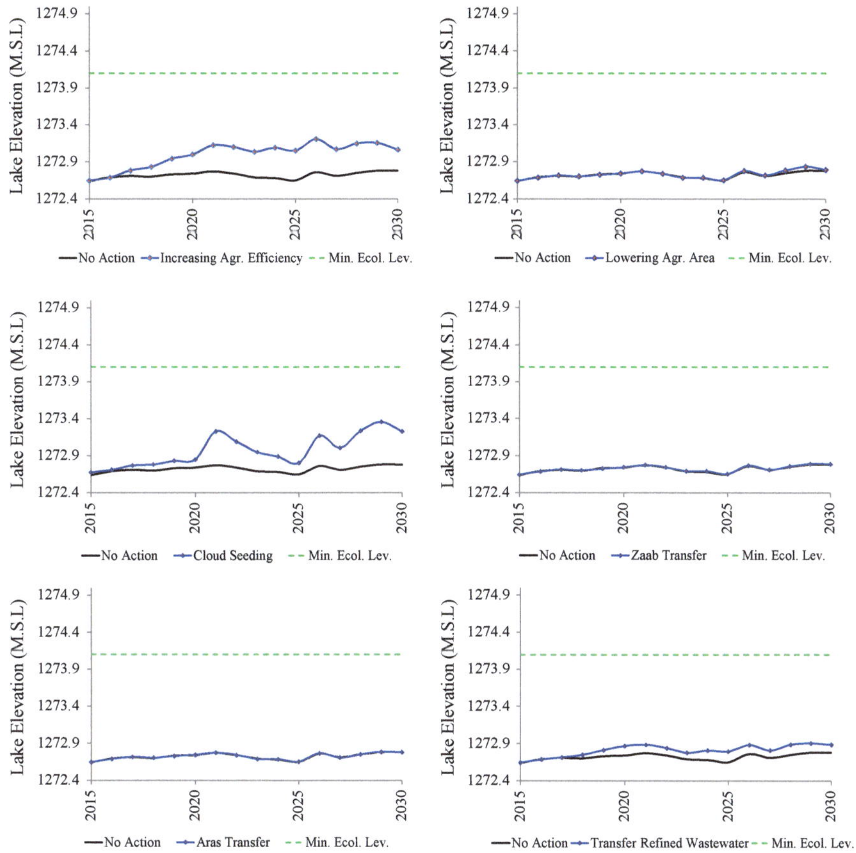

Fig. 7 Simulation of the individual effect of each plan on lake level by ULSD model

basin under this plan. This plan, if implemented, would result in negative socio-economic consequences.

It is seen in Fig. 7 that adding extra water to the basin by cloud seeding cannot revitalize the lake on its own, but it is beneficial. Cloud seeding project improves *GD* to 0.3, but it makes *RWSI* worse in the basin because of more water consumption as a consequence of having more water. As a note, it is assumed that this project raises precipitation 7% in the basin, though the uncertainties of this estimation must be acknowledged.

Inter-basin water transfers of Zaab and Aras are expected to have considerable favor for restoration of the Urmia Lake. However based on Fig. 7 it appears these projects cannot help the lake from drying up, because they partially contribute to the agricultural water consumption but not to restoration of the lake. Transferring water without any adjustment in water consumption habits and policies can backfire (Gohari et al. 2013). It must be also noted that some international water transfer plans would require additional levels of understanding, cooperation and diplomacy

(Islam and Susskind 2015). It is seen that stopping agricultural growth like other plans cannot restore the lake on its own. Also, it is no more effective than Zaab and Aras projects in contribution to sustainability indicators of *RWSI* and *GD*.

Zaab and Aras projects are proposed to transfer water to the basin and to supply water needs, but refined wastewater is proposed to be transferred straight to the Urmia Lake. Therefore, this project is expected to be more effective in saving the lake, but implementing this project is not much more effective than Zaab and Aras because of the low volume that it contributes to the lake.

4.2 Accumulative Effect of Plans

It is seen that a single plan could not restore the lake, therefore, a combination of plans is required. Hence, assuming all of the plans are in action with sufficient funds, then estimating the contribution of each plan is the objective of this paper. In the case of implementing all six plans, *RWSI* is estimated to be almost 0.6 and *GD* almost 0.4. Although *RWSI* indicates the basin is still highly stressed based on international standards, it is below average *RWSI* in the region, which could make it acceptable. In this scenario, the dynamics of the basin are totally different from the scenarios in which each plan is applied alone. Due to the combination of plans in this scenario, there is more water in the basin because of water transfers and cloud seeding, and consequently, smaller area needs to be taken out from farming. The contribution of each plan is plotted in Fig. 8.

Based on the results of this paper, implementing all of the plans could cause Urmia Lake to be restored by 2022 to minimum ecological level even though some

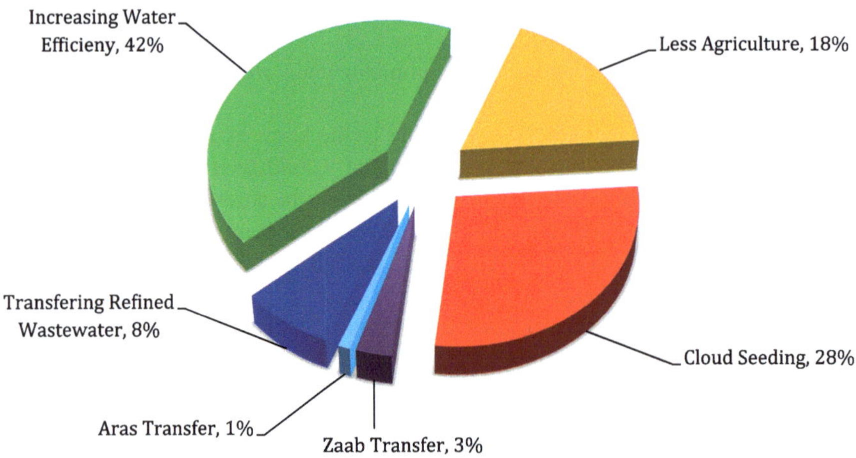

Fig. 8 Contribution of each plan (%) in restoring the Urmia Lake to reach its ecological level by 2022

plans are not completed by that time. It is seen that increasing water efficiency has the most important share in saving the Urmia Lake. Having 7% increase in annual precipitation by cloud seeding has a contribution about 28% in restoration of the Lake. Based on the dynamics in ULSD model, 8% decrease in agricultural area in 3 years and stopping agricultural growth—until Urmia Lake is reached its historical level (5 years after reaching minimum ecological level)—could have an 18% share in revitalizing Urmia Lake. Implementing all plans makes it unnecessary to lower agricultural area continuously. Refined water drained into the Urmia Lake is 8% effective in refilling the lake. Since water transfer projects are in action from 2020 and the lake is restoring in 2022, these 2 years' contributions in set of action plans provides a small impact of 4%. Economic, social and environmental externalities of each plan must be evaluated in detail, and these are the topics of next coming researches. Considering the costs of plans, cloud seeding and increasing irrigation efficiency are the most financially efficient ones. In addition, as there are uncertainties in the parameters and modeling, the results need sensitivity analysis. For this purpose, the Monte Carlo simulation is used to test the effects of the uncertain plans. The tolerance in the effect of plans (as shown in the last column of Table 1) is used in establishing uniform probability distribution functions. Finally, Fig. 9 represents the accumulative effect of plans on the lake level under uncertain condition in implementing the six plans.

However, since restoring the lake is a complex problem, more studies are needed. One important extension of this work is to compare alternative plans using multi-criteria analysis. Finding near-optimal combinations of cultivated land areas to restore the lake and also improve agricultural sustainability is vital. Conflict resolution mechanisms and experiments with stakeholder groups need to be added to the model and also conducted with real stakeholders in future research. However, the main contribution of the study is that it showed that the lake could be revived by using a combination of the restoration plans described above. This is in contradiction with the statement of some scholars (e.g. Kardovani 2014) who already rejected any hope for the lake. In addition, there is currently a debate over whether the resources within the basin should be used for restoration or water should be imported from the outside. In this case also, it is very clearly presented that both the

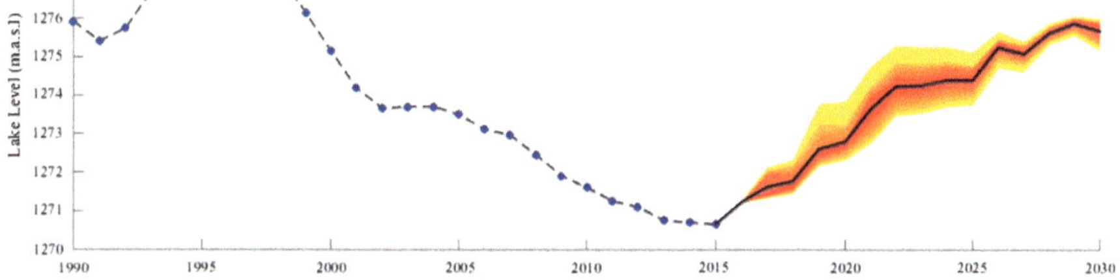

Fig. 9 Simulation of the accumulative effect of plans on lake level within 50% limits representing uncertainty

inner and outer resources are collectively needed for lake restoration. The results of this research bring hopes to reach consensus and cooperation among conflicting users which has positive (Zarghami et al. 2015) effect in water management.

5 Conclusions

The results of ULSD model support the possibility of Urmia Lake restoration. Based on the outcomes, increasing the water efficiency over a 10 year period is the most effective plan to restore the lake. In short term, the Zaab and Aras water transfer has the lowest contribution to lake recovery among the plans and would need high levels of cooperation which may be difficult to achieve. Also, it is shown that decreasing agricultural area by 8% within a 3 year period (2015–2018) and stopping any agricultural growth for 8 years (2018–2026) could make a contribution of about 18% to lake level restoration. Using the simulation model and providing its visualized results can help the stakeholders to see the effects of different policies and then it may help to foster the lake restoration process. The impact of climate change on the success of the plans is the topic of next forthcoming research.

Acknowledgements The authors are also thankful for comments and supports of Shafiqul Islam (Tufts University), James Wescoat (MIT), John Ikeda (World Bank), Seyed Mahdi Hashemian (MIT), Terrence Smith (Tufts University), Jory Hecht (Tufts University), Margaret Garcia (Tufts University) and Mohammad Jahandar (University of Tabriz). Some parts of the earlier version of manuscript is presented at the 33rd International Conference of the System Dynamics Society Cambridge, Massachusetts, USA, July 19–July 23, 2015, and here the effort of conference organizers are appreciated. The first author received a scholarship from the University of Tabriz to work as an affiliate researcher at the Tufts University and MIT.

References

Abbaspour, M., & Nazaridoust, A. (2007). Determination of environmental water requirements of Lake Urmia, Iran: An ecological approach. *International Journal of Environmental Studies, 64*(2), 161–169. doi:10.1080/00207230701238416.

Acharya, A., Piechota, T. C., Stephen, H., & Tootle, G. (2011). Modeled streamflow response under cloud seeding in the North Platte River watershed. *Journal of Hydrology, 409*(1–2), 305–314. doi:10.1016/j.jhydrol.2011.08.027.

Alipour, S. (2006). Hydrogeochemistry of seasonal variation of Urmia Salt Lake, Iran. *Saline Systems, 2*(9). doi:10.1186/1746-1448-2-9.

AghaKouchak, A., Norouzi, H., Madani, K., Mirchi, A., Azarderakhsh, M., Nazemi, A., et al. (2015). Aral Sea syndrome desiccates Lake Urmia: Call for action. *Journal of Great Lakes Research, 41*(1), 307–311. doi:10.1016/j.jglr.2014.12.007.

Ardakanian, R. (2005). Overview of water management in Iran. Water conservation, reuse and recycling. In *Proceeding of the Iranian-American Workshop* (pp. 153–172).

Arshadi, M., & Bagheri, A. (2014). A system dynamic approach to sustainability analysis in Karun River Basin, Iran. *Iran-Water Resources Research, 9*(1), 1–13.

Asem, A., Eimanifar, A., Djamali, M., Rios, P., & Wink, M. (2014). Biodiversity of the hypersaline Urmia Lake national park (NW Iran). *Diversity, 6*(1), 102–132. doi:10.3390/d6010102.

Barghouti, S. (2006). Case study of the Aral Sea water and environmental management project: An independent evaluation of the World Bank's support of regional programs. The World Bank (Report).

Belaney, H. F., & Criddle, W. D. (1950). *Determining water requirements in irrigated area from climatological irrigation data.* US Department of Agriculture, Soil Conservation Service.

CIWP: Conservation of Iranian Wetlands Project. (2008). *Integrated management plan for lake Urmia*. Tehran: Department of Environment.

Coe, M. T., & Foley, J. A. (2001). Human and natural impacts on the water resources of the Lake Chad basin. *Journal of Geophysical Research: Atmospheres, 106*(D4), 3349–3356. doi:10.1029/2000JD900587.

Cooper, E., Islam, I., & Susskind, L. (2015). Coping with uncertainty and feedback in the Nile Basin. *Water Diplomacy.* Retrieved October 13, 2016 from http://blog.waterdiplomacy.org/2015/10/coping-with-uncertainty-and-feedback-in-the-nile-basin/.

Curic, M., Janc, D., & Vuckovic, V. (2007). Cloud seeding impact on precipitation as revealed by cloud-resolving mesoscale model. *Meteorology and Atmospheric Physics, 95*(3), 179–193. doi:10.1007/s00703-006-0202-y.

DeFelice, T. P., Golden, J., Griffith, J., Woodley, W., Rosenfeld, D., Breed, D., et al. (2014). Extra area effects of cloud seeding—An updated assessment. *Atmospheric Research, 135–136,* 193–203. doi:10.1016/j.atmosres.2013.08.014.

Delju, A. H., Ceylan, A., Piguet, E., & Rebetez, M. (2013). Observed climate variability and change in Urmia Lake Basin, Iran. *Theoretical and Applied Climatology, 111*(1), 285–296. doi:10.1007/s00704-012-0651-9.

Fathian, F., Morid, S., & Kahya, E. (2014). Identification of trends in hydrological and climatic variables in Urmia Lake Basin, Iran. *Theoretical and Applied Climatology, 119*(3), 443–464. doi:10.1007/s00704-014-1120-4.

Forrester, J. W. (1973). *World dynamics* (2nd ed.). Cambridge, Massachusetts: Wright Allen Press.

Garcia, M., Portney, K., & Islam, S. (2016). A question driven socio-hydrological modeling process. *Hydrology and Earth System Sciences, 20,* 73–92. doi:10.5194/hess-20-73-2016.

Gohari, A., Eslamian, S., Mirchi, A., Abedi-Koupei, J., Massah Bavani, A., & Madani, K. (2013). Water transfer as a solution to water shortage: A fix that can backfire. *Journal of Hydrology, 491,* 23–39. doi:10.1016/j.jhydrol.2013.03.021.

Guo, H. C., Liu, L., Huang, G. H., Fuller, G. A., Zou, R., & Yin, Y. X. (2001). A system dynamics approach for regional environmental planning and management: A study for the Lake Erhai Basin. *Journal of Environmental Management, 61*(1), 93–111. doi:10.1006/jema.2000.0400.

Hassanzadeh, E., Zarghami, M., & Hassanzadeh, Y. (2012). Determining the main factors in declining the Urmia Lake level by using system dynamics modeling. *Water Resources Management, 26*(1), 129–145. doi:10.1007/s11269-011-9909-8.

Hurlimann, A., & Dolnicar, S. (2012). Newspaper coverage of water issues in Australia. *Water Research, 46*(19), 6497–6507. doi:10.1016/j.watres.2012.09.028.

Islam, S., & Susskind, L. (2015). Understanding the water crisis in Africa and the Middle East: How can science inform policy and practice? *Bulletin of the Atomic Scientists, 71*(2), 39–49.

Kardovani, P. (2014). Urmia Lake could not be restored. Interview by *Khabaronline.* Retrieved October 13, 2016 from http://www.khabaronline.ir/detail/346505/society/environment (in Farsi).

Khatami, S. (2013). Nonlinear chaotic and trend analyses of water level at Urmia Lake, Iran. M.Sc. Thesis Report: TVVR 13/5012, Lund, Sweden: Lund University. ISSN: 1101-9824.

Liu, Y., Guo, H. C., Yajuan, Y. U., Dai, Y. L., & Zhou, F. (2008). Ecological–economic modeling as a tool for watershed management: A case study of Lake Qionghai watershed, China. *Limnologica- Ecology and Management of Inland Waters, 38*(2), 89–104. doi:10.1016/j.limno.2007.11.001.

Madani, K. (2014). Water management in Iran: What is causing the looming crisis? *Journal of Environmental Science and Studies, 4*(4), 315–328. doi:10.1007/s13412-014-0182-z.

Merufinia, E., Aram, A., & Esmaeili, F. (2014). Saving the Lake Urmia: From slogan to reality (challenges and solutions). *Bulletin of Environment, Pharmacology and Life Sciences, 3*(3), 277–288.

Mirchi, A. (2013). System dynamics modeling as a quantitative–qualitative framework for water resources management: Insights for water quality policy in the great lakes region. PhD Dissertation, Houghton, USA: Michigan Technological University.

Nihoul, J. C., Zavialov, P. O., & Micklin, P. P. (Eds). (2004). *Dying and dead seas climate versus anthropic causes.* Dordrecht: Springer Science + Business Media.

Regional Council on Lake Urmia Basin Management. (2012). Drought risk management plan for Lake Urmia Basin. Technical Report, Tehran.

Rahmani, M. A., & Zarghami, M. (2013). A new approach to combine climate change projections by ordered weighting averaging operator; Applications to Northwestern Provinces of Iran. *Global and Planetary Change, 102*, 41–50. doi: 10.1016/j.gloplacha.2013.01.007.

Sahlke, G., & Jacobson, J. (2005). System dynamics modeling of transboundary system: The Bear River basin model. *Groundwater, 43*(5), 722–730. doi:10.1111/j.1745-6584.2005.00065.x.

Silverman, B. A. (2010). An evaluation of eleven operational cloud seeding programs in the watersheds of the Sierra Nevada Mountains. *Atmospheric Research, 97*(4), 526–539. doi:10.1016/j.atmosres.2010.06.013.

Simonovic, S. P. (2009). *Managing water resources, methods and tools for a systems approach.* London: UNESCO, Paris and Earthscan James & James.

Sterman, J. D. (2000). *Business dynamics: Systems thinking and modeling for a complex world.* Boston: McGraw-Hill.

Vrba, J., Hirata, R., Girman, J., Haie, N., Lipponen, A., Shah, T., et al. (2007). *Groundwater resources sustainability indicators*. Paris: UNESCO.

Vörösmarty, C. J., Douglas, E. M., Green, P. A., & Revenga, C. (2005). Geospatial indicators of emerging water stress: An application to Africa. *AMBIO: Journal of Human Environment, 34* (3), 230–236. doi:10.1579/0044-7447-34.3.230.

WRI (Water Resources Institute). (2013a). Land use change in Urmia Lake Basin. Tehran, (Report in Farsi).

WRI (Water Resources Institute). (2013b). Population studies in Urmia Lake Basin. (Report in Farsi).

WWAP (United Nations World Water Assessment Programme). (2016). *The United Nations World Water Development Report 2016: Water and Jobs*. Paris: UNESCO.

Zarghami, M. (2011). Effective watershed management: Case study of Urmia Lake, Iran. *Lake and Reservoir Management, 27*(1), 87–94. doi:10.1080/07438141.2010.541327.

Zarghami, M., & Alemohammad, S. H. (2015). Can water diplomacy enable a new future for the Urmia Lake? In *Tufts Water Diplomacy Program and Iranian Studies Group at MIT*.

Zarghami, M., Safari, F., Szidarovszky, F., & Islam, S. (2015). Nonlinear interval parameter programming combined with cooperative games: A tool for addressing uncertainty in water allocation using water diplomacy framework. *Water Resources Management, 29*(12), 4285–4303. doi:10.1007/s11269-015-1060-5.

A Decision Support System for Managing Water Resources in Real-Time Under Uncertainty

Emery A. Coppola and Suna Cinar

Abstract Over-pumping of groundwater resources is a serious problem world-wide. In addition to depleting this valuable water supply resource, hydraulically connected wetlands and surface water bodies are often impacted and even destroyed by over-pumping. Effectively managing groundwater resources in a way that satisfy human needs while preserving natural resources is a daunting problem that will only worsen with growing populations and climate change. What further complicates management of these systems is that even when pumping rates of wells are held fairly constant, their hydraulic effects are often highly transient due to variable weather and hydrologic conditions. Despite this, transient conditions are rarely if ever accounted for by management models due to the difficulties in separating pumping effects from natural factors like weather. To address this shortcoming, a conceptual real-time decision support system for managing complex groundwater/surface water systems affected by variable weather, hydrologic, and pumping conditions over space and time is presented in this chapter. For the hypothetical but realistic groundwater/surface water system presented here, the decision support system, based upon previous work by Coppola et al. (2003a, b, 2005a, b, c, 2007, 2014) consists of real-time data streams combined with artificial neural network (ANN) prediction models and formal optimization. Time variable response coefficients derived from ANN prediction models are used by an optimization model to maximize total groundwater pumping in a multi-layered aquifer system while protecting against aquifer over-draft, streamflow depletion, and dewatering of riparian areas. Optimization is performed for different management constraint sets for both the wet and dry seasons, resulting in significantly different groundwater pumping extraction solutions. Stochastic optimization is also performed for different precipitation forecast events to address corresponding

E.A. Coppola (✉)
NOAH LLC, Lawrenceville, NJ, USA
e-mail: ecoppolajr@gmail.com

S. Cinar
Wichita State University, Wichita, KS, USA
e-mail: cinarsuna@yahoo.com

A. Matsumoto (ed.), *Optimization and Dynamics with Their Applications*,
DOI 10.1007/978-981-10-4214-0_16

uncertainty associated with weather-dependent irrigation pumping demand. This data-driven support system can continuously adapt in real-time to existing and forecasted hydrological and weather conditions, as well as water demand, providing superior management solutions.

1 Introduction

Groundwater resources are being depleted by over-pumping in almost all parts of the world across the entire development spectrum, from economically disadvantaged to economically advanced nations (Brown 2011). In many instances, over-pumping not only mines the aquifer but degrades and even destroys hydraulically connected riparian areas and surface water bodies (Barlow and Leake 2012; Winter et al. 1998). There is no shortage of examples of over-pumped groundwater resources resulting in irreversible losses of ecologically and economically valuable areas (Brown 2011; Mays 2007).

Tucson, Arizona is a mid-sized city located in the southwestern United States. Relying exclusively upon groundwater in a desert valley, Tucson pumped the water table aquifer down hundreds of feet, destroying riparian corridors along once perennial rivers that are now dry except in the monsoon season. The city that was once inhabited by the Tohono O'odom Native Americans, famously known as "the water harvesters of the desert", now depends for its survival upon diverted Colorado River water, transported hundreds of miles from its source.

The Minquin Oasis, a once highly productive agricultural region in northwestern China, lost its vast system of lakes and wetlands by groundwater overpumping. By destroying the natural equilibrium, surrounding deserts encroached into this once lush oasis, transforming its fertile croplands into desert wasteland, which is now the source of the largest dust storms in Asia. Because of the dropping water table residents were forced to drill much deeper wells. The deeper and older groundwater contains higher dissolved concentrations of natural compounds like arsenic, causing cancer and other illnesses in residents and livestock to soar in the region.

Despite the abundance of tragic cautionary examples like Tucson and the Minquin Oasis, groundwater over-pumping continues unabated world-wide because of socioeconomic pressures, with climate change stresses and exploding populations imposing increasingly unsustainable demands on shrinking resources (Brown 2011). Because groundwater resources are finite, and in many areas are not replenished by nature at rates necessary to prevent aquifer over-draft, there is an urgent need for optimal water management. Achieving this will require better decision support systems than those currently used today.

The current state-of-the-art approach is to couple numerical groundwater models with optimization management models to identify appropriate pumping strategies (Karamouz et al. 2003; Mays 2007; Peralta and Kalwij 2012). Numerical groundwater models consist of equations that embody the physics of groundwater and when necessary surface water flow (i.e., conservation of mass and momentum).

Although theoretically capable of approximating the spatial and temporal variability of real-world groundwater/surface water systems, the models' predictive accuracy are inherently limited by simplifying physical and mathematical assumptions combined with incomplete characterization of complex hydrogeologic systems (Coppola et al. 2003a, 2014). Furthermore, numerical groundwater models are typically calibrated with limited historical data to select their boundary and initial conditions. These limitations not only reduce model robustness, but also precludes models from being initialized to real-time or even pseudo real-time conditions, further reducing their prediction and optimization capabilities.

ANNs provide a compelling alternative to numerical models. As shown by Coppola, et al. (2003a, 2014), they can achieve higher predictive accuracy than numerical groundwater models in complex hydrogeologic systems. ANNs can also be integrated with optimization management models to serve as the simulator of the real-world system (Coppola et al. 2007, 2014). Because ANNs are "data-driven", they are ideal for coupling with real-time data streams for continuous initialization to real-time conditions, further increasing their prediction and simulation accuracy, thereby improving the corresponding optimization solutions. Furthermore, ANNs models can also be continuously retrained using the real-time data streams.

In this chapter, the applications of automatic data collection and control systems, ANN models, and optimization to water resources are briefly presented. Following this, a hypothetical but realistic groundwater/surface water system with production wells is presented. From this, the corresponding hypothetical ANN prediction models and optimization management formulations, with their corresponding LINDO-generated solutions for optimal pumping strategies, are presented. A stochastic optimization analysis is then performed to account for uncertain pumping requirements due to weather-driven agricultural irrigation demand. Finally, the conclusions drawn from this study and the advantages of the real-time decision support system are summarized.

2 Data Collection Systems

Over the last several decades, there has been a digital data revolution in hydrology. While even large-scale regional water studies were once performed by deploying field personnel to manually collect data, automatic data loggers with sophisticated telemetry systems are replacing human labor. Today, many water utilities as well as large industrial and agricultural water users who operate their own water systems are installing automated data collection systems.

Field instruments automatically measure and record at any frequency of interest (e.g., every 5 s) a broad range of critical system information, including state variables like groundwater and surface water elevations, water temperature, and water quality, decision variables like pumping rates, and random variables like weather conditions. Telemetry relays the data in real-time to computer work stations for retrieval and processing as desired. Data collection systems have evolved into SCADA (i.e., Supervisory Control and Data Acquisition) systems, where data collection and operational variables are remotely controlled. Pumping rates of supply wells can be automatically adjusted by pre-set threshold values and/or interactively by operator preference in response to changing water demand and other system conditions like reservoir levels.

While stored data is sometimes used to retrospectively assess system performance and/or for regulatory compliance purposes, the data is largely under-utilized, and is rarely used for facilitating real-time decision making capability. Consequently, for the vast majority of systems, there is a lost opportunity to exploit the enormous untapped potential of SCADA for optimizing operational controls and strategies. As presented next, the data-driven structure of the ANN models is ideal for exploiting the enormous value of real-time data streams to their full potential.

3 ANN Prediction Models

ANNs are a form of artificial intelligence modeled after the brain in both their structure and operation (Poulton 2001). Because they "learn" directly from data, they are often called "data-driven models." The ANN models consist of nodes (i.e., brain neurons) assembled in distinct layers which are interconnected by transfer functions (i.e., brain synapses). ANN models generally consist of three layers, including the input layer, representing the predictor variables, the hidden layer, which receives the mathematically transformed values of the input values, and the output layer, which represents the final prediction values. During the learning process, representative data sets are processed through the ANN architecture, during which connection weights between nodes within the transfer functions are adaptively adjusted to minimize the prediction errors. As proven by Kolmogorov's Theorem, because of their mathematical structure, ANNs are highly adept at accurately modeling complex non-linear systems. A more detailed overview of the theoretical foundation of ANNs, their mathematical structure, and learning algorithms with related development issues and guidelines may be found in Coppola et al. (2014).

Coppola et al. (2003a, b, 2005a, b, c, 2007, 2014) proved the feasibility of developing accurate prediction models using easily measurable field data for a number of a range of important water resources applications. Applications include predicting groundwater elevations, surface water elevations, water demand, water distribution system pressures and storage tank water levels, and groundwater and surface water quality. A small subset of these ANN applications would constitute the prediction component of the conceptual real-time water management system presented in this chapter.

4 Optimization

Optimization has been applied to numerous water management problems, including aquifer overdraft, salt-water intrusion, contaminant remediation, and conjunctive groundwater/surface water management (Mays 2007; Peralta and Kalwij 2012). Although single and multi-objective optimization are typically performed with a numerical model simulator, ANN models have been used with much success. Coppola, et al. performed a multi-objective optimization for a real-world public supply wellfield where the conflicting objectives of maximizing groundwater pumping while minimizing vulnerability to contamination were optimized using ANN models developed from a numerical groundwater flow model (Coppola et al. 2005c). Coppola et al. also performed optimization for a real-world water distribution system for a mid-sized city using ANN models developed directly from hydraulic data collected from the actual system (Coppola et al. 2014).

5 Conceptual Study Area

As depicted in Fig. 1, the hypothetical watershed used in this example application is bounded by mountains.

A large river flows through the watershed, with its headwaters beginning at the mountain range to the east and discharging into the large lake on the western side of the watershed. Portions of the river are "losing reaches", where surface water seeps downward through the river bed and recharges the underlying water table aquifer. Most of the river, however, consists of "gaining reaches", where some groundwater

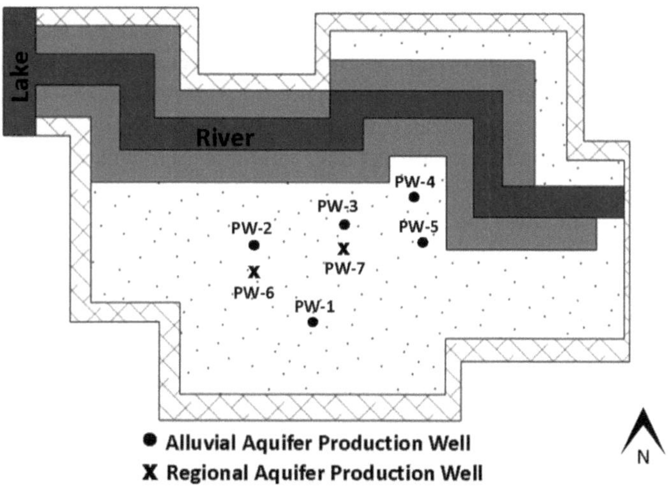

Fig. 1 Hypothetical watershed used for modeling analysis

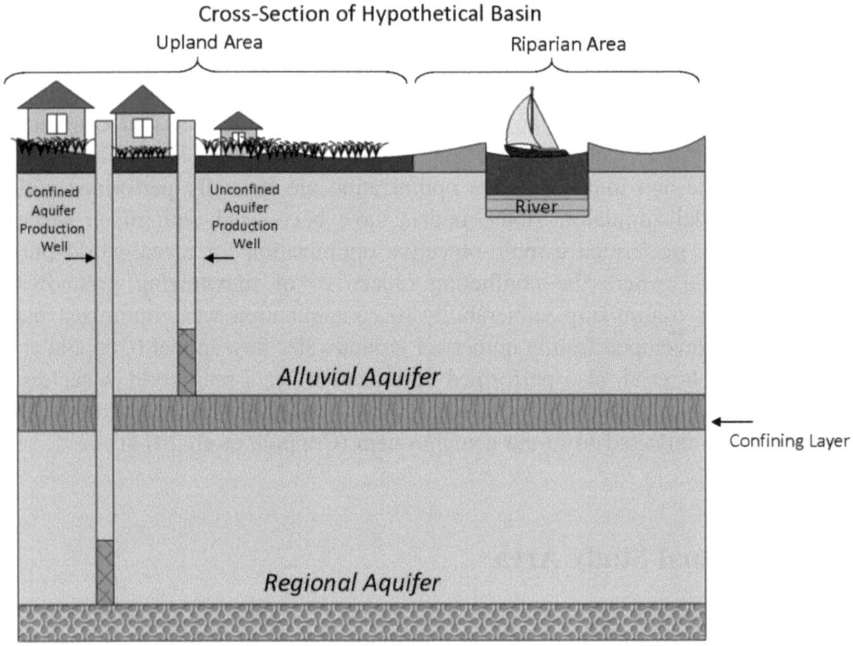

Fig. 2 Simplified cross-section of hypothetical watershed

from the water table aquifer discharges into the river. Along most of the river is a riparian corridor, which is a habitat for plants, fish, reptiles, mammals, and birds.

As depicted in Fig. 2, the subsurface portion of the groundwater/surface water system is a complex multi-layered aquifer, with the alluvial unconfined (i.e., water table) aquifer separated from the deeper regional confined aquifer by a low permeability clay layer. High capacity production wells are installed in both aquifers.

As is common to groundwater systems, each aquifer consists of different geologic media or lithologies that vary over space (i.e., heterogeneity). As depicted by Figs. 3 and 4 for the alluvial and regional aquifers, respectively, the sediments range from finer grained materials like silty sand that yield and transmit less water to coarser grained materials like sand and gravels which are highly prolific water bearing and transmission zones.

Seven high capacity production wells operate in the study area, with five wells pumping from the unconfined aquifer, and two deeper wells pumping from the confined aquifer. Because of the geologic heterogeneity and variable boundary conditions, both aquifers respond differently over space to pumping stresses. Similarly, river reaches (i.e., discrete sections) also respond differently to the pumping stresses of the individual production wells, depending upon their proximity and surrounding hydrogeology. Adding to system complexity are the time-varying hydrologic and weather conditions, particularly distinct by season, which produce temporally varying pumping responses in the system.

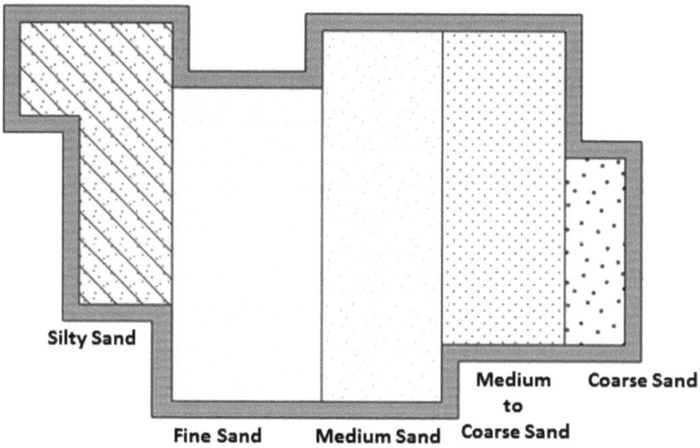

Fig. 3 Alluvial aquifer lithology

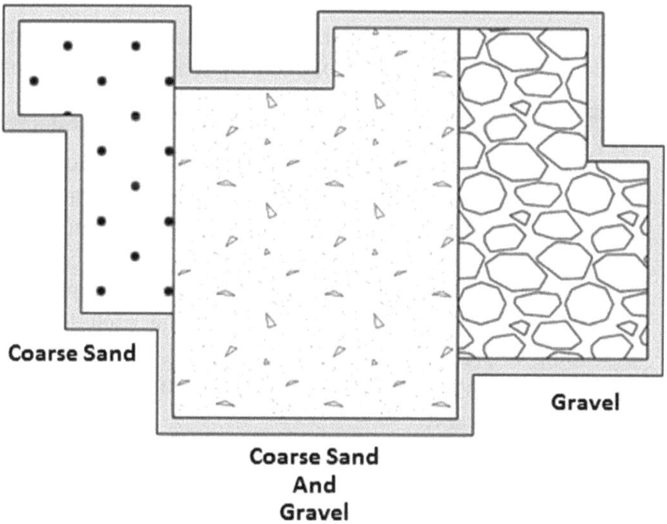

Fig. 4 Regional aquifer lithology

6 Conceptual Real-Time Management System

For this real-time management system, field data consisting of groundwater elevations, surface water flows, precipitation, and pumping rates are collected to develop ANN prediction models. The ANNs predict in real-time groundwater elevations in both the unconfined and confined aquifers, including riparian areas, and surface water flows in the river in response to variable weather, hydrologic, and pumping conditions.

Symbolic functional representation of the ANN models is presented below. The models predict groundwater elevations and surface water flows as a function of initial hydraulic conditions (i.e., groundwater elevation and streamflow rates), weather conditions, and pumping rates of the production wells.

$$GWLU_{i,t} = f\left(GWLU_{i,t-1}, Q_{m,t}, Q_{n,t}, P_t, T_t, SWF_{j,t-1}\right) \tag{1}$$

$$SWF_{j,t} = f\left(GWLU_{i,t-1}, Q_{m,t,}, Q_{n,t}, P_t, T_t, SWF_{j,t-1}\right) \tag{2}$$

$$GWLC_{k,t} = f(GWLC_{k,t-1}, Q_{m,t}, Q_{n,t}) \tag{3}$$

where

$GWLU_{i,t}$ the vector element representing the final groundwater elevations in the unconfined aquifer at each location *i* at the end of prediction period *t*;

$GWLU_{i,t-1}$ the vector element representing the initial groundwater elevations in the unconfined aquifer at each location *i* at the beginning of prediction period *t*;

$Q_{m,t}$ the vector element representing the measured pumping rates of the production wells in the unconfined aquifer at each location *m* corresponding to prediction period *t*;

$GWLC_{k,t}$ the vector element representing the final groundwater elevations in the confined aquifer at each location *k* at the end of prediction period *t*;

$GWLC_{k,t-1}$ the vector element representing the initial groundwater elevations in the confined aquifer at each location *k* at the beginning of prediction period *t*;

$Q_{n,t}$ the vector element representing the measured pumping rates of the production wells in the confined aquifer at each location *n* corresponding to prediction period *t*;

P_t a representative precipitation value for the study area corresponding to prediction period *t*;

T_t a representative air temperature value for the study area corresponding to prediction period *t*;

$SWF_{j,t-1}$ the vector element representing the initial surface water flow rate in the river at each location *j* at the beginning of prediction period *t*;

$SWF_{j,t}$ the vector element representing the final surface water flow rate in the river at each location *j* at the end of prediction period *t*;

Each of the ANN prediction models include some subset of the state variables of interest, namely groundwater elevations and surface water flow rates, random weather variables like precipitation and temperature values, and finally, the pumping rates of the wells, which constitute the decision control variables we seek to optimize. The initial state values, as well as the combination of weather and pumping conditions, collectively determine the final values of the state variables at the end of the prediction period.

As demonstrated by Coppola et al. (2003a, b, 2005a, b, c, 2007, 2014), a fundamental understanding of the hydrogeological system combined with sensitivity analyses is important for converging to the critical set of prediction variables for each ANN model. For this system, it is assumed that the confined aquifer is not significantly affected by precipitation and temperature over the length of the prediction period. Accordingly, these variables would be excluded from the ANN models used to predict groundwater elevations in this deeper aquifer.

These inter-variable functional relationships help demonstrate how ANN models can be initialized to real-time conditions. They also illustrate how ANN models can be developed to simulate any range of conditions for which historical data exists, and equally important, differentiate weather and/or hydrologic conditions from pumping effects on critical system states like groundwater elevations and river flow rates. This is a significant advantage over traditional physics-based models, the de facto method for simulating and optimizing pumping in complex groundwater/surface water systems. Despite the significant influence of weather and hydrologic conditions on these systems and how they respond to pumping, these natural time-varying factors are typically ignored.

As stated by the United States Geological Survey in their 2012 report entitled "*Streamflow Depletion by Wells—Understanding and Managing the Effects of Groundwater Pumping on Streamflow*" (Mays 2007):

> Theoretically, response functions could be determined by monitoring changes in streamflow that result from pumping at a particular well, but this approach is often not technically feasible because of difficulty in separating depletion changes from streamflow responses to other changes, such as those driven by climate. In practice, response functions are determined by using analytical or numerical models.

ANN models bridge the gap between what is theoretically possible and what is achievable by learning directly from real-world data, and functionally mapping a combination of weather, hydrologic, and pumping variables to distinct system states at specific locations. Because of this, with sufficient data, system responses to any combination of conditions and stresses can be accurately predicted.

The set of ANN equations constitutes a highly efficient representation of the physical system behavior of interest. These equations can be explicitly embedded into the optimization program or used to generate response coefficients for the optimization model as part of the constraint set and/or objective function. The next section describes the general optimization formulation for this problem.

7 Management Formulation

In this water management problem, at any given time, there is a finite volume of water within the watershed that varies by season, day, and even time of day. As groundwater is extracted, loss of aquifer storage will reduce groundwater elevations and potentially surface water flows. Depending upon their severity, a combination of adverse environmental and economic impacts may occur.

Environmental impacts include drying of river reaches and riparian areas which destroy habitats for flora and fauna. Beyond less easily quantifiable loses like aesthetics and overall quality of life, environmental loses can also translate directly into economic costs, like loss of aquatic species for harvesting and diminished eco-tourism. More easily measurable economic costs included higher costs to drill deeper wells to reach the dropping water table aquifer, higher electricity costs associated with lifting the dropping groundwater surface, and by extension, more powerful and expensive pumps.

For this management problem, the decision makers want to maximize total groundwater pumping while minimizing potential negative impacts, which include:

- Excessive drawdown in the unconfined aquifer;
- Excessive drawdown in the confined aquifer;
- Excessive drawdown in the riparian corridor;
- Streamflow depletion of the river.

The objective is to maximize the combined groundwater pumping of the production wells without violating the groundwater and surface water management constraints imposed on the groundwater/surface water system. These constraints are formulated to help ensure that over-depletion of groundwater and surface water does not occur.

A single linear objective optimization management formulation was devised as follows:

Objective Function

$$Maximize\, Q_1 + Q_2 + Q_3 + Q_4 + Q_5 + Q_6 + Q_7 \tag{4}$$

Linear Constraint Types

$$RC_{u1,1}Q_1 + RC_{u1,2}Q_2 + RC_{u1,3}Q_3 + RC_{u1,4}Q_4 + RC_{u1,5}Q_5 + RCu_{1,6}Q_6 + RC_{u1,7}Q_7 \le DD_{u1} \tag{5}$$

$$RC_{c1,1}Q_1 + RC_{c1,2}Q_2 + RC_{c1,3}Q_3 + RC_{c1,4}Q_4 + RC_{c1,5}Q_5 + RCc_{1,6}Q_6 + RC_{c1,7}Q_7 \le DD_{c1} \tag{6}$$

$$RC_{r1,1}Q_1 + RC_{r1,2}Q_2 + RC_{r1,3}Q_3 + RC_{r1,4}Q_4 + RC_{r1,5}Q_5 + RCc_{r,6}Q_6 + RC_{r1,7}Q_7 \le DD_{r1} \tag{7}$$

$$RC_{s1,1}Q_1 + RC_{s1,2}Q_2 + RC_{s1,3}Q_3 + RC_{s1,4}Q_4 + RC_{s1,5}Q_5 + RCc_{s,6}Q_6 + RC_{s1,7}Q_7 \le SFD_{s1} \tag{8}$$

$$0 \le Q_i \le Q_{upperlimit} \quad \mathrm{i=1,7} \tag{9}$$

Equation 4 is the objective function used to maximize total pumping of the seven production wells subject to the imposed constraints, where Q_i represents the pumping rate of public supply well P_i in cubic feet/second (cfs). Equations 5, 6, and 7 are example constraints for limiting drawdown (DD) in the unconfined aquifer, the confined aquifer, and riparian areas (which are part of the unconfined aquifer), respectively, in feet. Equation 8 is an example constraint for limiting streamflow depletion (SFD) in the river along a particular reach in cfs. Equation 9 bounds the pumping rate of each production well between 0 and some maximum value, typically equal to the maximum sustainable pumping rate of the well.

The response coefficients in each constraint, generically symbolized by *RC*, quantifies how much a system state value changes at a particular location per unit stress of the corresponding production well (i.e., decision or control variable). For example, $RC_{r1,3}$ in Eq. 7 is the response coefficient for production well P-3 (pumping rate designated as Q_3) pertaining to drawdown at location 1 of the riparian area. In quantitative terms, a drawdown response coefficient of 3.50 signifies that each unit pumping rate (e.g., one cubic feet per second) by P-3 induces 3.5 feet of drawdown (i.e., groundwater elevation decline) at location 1 of the riparian area. Similarly, a streamflow depletion response coefficient of 0.52 for a particular production well signifies that for each unit pumping rate, it induces a loss in the river flow or discharge rate of 0.52 cfs at the stream location of interest. The higher the response coefficient, the more the particular production well induces a response in the state variable of interest at the corresponding monitoring/management location.

The general process is as follows. ANN models trained with real-world data generate the response coefficient values for each decision variable for all responses and locations of interest. The response coefficients are generated in real-time to account for existing hydrological and weather conditions (e.g., wet period). The generated response coefficients are substituted into the optimization model, which is then solved to compute the optimal pumping rates for all seven production wells. Response coefficients are re-computed and updated as necessary to reflect changing conditions for each subsequent optimization management period.

For the analysis presented here, different combinations of upper bound constraint values were assigned to the optimization formulation to reflect a possible range of decision-making preferences. For example, in some cases, limiting drawdown values in the aquifers would not be considered as important as protecting the river against depletion.

Because of varying but generally present hydraulic interconnections between aquifers, the river, and riparian areas, the optimization solution in minimizing impacts to one resource will often reduce impacts to other resources. An exception, as shown later, is during the wet season, when groundwater pumping has minimal hydraulic effects on river flows. In contrast, during the dry season, surface water is much more to vulnerable to groundwater pumping, which if not properly managed, can deplete river flows.

As will be shown, these significant variations in system responses and corresponding optimization solutions under different hydrologic conditions (e.g., wet versus dry) demonstrate the utility and value of a dynamic decision support system that accurately reflect real-time conditions.

8 Computed Optimal Solutions

Table 1 presents the assigned constraint values for the ten optimization scenarios.

The ten optimization scenarios provide a wide range of possible upper constraint values for drawdown in the two aquifers and the riparian area, as well as depletion of river flow. The only upper bound constraint value that does not change is the upper pumping rate of each of the seven production wells.

Table 2 shows the corresponding computed optimal solutions for the ten different scenarios for both the dry and wet seasons generated by the optimization software program LINDO, where Q_i represents the pumping rate of production well i.

Figure 5 is a bar graph depicting the different optimization solutions for the ten scenarios for both the dry and wet seasons.

For most scenarios, the optimal solution for the dry season produced lower total pumping than the corresponding solution for the wet season. For scenarios where the upper bound constraint values were low (i.e., highly protective), the optimal pumping solutions were identical for both wet and dry seasons. There are no scenarios where the dry season had total higher optimal pumping rates than the wet season.

In addition to seasonal influences, time-constant hydrogeological conditions also affected computed optimal solutions. Deeper production wells P-6 and P-7 generally pump at higher rates than the shallow production wells, as the unconfined wells are more hydraulically connected to the river and riparian areas, and therefore, impact these resources more.

Table 1 Constraint sets for the ten scenarios for dry and wet seasons

Constraint	1	2	3	4	5	6	7	8	9	10
Unconfined drawdown (ft.)	≤ 100	≤ 65	≤ 65	≤ 65	≤ 65	≤ 65	≤ 100	≤ 100	≤ 65	≤ 35
Confined drawdown (ft.)	≤ 80	≤ 60	≤ 60	≤ 60	≤ 60	≤ 80	≤ 60	≤ 80	≤ 60	≤ 30
Wetlands drawdown (ft.)	≤ 50	≤ 50	≤ 5	≤ 50	≤ 5	≤ 50	≤ 50	≤ 50	≤ 10	≤ 5
Streamflow depletion (cfs)	≤ 5	≤ 5	≤ 5	≤ 2	≤ 2	≤ 5	≤ 5	≤ 2	≤ 5	≤ 2
Maximum pumping rate (cfs)	≤ 20	≤ 20	≤ 20	≤ 20	≤ 20	≤ 20	≤ 20	≤ 20	≤ 20	≤ 20

Table 2 Computed optimal pumping solutions for the ten scenarios for dry and wet seasons (in cfs)

Solutions	Dry season scenarios									
Q1	0	0	0	0	0	0	0	0	0	0
Q2	0	0	0	0	0	0	0	0	0	0
Q3	0	0	0	0	0	0	0	0	0	0
Q4	0	0	0.42	0	0.42	0	0	0	0.84	0.42
Q5	3.48	4.41	0.6	0.1	0.6	3.48	4.41	0	1.21	0.6
Q6	15.8	11.46	3.11	12.4	3.11	15.8	11.46	20	6.22	3.11
Q7	17.9	12.91	0	14.26	0	17.9	12.91	8.11	0	0
Total Q (gpm)	37.27	28.79	4.14	26.77	4.14	37.27	28.79	28.11	8.29	4.14
Solutions	Wet season scenarios									
Q1	18.28	3.96	0	8.8	0	1.84	18.66	0	0	0
Q2	0	4.73	0	9.17	0	2.31	10.77	18.42	0	0
Q3	0	6.85	0	0	0	0.19	0	0	0	0
Q4	0	6.15	0.42	0	0.42	2.69	5.49	0	0.84	0.42
Q5	4.85	7.74	0.6	0	0.6	5.65	4.97	0	1.21	0.6
Q6	11.92	4.08	3.31	10.43	3.31	13.04	0	10	6.22	3.11
Q7	6.11	4.8	0	0	0	15.15	0	0	0	0
Total Q (gpm)	41.19	38.42	4.14	28.41	4.14	40.89	39.9	28.42	8.29	4.14

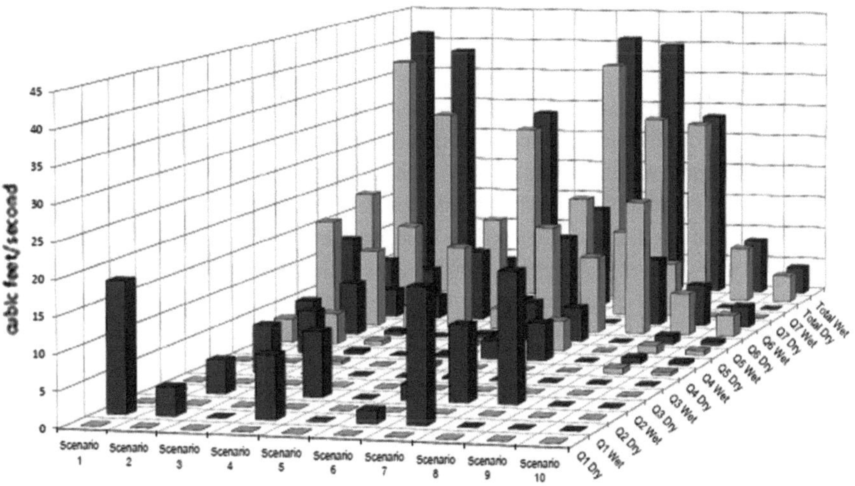

Fig. 5 Optimal solutions for dry and wet seasons for ten scenarios

9 Stochastic Analysis

The above optimization analysis considered distinct dry and wet seasons, the effects of which can significantly change the relative pumping effects of individual wells on components of the hydrogeologic system like river flow. What this analysis did not consider is the stochastic component where pumping demand can change in response to changing weather conditions. In this section, it is assumed that two production wells, P-5 and P-7, are used exclusively for irrigating agricultural areas. In a properly managed area, groundwater pumping for irrigation is modified as a function of weather conditions, where higher precipitation periods require less irrigation than during lower precipitation periods.

To account for time-varying weather driven irrigation demand, stochastic groundwater pumping optimization was performed in accordance with precipitation forecasts and their associated uncertainty. In this analysis, it is assumed that weather forecasts have historically been shown to be accurate to within plus or minus 0.2 inches for a 90% confidence interval within the range of forecasts provided in this analysis. In reality, the range of the lower and upper bound 90% confidence interval would likely change during different seasons and for different ranges of values of precipitation forecasts.

Irrigation demand changes in direct linear proportion to precipitation for a given management period. When no precipitation occurs over this period, the irrigation demand is 100% of what the crops require, assumed equivalent to 14 cfs for the agricultural management area. It was also assumed that if precipitation over the optimization management period is 2.4 inches or more in the agricultural area, no groundwater irrigation is required as rainfall satisfies total crop water demand. Any precipitation less than this constitutes a crop water deficit condition that requires some linearly proportional contribution from groundwater pumping.

The weather forecast for the management period was used to predict expected agricultural groundwater pumping demand by computing the expected percent irrigation demand required with a 90% confidence interval. To do this, the lower and upper 90% statistical bound estimates of the forecasted precipitation as well as the actual forecasted value are used to compute the percent irrigation demand for each. The optimization management problem was then solved separately for each of these three values by constraining the total combined pumping of production wells P-5 and P-7 equal to the corresponding stochastically estimated irrigation pumping demand.

Table 3 depicts the three different irrigation water demand forecasts, representing a lower than average, average, and higher than average precipitation forecasts for the area during the wet season. The total irrigation demand represents values, obtained in accordance with the precipitation forecasts and their corresponding 90% statistical bounds, derived from the above described linear relationship. For example, for the lower than average precipitation forecast event, in accordance with the forecasted precipitation, the expected irrigation water demand

Table 3 Stochastic optimization solutions for expected irrigation demand in accordance with precipitation forecasts during the wet season for optimization scenario 2

	Forecast lower than average			Average forecast			Forecast higher than average		
	Lower bounds (90% confidence interval)		Upper bounds (90% confidence interval)	Lower bounds (90% confidence interval)		Upper bounds (90% confidence interval)	Lower bounds (90% confidence interval)		Upper bounds (90% confidence interval)
Q5 + Q7 =	10.03	8.87	7.7	8.17	7	5.83	3.03	1.87	0.7
Q1	4.16	4.25	4.33	4.3	4.37	4.45	4.63	4.07	4.77
Q2	4.88	4.94	5	4.98	5.01	5.04	5.11	5.13	5.16
Q3	7.11	7.23	7.33	7.3	7.38	7.46	7.66	7.75	7.83
Q4	7.11	7.54	8.06	7.8	8.55	9.31	11.13	11.88	12.64
Q5	7.95	8.06	7.7	8.12	7	5.83	3.03	1.87	0.7
Q6	4.87	5.22	5.46	5.43	5.5	5.56	5.71	5.77	5.83
Q7	2.06	0.8	0	0.048	0	0	0	0	0
Total pumping	38.17	38.06	37.93	38	37.84	37.67	37.28	37.12	36.96

to meet total crop water demand is 8.87 cfs, with lower and upper 90% statistical bounds of 10.03 cfs, and 7.7 cfs, respectively, corresponding to their estimated precipitation values.

These irrigation demand values were assigned to an equality constraint for combined pumping of the two irrigation production wells, which was added to the constraint set used for optimization scenario 2, presented in Table 1. The resulting optimization solutions for this stochastic analysis are included in Table 3.

As shown, the combined groundwater pumping rates for the seven production wells do not significantly change between the lower than average, average and higher than average precipitation forecast events. What often does change, however, is the distribution of the pumping rates among production wells. For the higher precipitation forecast periods, when the expected total required irrigation pumping via production wells P-5 and P-7 is lower, higher pumping rates are distributed to other production wells. For example, the pumping rate of P-4 for the lower than average precipitation forecast event almost doubles for the higher than average precipitation forecast event.

From a water management perspective, for higher than average precipitation periods, higher groundwater pumping can be allocated to the industrial and public supply sectors. This surplus water extraction can be stored in reservoirs for future use during periods when demand by different sectors may be higher and/or for emergency storage for use during prolonged low precipitation periods like drought. In contrast, during lower than average precipitation periods, less pumping can be diverted from irrigation to the other sectors.

10 Conclusions

The conceptual water management system presented in this chapter, based upon previous work by Coppola et al. (2003a, b, 2005a, b, c, 2007, 2014) combines continuous data streams with the real-time accuracy of data-driven ANN prediction models coupled to optimization management models. In this hypothetical application, optimal pumping rates were identified to reflect real-time conditions, representing dry and wet seasons, for a complex multi-layered groundwater/surface water system.

The solutions are consistent with the hydraulics of the hydrogeologic system. During the dry season, the higher potential for river flow depletion by groundwater pumping produced lower optimal pumping rates for the unconfined production wells. That is, because there is less water in the hydrologic system during the dry season, pumping hydraulically captures more river water.

In contrast, during the wet season, snowmelt from neighboring mountains would produce significantly higher surface water flows. Streams and sections of the river that are often dry or with low flow would be flowing full, also resulting in higher bank storage and additional surface water accumulation in wetland areas, as well as higher groundwater recharge into the unconfined aquifer. There would also be increased groundwater flux into the watershed via mountain front recharge. These additional water sources in the watershed provide more water for the production wells to draw from, decreasing their relative pumping effects on the river during the wet season. As a result, the streamflow depletion constraint in the wet season has little or no effect on the optimization solutions.

The optimization results demonstrate how sensitive the optimal pumping solutions are to dry and wet seasons. They also show how a failure to represent real-time conditions can result in identification and implementation of inappropriate and even dangerous groundwater pumping strategies, putting water and ecological resources at risk. Although not shown here, many if not most watersheds have significant intra-daily and even inter-daily variability in hydrologic conditions, which could change the optimal pumping rates in very short time periods.

To further improve decision making capability, stochastic analysis was performed to factor precipitation forecasts and their uncertainty into optimized groundwater pumping strategies. Stochastic optimization as presented here would allow decision makers to more appropriately allocate groundwater pumping among different sectors in real-time, while also providing more effective longer term water management strategies.

In contrast to the ANN-based prediction system presented here, physics-based numerical groundwater models generally lack the capability to accurately reflect real-time hydrologic and weather conditions. Because of this limitation, there is the increased possibility of over-pumping groundwater, particularly during drought periods, which are becoming more prevalent with climate change. At the other extreme is under-pumping during wetter or higher precipitation conditions, which can result in lost opportunities for increased water storage as insurance to help mitigate shortages during high demand and/or drought conditions.

In conclusion, the conceptual real-time decision support system presented here provides numerous important advantages and benefits. This includes maximizing the value of automated data collection systems, providing a more transparent data-driven modeling process that, unlike traditional physics-based models, can accurately differentiate between pumping and natural factors like weather, and computing more appropriate real-time and longer-term water management strategies. As water demand continues to increase world-wide, imposing more stress on dwindling groundwater and surface water resources, the type of dynamic decision support system presented here that continuously adapts to real-time conditions will be essential.

Acknowledgements Nicole Chamoun for preparation of the study area figures, Diane Trube for preparation of the tables and editing, and Anthony Verdi for assistance with manuscript preparation.

References

Barlow, P. M., & Leake, S. A. (2012). Streamflow depletion by wells—understanding and managing the effects of groundwater pumping on streamflow. Circular 1376, United States Geological Survey, Reston, Virginia.

Brown, L. R. (2011). *World on the edge*. New York: W. W. Norton & Company.

Coppola, E., Szidarovszky, F., Poulton, M., & Charles, E. (2003a). Artificial neural network approach for predicting transient water levels in a multilayered groundwater system under variable state, pumping, and climate conditions. *Journal of Hydrologic Engineering, 8*(6), 348–359.

Coppola, E., Poulton, M., Charles, E., Dustman, J., & Szidarovszky, F. (2003b). Application of artificial neural networks to complex groundwater management problems. *Journal of Natural Resources Research, 12*(4), 303–320.

Coppola, E., Rana, A., Poulton, M., Szidarovszky, F., & Uhl, V. (2005a). A neural network model for predicting water table elevations. *Journal of Ground Water, 43*(2), 231–241.

Coppola, E., McLane, C., Poulton, M., Szidarovszky, F., & Magelky, R. (2005b). Predicting conductance due to upcoming using neural networks. *Journal of Ground Water, 43*(6), 827–836.

Coppola, E., Szidarovszky, F., & Poulton, M. (2005c). Application of artificial neural networks to complex groundwater prediction and management problems. *Journal of Southwest Hydrology, 4*(3).

Coppola, E., Szidarovszky, F., Davis, D., Spayd, S., Poulton, M., & Roman, E. (2007). Multiobjective analysis of a public wellfield using artificial neural networks. *Journal of Ground Water, 45*(1), 53–61.

Coppola, E., Szidarovszky, A., & Szidarovszky, F. (2014). Artificial neural network based modeling of hydrologic processes. In *Handbook of engineering hydrology*. CRC Press.

Karamouz, M., Szidarovszky, F., & Zahraie, B. (2003). *Water resources systems analysis*. Boca Raton, Florida: Lewis Publishers.

Mays, L. W. (2007). *Water resources sustainability*. New York: The McGraw Hill Companies.

Peralta, R. C., & Kalwij, I. M. (2012). *Groundwater optimization handbook*. Boca Raton, Florida: CRC Press.

Poulton, M. (Ed.). (2001). *Computational neural networks for geophysical data processing*. Amsterdam: Pergamon, 335 p.

Winter, T. C., Harvey, J. W., Franke, O. L., & Alley, W. M. (1998). Groundwater and surface water, a single resource. U.S. Geological Survey Circular 1139, Denver, Colorado.

MIX
Papier aus verantwortungsvollen Quellen
Paper from responsible sources
FSC® C105338

If you have any concerns about our products,
you can contact us on
ProductSafety@springernature.com

In case Publisher is established outside the EU,
the EU authorized representative is:
Springer Nature Customer Service Center GmbH
Europaplatz 3, 69115 Heidelberg, Germany

Printed by Libri Plureos GmbH
in Hamburg, Germany